Serverless Analytics with Amazon Athena

Query structured, unstructured, or semi-structured
data in seconds without setting up any infrastructure

Anthony Virtuoso

Mert Turkay Hocanin

Aaron Wishnick

BIRMINGHAM—MUMBAI

Serverless Analytics with Amazon Athena

Group Product Manager: Kunal Parikh

Publishing Product Manager: Devika Battike

Senior Editor: David Sugarman

Content Development Editor: Joseph Sunil

Technical Editor: Rahul Limbachiya

Copy Editor: Safis Editing

Project Coordinator: Aparna Nair

Proofreader: Safis Editing

Indexer: Tejal Soni

Production Designer: Shankar Kalbhor

First published: November 2021

Production reference: 1131021

Published by Packt Publishing Ltd.
Livery Place
35 Livery Street
Birmingham
B3 2PB, UK.

ISBN 978-1-80056-234-9

www.packt.com

*To my wife, Cristina, thank you for the support and understanding
as I spent late nights and early mornings working on this book. I also
appreciate all the laughs we had over my terrible spelling. For my sons,
Luca and Massimo, who worked on their own pop-up books alongside me;
I'll be first in line for an advanced copy of your books.*

– Anthony Virtuoso

*I dedicate this book to my wife, Subrina, who has been incredibly
supportive, and our son, Tristan, who was born while writing this book.
Without the both of you and the encouragement and love you gave me,
this book would not have been possible. I also want to thank my parents,
siblings, and everyone else who helped make this possible.*

– Mert Turkay Hocanin

Foreword

Creating a data strategy is a top priority for leading organizations. That's because with any major initiative, from creating new experiences to building new revenue streams, leaders must be able to quickly gather insights and get to the truth. Data-driven organizations seek the truth by treating data like an organizational asset, no longer the property of individual departments. They set up processes to collect and store valuable data. Their data is democratized, meaning it's available to the right people and systems that need it. And their data is used to build new and innovative products that use data and **machine learning (ML)** to deliver new customer experiences.

AWS offers the broadest and deepest set of services for analytics and ML, and Amazon Athena is a key pillar of our offerings. Amazon Athena is a serverless analytics service that enables customers to use standard SQL to analyze all the data in their Amazon S3 data lakes, their data warehouses, and their transactional databases, as well as data that lives on-premises, in SaaS applications, and in other clouds. In other words, with Athena, you can query all your data from a single place using a language familiar to most analysts, using any business intelligence or ML tools you'd like. It's really all about having all your data at your fingertips.

I am incredibly lucky to have worked on creating and launching virtually all of the analytics offerings from AWS over the past decade. I was part of the team that created the original vision for Athena and launched the service in 2016. We created Athena because customers wanted a way to query all their data, both the structured data from databases as well as the semi-structured and unstructured data in their data lakes and other data sources, without having to manage infrastructure or give up SQL or the standard tools they were already using. We launched Athena at re:Invent 2016 and have been iterating on and improving the service ever since.

Mert, Aaron, and Anthony were founding members of the Amazon Athena team and have played pivotal roles in defining, building, and evolving the service. They are deeply passionate engineers who love helping customers succeed with Athena and with analytics overall. At AWS, the vast majority of our roadmap is driven by working closely with our customers, understanding their requests and priorities and bringing them into our services. Mert, Aaron, and Anthony are customer-obsessed, always looking for ways to help customers get more from Athena, and they have an innate ability to teach and bring people along. I'm so grateful they chose to write this book to share their expertise with all of us.

This book, like Amazon Athena, is designed to get you up and running with queries with minimal upfront setup and work. You'll progress from running simple queries to building sophisticated, automated pipelines to work with near-real-time event data, queries to external data sources, custom functions, and more, all while learning from Mert, Aaron, and Anthony's experience working with real-world customer scenarios.

I highly recommend this book to any new or existing customers looking to transform their business with data and with Amazon Athena.

Rahul Pathak, VP, AWS Analytics

Contributors

About the authors.

Anthony Virtuoso works as a principal engineer at Amazon and holds multiple patents in distributed systems, software-defined networks, and security. In his 8 years at Amazon, he has helped launch several Amazon web services, the most recent of which was Amazon Managed Blockchain. As one of the original authors of Athena Query Federation, you'll often find him lurking on the Athena Federation GitHub repository answering questions and shipping bug fixes. When not at work, Anthony obsesses over a different set of customers, namely his wife and two little boys, aged 2 and 5. His kids enjoy doing science experiments with their dad, such as 3D printing toys, building with LEGO, or searching the local pond for tardigrades.

Mert Turkay Hocanin is a principal big data architect at Amazon Web Services within the AWS Glue and AWS Lake Formation services and has previously worked for several other services, including Amazon Athena, Amazon EMR, and Amazon Managed Blockchain. During his time at AWS, he has worked with several Fortune 500 companies on some of the largest data lakes in the world and was involved with the launching of three Amazon web services. Prior to being a big data architect, he was a senior software developer within Amazon's retail systems organization, building one of the earliest data lakes in the company in 2013. When he is not helping customers build data lakes, he enjoys spending time with his wife, Subrina, and son, Tristan, and exploring New York City.

Aaron Wishnick works as a senior software engineer at Amazon, where he has been for 7 years. During that time, he has worked on Amazon's payment systems and financial intelligence systems, as well as working for AWS on Athena and AWS Proton. When not at work, Aaron and his fiance, Alyssa, are on a quest to determine just how much dog fur is too much, with their husky and malamute, Mina and Wally.

About the reviewers

Seth Denney is a software engineer who has spent most of his career in big data analytics, building infrastructure and query engines to support a wide variety of use cases at companies including Amazon and Google. While on the AWS Athena team, he was intimately involved with the Lake Formation and Query Federation projects, to name a few.

Janak Agarwal has been the product manager for Amazon Athena since he joined AWS in December 2018. Prior to joining AWS, Janak was at Microsoft for 9+ years, where he led a team of engineers for Microsoft Office 365. He also co-founded CourseKart, an e-learning platform in India, and TaskUnite, a medical technology company in the US. Janak holds a master's in electrical engineering from USC and an MBA from the Wharton School.

Table of Contents

3

Key Features, Query Types, and Functions

Section 2: Building and Connecting to Your Data Lake

4

Metastores, Data Sources, and Data Lakes

5
Securing Your Data

6
AWS Glue and AWS Lake Formation

Section 3: Using Amazon Athena

7
Ad Hoc Analytics

8
Querying Unstructured and Semi-Structured Data

9
Serverless ETL Pipelines

10

Building Applications with Amazon Athena

11

Operational Excellence – Monitoring, Optimization, and Troubleshooting

Section 4: Advanced Topics

12
Athena Query Federation

13
Athena UDFs and ML

14
Lake Formation – Advanced Topics

Other Books You May Enjoy

Index

Preface

Amazon Athena is an interactive query service that makes it easy to analyze data in Amazon S3 using standard SQL, without needing to manage any infrastructure.

This book begins with an overview of the serverless analytics experience offered by Athena and teaches you how to build and tune an S3 data lake using Athena, including how to structure your tables using open source file formats such as Parquet. You'll learn how to build, secure, and connect to a data lake with Athena and Lake Formation. Next, you'll cover key tasks such as ad hoc data analysis, working with ETL pipelines, monitoring and alerting KPI breaches using CloudWatch Metrics, running customizable connectors with AWS Lambda, and more. Moving ahead, you'll work through easy integrations, troubleshooting and tuning common Athena issues, and the most common reasons for query failure, as well as reviewing tips for diagnosing and correcting failing queries in your pursuit of operational excellence. Finally, you'll explore advanced concepts such as Athena Query Federation and Athena ML to generate powerful insights without needing to touch a single server.

By the end of this book, you'll be able to build and use a data lake with Amazon Athena to add data-driven features to your app and perform the kind of ad hoc data analysis that often precedes many of today's ML modeling exercises.

Who this book is for

BI analysts, application developers, and system administrators who are looking to generate insights from an ever-growing sea of data while controlling costs and limiting operational burdens will find this book helpful. Basic SQL knowledge is expected to make the most out of this book.

What this book covers

Chapter 1, Your First Query, is all about orienting you to the serverless analytics experience offered by Amazon Athena. For now, we will simplify things in order to run your first queries and demonstrate why so many people choose Amazon Athena for their workloads. This will help establish your mental model for the deeper discussions, features, and examples of later sections.

Chapter 2, Introduction to Amazon Athena, continues your introduction to Athena by discussing the service's capabilities, scalability, and pricing. You'll learn when to use Amazon Athena and how to estimate the performance and costs of your workloads before building them on Athena. We'll also take a look behind the scenes to see how Athena uses PrestoDB, an open source SQL engine from Facebook, to process your queries.

Chapter 3, Key Features, Query Types, and Functions, concludes our introduction to Amazon Athena by exploring built-in features you can use to make your reports or application more powerful. This includes approximate query techniques to speed up analysis of large datasets and **Create Table As Select (CTAS)** statements for running queries that generate significant amounts of result data.

Chapter 4, Metastores, Data Sources, and Data Lakes, teaches you what a metastore is and what they contain. We will introduce Apache Hive and AWS Glue Data Catalog implementations of a metastore. We'll then learn how to create tables through Athena or discover datasets in S3 using AWS Glue crawlers. We then focus on a typical data lake architecture, which contains three different stages for data.

Chapter 5, Securing Your Data, covers the various methods that can be employed to secure your data and ensure it can only be viewed by those that have permission to do so.

Chapter 6, AWS Glue and AWS Lake Formation, demonstrates step by step how to build a secure data lake in Lake Formation and how Athena interacts with Lake Formation to keep data safe.

Chapter 7, Ad Hoc Analytics, focuses on how you can use Athena to quickly get to know your data, look for patterns, find outliers, and generally surface insights that will help you get the most from your data.

Chapter 8, Querying Unstructured and Semi-Structured Data, shows how Amazon Athena combines a traditional query engine, and its requirement for an upfront schema, with extensions that allow it to handle data that contains varying or no schema.

Chapter 9, Serverless ETL Pipelines, continue with the theme of controlling chaos by using automation to normalize newly arrived data through a process known as **extract, transform, load (ETL)**.

Chapter 10, Building Applications with Amazon Athena, tells you what to do when integrating Amazon Athena into your applications. How will the application make Athena calls? How should credentials be stored? Should you use JDBC, ODBC, or Athena's SDK? What are the best practices on setting up connectivity between your application and Athena and the security considerations? Lastly, what is the best way for me to store my data on S3 to optimize speed and cost? This chapter will answer all these questions and give examples – including working code – to get you started integrating with Athena fast, easily, and in a secure way.

Chapter 11, Operational Excellence – Maintenance, Optimization, and Troubleshooting, focuses on operational excellence by looking at what could go wrong when using Athena in a production environment. We'll learn how to monitor and alert KPI breaches – such as queue dwell times – using CloudWatch metrics so you can avoid surprises. You'll also see how to optimize your data and queries to avoid problems before they happen. We'll then look at how the layout of data stored in S3 can have a significant impact on both cost and performance. Lastly, we will look at the most common reasons for query failure and review tips to help diagnose and correct failing queries.

Chapter 12, Athena Query Federation, is all about getting the most out of Amazon Athena by using Athena's Query Federation capabilities to expand beyond queries over data in S3. We will illustrate how Query Federation allows you to combine data from multiple sources (for example, S3 and Elasticsearch) to provide a single source of truth for your queries. Then we will peel back the hood and explain how Amazon Athena uses AWS Lambda to run customizable connectors. We will even write our own connector in order to show you how easy it is to customize Athena with your own code.

Chapter 13, Athena UDFs and ML, continues the theme of enhancing Amazon Athena with our own functionality by adding our own user-defined functions and machine learning models. These capabilities allow us to do everything from applying ML inference to identify suspicious records in our dataset to converting port numbers in a VPC flow log to the common name for that port (for example, HTTP). In all of these examples, we add our own logic to Athena's row-level processing without the need to run any servers of our own.

Chapter 14, Lake Formation – Advanced Topics, covers some of the advanced features that Lake Formation brings to the table, and explores various use cases that are enabled by these features.

To get the most out of this book

To work on the technologies in this book, you will need a computer with a Chrome, Safari, or Microsoft Edge browser installed and AWS CLI version 2 installed.

If you are using the digital version of this book, we advise you to type the code yourself or access the code from the book's GitHub repository (a link is available in the next section). Doing so will help you avoid any potential errors related to the copying and pasting of code.

Please ensure that you close any outstanding AWS instances after you are done working on them so that you don't incur unnecessary expenses.

Download the example code files

You can download the example code files for this book from GitHub at `https://github.com/PacktPublishing/Serverless-Analytics-with-Amazon-Athena`. If there's an update to the code, it will be updated in the GitHub repository.

We also have other code bundles from our rich catalog of books and videos available at `https://github.com/PacktPublishing/`. Check them out!

Download the color images

We also provide a PDF file that has color images of the screenshots and diagrams used in this book. You can download it here: `http://www.packtpub.com/sites/default/files/downloads/9781800562349_ColorImages.pdf`.

Conventions used

There are a number of text conventions used throughout this book.

`Code in text`: Indicates code words in text, database table names, folder names, filenames, file extensions, pathnames, dummy URLs, user input, and Twitter handles. Here is an example: "We simply specify a `SYSTEM_TIME` that Athena will use to set the read point in the transaction log."

A block of code is set as follows:

```
try:
    sink.writeFrame(new_and_updated_impressions_dataframe)
    glueContext.commit_transaction(txid1)
```

```
except:
    glueContext.abort_transaction(txid1)
```

When we wish to draw your attention to a particular part of a code block, the relevant lines or items are set in bold:

```
"inventory_id","item_name","available_count"
"1","A simple widget","5"
"2","A more advanced widget","10"
"3","The most advanced widget","1"
"4","A premium widget","0"
"5","A gold plated widget","9"
```

Bold: Indicates a new term, an important word, or words that you see onscreen. For instance, words in menus or dialog boxes appear in **bold**. Here is an example: "Administrators can set a workgroup to encrypt query results. In the workgroup settings, set query results to be encrypted using SSE-KMS, CSE-KMS, or SSE-S3 and check the **Override client-side** settings."

> **Tips or important notes**
> Appear like this.

Get in touch

Feedback from our readers is always welcome.

General feedback: If you have questions about any aspect of this book, email us at customercare@packtpub.com and mention the book title in the subject of your message.

Errata: Although we have taken every care to ensure the accuracy of our content, mistakes do happen. If you have found a mistake in this book, we would be grateful if you would report this to us. Please visit www.packtpub.com/support/errata and fill in the form.

Piracy: If you come across any illegal copies of our works in any form on the internet, we would be grateful if you would provide us with the location address or website name. Please contact us at copyright@packt.com with a link to the material.

If you are interested in becoming an author: If there is a topic that you have expertise in and you are interested in either writing or contributing to a book, please visit `authors.packtpub.com`.

Share Your Thoughts

Once you've read *Serverless Analytics with Amazon Athena*, we'd love to hear your thoughts! Scan the QR code below to go straight to the Amazon review page for this book and share your feedback.

https://packt.link/r/1-800-56234-9

Your review is important to us and the tech community and will help us make sure we're delivering excellent quality content.

Section 1: Fundamentals Of Amazon Athena

In this section, you will run your first Athena queries and establish an understanding of key Athena concepts that will be put into practice in later sections.

This section consists of the following chapters:

- *Chapter 1, Your First Query*
- *Chapter 2, Introduction to Amazon Athena*
- *Chapter 3, Key Features, Query Types, and Functions*

1
Your First Query

This chapter is all about introducing you to the serverless analytics experience offered by **Amazon Athena**. Data is one of the most valuable assets you and your company generate. In recent years, we have seen a revolution in data retention, where companies are capturing all manner of data that was once ignored. Everything from logs to clickstream data, to support tickets are now routinely kept for years. Interestingly, the data itself is not what is valuable. Instead, the insights that are buried in that mountain of data are what we are after. Certainly, increased awareness and retention have made the information we need to power our businesses, applications, and decisions *more available* but the explosion in data sizes has made the insights we seek *less accessible*. What could once fit nicely in a traditional RDBMS, such as Oracle, now requires a distributed filesystem such as HDFS and an accompanying **Massively Parallel Processing (MPP)** engine such as Spark to run even the most basic of queries in a timely fashion.

Enter Amazon Athena. Unlike traditional analytics engines, Amazon Athena is a fully managed offering. You will never have to set up any servers or tune cryptic settings to get your queries running. This allows you to focus on what is most important: using data to generating insights that drive your business. You can just focus on getting the most out of your data. This ease of use is precisely why this first chapter is all about getting hands-on and running your first query. Whether you are a seasoned analytics veteran or a newcomer to the space, this chapter will give you the knowledge you need to be running your first Athena query in less than 30 minutes. For now, we will simplify things to demonstrate why so many people choose Amazon Athena for their workloads. This will help establish your mental model for the deeper discussions, features, and examples of later sections.

In this chapter, we will cover the following topics:

- What is Amazon Athena?
- Obtaining and preparing sample data
- Running your first query

Technical requirements

Wherever possible, we will provide samples or instructions to guide you through the setup. However, to complete the activities in this chapter, you will need to ensure you have the following prerequisites available. Our command-line examples will be executed using **Ubuntu**, but most flavors of **Linux** should also work without modification.

You will need internet access to GitHub, S3, and the AWS Console.

You will also require a computer with the following installed:

- Chrome, Safari, or Microsoft Edge
- The AWS CLI

In addition, this chapter requires you to have an **AWS account** and accompanying IAM user (or role) with sufficient privileges to complete the activities in this chapter. Throughout this book, we will provide detailed IAM policies that attempt to honor the age-old best practice of "least privilege." For simplicity, you can always run through these exercises with a user that has full access, but we recommend that you use scoped-down IAM policies to avoid making costly mistakes and to learn more about how to best use IAM to secure your applications and data. You can find the suggested IAM policy for this chapter in this book's accompanying GitHub repository, listed as `chapter_1/iam_policy_chapter_1.json`:

`https://github.com/PacktPublishing/Serverless-Analytics-with-Amazon-Athena/tree/main/chapter_1`

This policy includes the following:

- Read and Write access to one S3 bucket using the following actions:
  - `s3:PutObject`: Used to upload data and also for Athena to write query results.
  - `s3:GetObject`: Used by Athena to read data.
  - `s3:ListBucketMultipartUploads`: Used by Athena to write query results.

- `s3:AbortMultipartUpload`: Used by Athena to write query results.

- `s3:ListBucketVersions`

- `s3:CreateBucket`: Used by you if you don't already have a bucket you can use.

- `s3:ListBucket`: Used by Athena to read data.

- `s3:DeleteObject`: Used to clean up if you made a mistake or would like to reattempt an exercise from scratch.

- `s3:ListMultipartUploadParts`: Used by Athena to write a result.

- `s3:ListAllMyBuckets`: Used by Athena to ensure you own the results bucket.

- `s3:ListJobs`: Used by Athena to write results.

- Read and Write access to one Glue Data Catalog database, using the following actions:

  - `glue:DeleteDatabase`: Used to clean up if you made a mistake or would like to reattempt an exercise from scratch.

  - `glue:GetPartitions`: Used by Athena to query your data in S3.

  - `glue:UpdateTable`: Used when we import our sample data.

  - `glue:DeleteTable`: Used to clean up if you made a mistake or would like to reattempt an exercise from scratch.

  - `glue:CreatePartition`: Used when we import our sample data.

  - `glue:UpdatePartition`: Used when we import our sample data.

  - `glue:UpdateDatabase`: Used when we import our sample data.

  - `glue:CreateTable`: Used when we import our sample data.

  - `glue:GetTables`: Used by Athena to query your data in S3.

  - `glue:BatchGetPartition`: Used by Athena to query your data in S3.

  - `glue:GetDatabases`: Used by Athena to query your data in S3.

  - `glue:GetTable`: Used by Athena to query your data in S3.

  - `glue:GetDatabase`: Used by Athena to query your data in S3.

  - `glue:GetPartition`: Used by Athena to query your data in S3.

  - `glue:CreateDatabase`: Used to create a database if you don't already have one you can use.

- glue:DeletePartition: Used to clean up if you made a mistake or would like to reattempt an exercise from scratch.

- Access to run Athena queries.

> **Important Note**
> We recommend against using Firefox with the Amazon Athena console as we have found, and reported, a bug associated with switching between certain elements in the UX.

What is Amazon Athena?

Amazon Athena is a query service that allows you to run standard SQL over data stored in a variety of sources and formats. As you will see later in this chapter, Athena is serverless, so there is no infrastructure to set up or manage. You simply pay $5 per TB scanned for the queries you run without needing to worry about idle resources or scaling.

> **Note**
> AWS has a habit of reducing prices over time. For the latest Athena pricing, please consult the Amazon Athena product page at https://aws. amazon.com/athena/pricing/?nc=sn&loc=3.

Athena is based on **Presto** (https://prestodb.io/), a distributed SQL engine that's open sourced by **Facebook**. It supports ANSI SQL, as well as Presto SQL features ranging from geospatial functions to rough query extensions, which allow you to run approximating queries, with statistically bound errors, over large datasets in only a fraction of the time. Athena's commitment to open source also provides an interesting avenue to avoid lock-in concerns because you always have the option to download and manage your own Presto deployment from GitHub. Of course, you will lose many of Athena's enhancements and must manage the infrastructure yourself, but you can take comfort in knowing you are not beholden to potentially punitive licensing agreements as you might be with other vendors.

While Athena's roots are open source, the team at AWS have added several enterprise features to the service, including the following:

- Federated Identity via SAML and Active Directory support

- Table, column, and even row-level access control via Lake Formation

- Workload classification and grouping for cost control via WorkGroups

- Automated regression testing to take the pain out of upgrades

Later chapters will cover these topics in greater detail. If you feel compelled to do so, you can use the table of contents to skip directly to those chapters and learn more.

Let's look at some use cases for Athena.

Use cases

Amazon Athena supports a wide range of use cases and we have personally used it for several different patterns. Thanks to Athena's ease of use, it is extremely common to leverage Athena for ad hoc analysis and data exploration.

Later in this book, you will use Athena from within a **Jupyter notebook** for machine learning. Similarly, many analysts enjoy using Athena directly from **BI Tools** such as **Looker** and **Tableau**, courtesy of Athena's JDBC driver. Athena's robust SQL dialect and asynchronous API model also enables application developers to build analytics right into their applications, enabling features that would not previously have been practical due to scale or operational burden. In many cases, you can replace RDBMS-driven features with Athena at a fraction of the cost and lower operational burden.

Another emerging use case for Athena is in the ETL space. While Athena advertises itself as being an engine that avoids the need for ETL by being able to query the data in place, as it is, we have seen the benefits of replacing existing or building new ETL pipelines using Athena where cost and capacity management are key factors. Athena will not necessarily achieve the same scale or performance as Spark, for example, but if your ETL jobs do not require multi-TB joins, you might find Athena to be an interesting option.

Separation of storage and compute

If you are new to serverless analytics, you may be wondering where your data is stored. Amazon Athena builds on the concept of *Separation of Storage and Compute* to decouple the computational resources (for example, CPU, memory, network) that do the heavy lifting of executing your SQL queries from the responsibility of keeping your data safe and available. In short, this means Athena itself does not store your data. Instead, you are free to choose from several data stores with customers increasingly pairing with DynamoDB to rapidly mutate data with S3 for their bulk data. With Athena, you can easily write a query that spans both data stores.

Amazon's **Simple Storage Service**, or **S3** for short, is easily the most recommended data store to use with Athena. When Athena launched in 2016, S3 was the first data store it supported. Unsurprisingly, Athena has been optimized to take advantage of S3's unique ability to deliver exabyte scale and throughput while still providing eleven nines (99.999999999%) of durability. In addition to effortless scaling from a few gigabytes of data up to many petabytes, S3 offers some of the lowest prices for performance that you can find. Depending on your replication requirements, storing 1 GB of data for a month will cost you between $0.01 and $0.023. Even the most cost-efficient enterprise hard drives cost around $0.21 per GB before you add on redundancy, the power to run them, or a server and data center to house them. As with most AWS services, you should consult S3's pricing page (`https://aws.amazon.com/s3/pricing/`) for the latest details since AWS has cut their prices more than 70 times in the last decade.

Metastore

In addition to accessing the raw 1s and 0s that represent your data, Athena also requires metadata that helps its SQL engine understand how to interpret the data you have stored in S3 or elsewhere. This supplemental information helps Athena map collections of files, or objects in the case of S3, to SQL constructs such as tables, columns, and rows. The repository for this data, about your data, is often called a metastore. Athena works with Hive-compliant metastores, including AWS's Glue Data Catalog service. In later chapters, we will look at AWS Glue Data Catalog in more detail, as well as how you can attach Athena to your own metastore, even a homegrown one. For now, all you need to know is that Athena requires the use of a metastore to discover key attributes of the data you wish to query. The most common pieces of information that are kept in the Metastore include the following:

- A list of tables that exist

- The storage location of each table (for example, the S3 path or DynamoDB table name)

- The format of the files or objects that comprise the table (for example, CSV, Parquet, JSON)

- The column names and data types in each table (for example, inventory column is an integer, while revenue is a decimal (10,2))

Now that we have a good overview of Amazon Athena, let's look at how to use it in practice.

Obtaining and preparing sample data

Before we can start running our first query, we will need some data that we would like to analyze. Throughout this book, we will try to make use of open datasets that you can freely access but that also contain interesting information that may mirror your real-world datasets. In this chapter, we will be making use of the **NYC Taxi & Limousine Commission's (TLC's)** Trip Record Data for New York City's iconic yellow taxis. Yellow taxis have been recording and providing ride data to TLC since 2009. Yellow taxis are traditionally hailed by signaling to a driver who is on duty and seeking a passenger (also known as a street hail). In recent years, yellow taxis have also started to use their own ride-hailing apps such as Curb and Arro to keep pace with emerging ride-hailing technologies from Uber and Lyft. However, yellow taxis remain the only vehicles permitted to respond to street hails from passengers in NYC. For that reason, the dataset often has interesting patterns that can be correlated with other events in the city, such as a concert or inclement weather.

Our exercise will focus on just one of the many datasets offered by the TLC. The yellow taxis data includes the following fields:

- `VendorID`: A code indicating the TPEP provider that provided the record. 1= Creative Mobile Technologies, LLC; 2= VeriFone Inc.
- `tpep_pickup_datetime`: The date and time when the meter was engaged.
- `tpep_dropoff_datetime`: The date and time when the meter was disengaged.
- `Passenger_count`: The number of passengers in the vehicle.
- `Trip_distance`: The elapsed trip distance in miles reported by the taximeter.
- `RateCodeID`: The final rate code in effect at the end of the trip. 1= Standard rate, 2= JFK, 3= Newark, 4= Nassau or Westchester, 5= Negotiated fare, 6= Group ride.
- `Store_and_fwd_flag`: This flag indicates whether the trip record was held in the vehicle's memory before being sent to the vendor, also known as "store and forward," because the vehicle did not have a connection to the server. Y= store and forward trip, while N= not a store and forward trip.
- `pulocationid`: Location where the meter was engaged.
- `dolocationid`: Location where the meter was disengaged.
- `Payment_type`: A numeric code signifying how the passenger paid for the trip. 1= Credit card, 2= Cash, 3= No charge, 4= Dispute, 5= Unknown, 6= Voided trip.
- `Fare_amount`: The time-and-distance fare calculated by the meter.

- `Extra`: Miscellaneous extras and surcharges. Currently, this only includes the $0.50 and $1 rush hour and overnight charges.

- `MTA_tax`: $0.50 MTA tax that is automatically triggered based on the metered rate in use.

- `Improvement_surcharge`: $0.30 improvement surcharge assessed trips at the flag drop. The improvement surcharge began being levied in 2015.

- `Tip_amount`: This field is automatically populated for credit card tips. Cash tips are not included.

- `Tolls_amount`: Total amount of all tolls paid in a trip.

- `Total_amount`: The total amount charged to passengers. Does not include cash tips.

- `congestion_surcharge`: Amount surcharges associated with time/traffic fees imposed by the city.

This dataset is easy to obtain and is relatively interesting to run analytics against. The inconsistency in field naming is difficult to overlook but we will normalize using a mixture of camel case and underscore conventions later:

1. Our first step is to download the Trip Record Data for June 2020. You can obtain this directly from the NYC TLC's website (`https://www1.nyc.gov/site/ tlc/about/tlc-trip-record-data.page`) or our GitHub repository using the following command:

   ```
   wget https://github.com/PacktPublishing/Serverless-
   Analytics-with-Amazon-Athena/raw/main/chapter_1/yellow_
   tripdata_2020-06.csv.gz
   ```

   If you choose to download it from the NYC TLC directly, please gzip the file before proceeding to the next step.

2. Now that we have some data, we can add it to our data lake by uploading it to Amazon S3. To do this, we must create an S3 bucket. If you already have an S3 bucket that you plan to use, you can skip creating a new bucket. However, we do encourage you to avoid completing these exercises in accounts that house production workloads. As a best practice, all experimentation and learning should be done in isolation.

3. Once you have picked a bucket name and the region that you would like to use for these exercises, you can run the following command:

```
aws s3api create-bucket \
--bucket packt-serverless-analytics \
--region us-east-1
```

> **Important Note**
>
> Be sure to substitute your bucket name and region. You can also create buckets directly from the AWS Console by logging in and navigating to S3 from the service list. Later in this chapter, we will use the AWS Console to edit and run our Athena queries. For simple operations, using the AWS CLI can be faster and easier to see what is happening since the AWS Console can hide multi-step operations behind a single button.

4. Now that our bucket is ready, we can upload the data we would like to query. In addition to the bucket, we will want to put our data into a subfolder to keep things organized as we proceed through later exercises. We have an entire chapter dedicated to organizing and optimizing the layout of your data in S3. For now, let's just upload the data to a subfolder called `tables/nyc_taxi` using the following AWS CLI command. Be sure to replace the bucket name, `packt-serverless-analytics`, in the following example command with the name of your bucket:

```
aws s3 cp ./yellow_tripdata_2020-06.csv.gz \
s3://packt-serverless-analytics/tables/nyc_taxi/yellow_
tripdata_2020-06.csv.gz
```

This command may take a few moments to complete since it needs to upload our roughly 10 MB file over the internet to Amazon S3. If you get a permission error or message about access being denied, double-check you used the right bucket name.

5. If the command seems to have finished running without issue, you can use the following command to confirm the file is where we expect. Be sure to replace the example bucket with your actual bucket name:

```
aws s3 ls s3://packt-serverless-analytics/tables/nyc_
taxi/
```

6. Now that we have confirmed our sample data is where we expect, we need to add this data to our Metastore, as described in the *What is Amazon Athena?* section. To do this, we will use AWS Glue Data Catalog as our Metastore by creating a database to house our table. Remember that Data Catalog will not store our data, just details about where engines such as Athena can find it (for example, S3) and what format was used to store the data (for example, CSV). Unlike Amazon S3, multiple accounts can have databases and tables with the same name so that you can use the following commands as-is, without the need to rename anything. If you already have a database that you would like to use, you can skip creating a new database, but be sure to substitute your database name into subsequent commands; otherwise, they will fail:

```
aws glue create-database \
--database-input "{\"Name\":\"packt_serverless_
analytics\"}" \
--region us-east-1
```

Now that both our data and Metastore are ready, we can define our table right from Athena itself by running our first query.

Running your first query

Athena supports both **Data Definition Language** (**DDL**) and **Data Manipulation Language** (**DML**) queries. Queries where you SELECT data from a table are a common example of DML queries. Our first meaningful Athena query will be a DDL query that creates, or defines, our NYC Taxis data table:

1. Let's begin by ensuring our AWS account and IAM user/role are ready to use Athena. To do that, navigate to the Athena query editor in the AWS Console: https://console.aws.amazon.com/athena/home.

    Be sure to use the same region that you uploaded your data and created your database in.

2. If this is your first time using Athena, you will likely be met by a screen like the following. Luckily, Athena is telling us that *"Before you run your first query, you need to set up a query result location in Amazon S3..."*. Since Athena writes the results of all queries to S3, even DDL queries, we will need to configure this setting before we can proceed. To do so, click on the highlighted text in the AWS Console that's shown in the following screenshot:

Figure 1.1 – The prompt for setting the query result's location upon your first visit to Athena

3. After clicking on the modal's link, you will see the following prompt so that you can set your query result's location. You can use the same S3 bucket we used to upload our sample data, with `results` being used as the name of the folder that Athena will write query results to within that bucket. Be sure your location ends with a "/" to avoid errors:

Settings

Settings apply by default to all new queries. Learn more

Workgroup: primary

Query result location	s3://YOUR_BUCKET_NAME]results/ ❶
	Example: s3://query-results-bucket/folder/
Encrypt query results	☐ ❶
Autocomplete	☐ ❶

Cancel   Save

Figure 1.2 – Athena's settings prompt for the query result's location

Next, let's learn how to create a table.

Creating your first table

It is now time to run our first Athena query. The following DDL query asks Athena to create a new table called `nyc_taxi` in the `packt_serverless_analytics` database, which is stored in the AWS Glue Data Catalog. The query also specifies the schema (columns), file format, and storage location of the table. For now, the other nuances of this create query are unimportant. You may find it easier to copy `create table` from the `create_nyc_taxi.sql` (`http://bit.ly/3mXj3K0`) file in the `chapter_1` folder of this book's GitHub repository. Paste it into Athena's query editor, change `LOCATION` so that it matches your bucket name, and click **Run query**. It should complete in a few seconds:

```
CREATE EXTERNAL TABLE 'packt_serverless_analytics'.'nyc_taxi'(
  'vendorid' bigint,
  'tpep_pickup_datetime' string,
  'tpep_dropoff_datetime' string,
  'passenger_count' bigint,
  'trip_distance' double,
  'ratecodeid' bigint,
  'store_and_fwd_flag' string,
  'pulocationid' bigint,
  'dolocationid' bigint,
  'payment_type' bigint,
  'fare_amount' double,
  'extra' double,
  'mta_tax' double,
  'tip_amount' double,
  'tolls_amount' double,
  'improvement_surcharge' double,
  'total_amount' double,
  'congestion_surcharge' double)
ROW FORMAT DELIMITED
  FIELDS TERMINATED BY ','
STORED AS INPUTFORMAT
  'org.apache.hadoop.mapred.TextInputFormat'
OUTPUTFORMAT
  'org.apache.hadoop.hive.ql.io.HiveIgnoreKeyTextOutputFormat'
LOCATION
```

```
    's3://<YOUR_BUCKET_NAME>/tables/nyc_taxi/'
TBLPROPERTIES (
  'areColumnsQuoted'='false',
  'columnsOrdered'='true',
  'compressionType'='gzip',
  'delimiter'=',',
  'skip.header.line.count'='1',
  'typeOfData'='file')
```

Once your table creation DDL query completes, the left navigation pane of the Athena console will refresh with the definition of your new table. If you have other databases and tables, you may need to choose your database from the dropdown before your new table will appear.

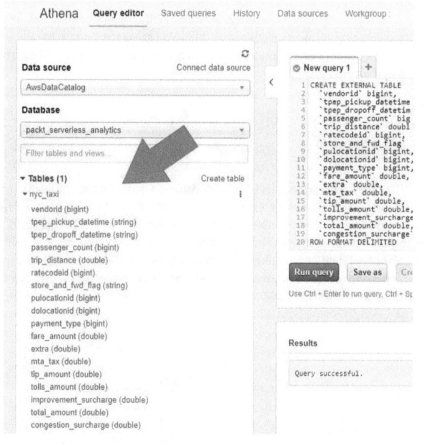

Figure 1.3 – Athena's Database navigator will show the schema of your newly created table

At this point, the significance of the query we just ran may not be entirely apparent, but rest assured we will go deeper into why serverless DDL queries are a powerful thing. Oh, and did we mention that Athena does not charge for DDL queries?

Running your first analytics queries

When working with a new or unfamiliar set of data, it can be helpful to view a sample of the rows before exploring the dataset in more meaningful ways. This allows you to understand the schema of your dataset, including verifying that the schema (for example, column names) match the values and types. There are a few ways to do this, including the following limit query:

```
SELECT * from packt_serverless_analytics.nyc_taxi limit 100
```

This works fine in most cases, but we can do better. Many query engines, Athena included, will end up returning all 100 rows requested in the preceding query from the same S3 object. If your dataset contains many objects or files, you are getting an extremely narrow view of the table. For that reason, I prefer using the following query to view data from a broader portion of the dataset:

```
SELECT *
FROM packt_serverless_analytics.nyc_taxi TABLESAMPLE BERNOULLI
(1)
limit 100
```

This query is like the earlier limit query but uses Athena's TABLESAMPLE feature to obtain our 100 requested rows using BERNOULLI sampling. When a table is sampled using the Bernoulli method, all the objects of the table may be scanned as opposed to likely stopping after the first object. This is because the probability of a row being included in the result is independent of any other row reducing the significance of the object scan order. In the following screenshot, we can see some of the rows that were returned using TABLESAMPLE with the BERNOULLI method:

	vendorid	tpep_pickup_datetime	tpep_dropoff_datetime	passenger_count	trip_distance	ratecodeid	store_and_fwd_flag	pul
1	2	2020-06-01 00:08:05	2020-06-01 00:21:33	1	3.41	1	N	
2	1	2020-06-01 01:41:39	2020-06-01 02:01:44	1	7.6	1	N	
3	2	2020-06-01 03:00:53	2020-06-01 03:09:52	1	2.13	1	N	
4	1	2020-06-01 03:08:27	2020-06-01 03:10:38	1	0.7	1	N	
5	2	2020-06-01 06:00:08	2020-06-01 06:01:43	2	0.61	1	N	
6	2	2020-06-01 06:42:11	2020-06-01 06:54:49	1	3.41	1	N	
7	1	2020-06-01 06:39:50	2020-06-01 06:48:05	1	3.7	1	N	
8	1	2020-06-01 06:25:16	2020-06-01 06:34:37	1	3.4	1	N	
9	2	2020-06-01 06:35:36	2020-06-01 06:46:24	1	2.03	1	N	
10	2	2020-06-01 06:35:49	2020-06-01 06:43:31	1	2.2	1	N	
11	1	2020-06-01 07:12:52	2020-06-01 07:16:14	1	1.0	1	N	
12	1	2020-06-01 07:57:04	2020-06-01 08:11:15	1	7.7	1	N	
13	1	2020-06-01 07:12:45	2020-06-01 07:18:15	1	1.0	1	N	
14	1	2020-06-01 07:27:43	2020-06-01 07:29:53	1	0.8	1	N	

Figure 1.4 – Results of executing TAMPLESAMPLE against our nyc_taxi table

While that query allowed us to confirm that Athena can indeed access our data and that the schema appears to match the data itself, we have not extracted any real insights from the data. For this, we will run our first real analytics query by generating a histogram of ride durations and distances. Our goal here is to learn how much time people are typically spending in taxis, but we'll also be able to gain insights into the quality of our data. The following query uses Athena's `numeric_histogram` function to approximate the distribution with 10 buckets according to the difference between `tpep_pickup_datetime` and `tpep_dropoff_datetime`. Since the dataset stores datetimes as strings, we are using the `date_parse` function to convert the values into actual timestamps that we can then use with Athena's `date_diff` function to generate the ride durations as minutes. Lastly, the query uses a CROSS JOIN with UNEST to turn the histogram into rows and columns. Normally, the `numeric_histogram` function returns a map containing the histogram, but this can be difficult to read. UNEST helps us turn it into a more intuitive tabular format. Do not worry about remembering all these functions and SQL techniques right now. Athena frequently adds new capabilities, and you can always consult a reference.

You can copy the following code from GitHub at `http://bit.ly/2Jm6o5v`:

```
SELECT ride_minutes, number_rides
    FROM (SELECT numeric_histogram(10,
        date_diff('minute',
            date_parse(tpep_pickup_datetime,'%Y-%m-%d %H:%i:%s'),
```

```
        date_parse(tpep_dropoff_datetime, '%Y-%m-%d %H:%i:%s')
        )
    )
FROM packt_serverless_analytics.nyc_taxi ) AS x (ride_
histogram)
CROSS JOIN
    UNNEST(ride_histogram) AS t (ride_minutes, number_rides);
```

Once you run the query, the results will look as follows. You can experiment with the number of buckets that are generated by adjusting the parameters of the numeric_ histogram function. Generating 100 or even 1,000 buckets can uncover patterns that were hidden with fewer buckets. Even with just 10 buckets, we can already see a strong correlation between the distance and the number of rides. I was surprised to see that such a large portion of the yellow cab rides lasted less than 7 minutes. From this query, we can also see some likely data quality issues in the dataset. Unless one of the June 2020 rides happened in a time-traveling DeLorean, we likely have an erroneous record. Less obvious is the fact that several hundred rides claim to have lasted longer than 24 hours:

ride_minutes	number_rides
-531231.0	1.0
6.328061	356705.0
16.226482	123621.0
29.021547	56296.0
51.916306	12283.0
518.5143	70.0
995.4878	41.0
1408.4696	741.0
2687.0	1.0
4497.0	1.0

Figure 1.5 – Ride duration histogram results

Let's try one more histogram query, but this time, we will target the trip distance of the rides that took less than 7 minutes. The following code block contains the modified histogram query you can run to understand that bucket of rides. You can download it from GitHub at http://bit.ly/3hkggJl:

```
SELECT trip_distance, number_rides
FROM
    (SELECT numeric_histogram(5,trip_distance)
        FROM packt_serverless_analytics.nyc_taxi
    WHERE date_diff('minute',
        date_parse(tpep_pickup_datetime,'%Y-%m-%d %H:%i:%s'),
        date_parse(tpep_dropoff_datetime, '%Y-%m-%d %H:%i:%s')
        ) <= 6.328061
    ) AS x (ride_histogram)
CROSS JOIN UNNEST(ride_histogram) AS t (trip_distance , number_
rides);
```

Considering that the average person can walk a mile in 15 minutes, New Yorkers must be in a serious hurry to opt for taxi rides instead of a 15-minute walk!

trip_distance (miles)	number_rides
0.2919	31882.0
0.8105	98200.0
1.4027	56680.0
2.5416	5788.0
56.5666	3.0

Figure 1.6 – Ride distance histogram results

With that, we've been through the basics of AWS Athena. Let's conclude by providing a recap of what we've learned.

Summary

In this chapter, you saw just how easy it is to get started running queries with Athena. We obtained sample data from the NYC TLC, used it to create a table in our S3-based data lake, and ran some analytics queries to understand the insights contained in that data. Since Athena is serverless, we spent absolutely no time setting up any infrastructure or software. Incredibly, all the operations we ran in this chapter cost less than $0.00135. Without the *serverless* aspect of Athena, we would have found ourselves purchasing many thousands of dollars of hardware or hundreds of dollars in cloud resources to run these basic exercises.

While the main goals of this chapter were to orient you to the uniquely serverless experience of using Amazon Athena, there are a few concepts worth remembering as you continue reading. The first is the role of the Metastore. We saw that uploading our data to S3 was not enough for Athena to query the data. We also needed to register the location, schema, and file format as a table in AWS Glue Data Catalog. Once our table was defined, it became queryable from Athena. *Chapter 3, Key Features, Query Types, and Functions*, will cover this topic in greater depth.

The next important thing we saw was the feature-rich SQL dialect we used in our basic analytics queries. Since Athena utilizes a customized variant of Presto, you can refer to Presto's documentation (`https://prestodb.io/docs/current/`) as a supplement for Athena's documentation.

Chapter 2, Introduction to Amazon Athena, will go deeper into Athena's capabilities and open source roots so that you can understand when to use Athena, as well as how you can gain deeper insight into specific behaviors of the service.

2
Introduction to Amazon Athena

The previous chapter walked you through your first, hands-on experience with serverless analytics using **Amazon Athena**. This chapter will continue that introduction by discussing Athena's capabilities, scalability, and pricing in more detail. In the past, vendors such as Oracle and Microsoft produced mostly one-size-fits-all analytics engines and RDBMSes. Bucking the historical norms, AWS has championed a *fit for purpose* database and analytics strategy. By optimizing for specific use cases, the analytics engines' very architecture could exploit nuances of the workload for which they were intended, thereby delivering an all-around better product. For example, Redshift, EMR, Glue, Athena, and Timestream all offer related but differentiated capabilities with their own unique advantages and trade-offs. The knowledge you will gain in this chapter provides a broad-based understanding of what functionality Athena offers as well as a set of criteria to help you determine whether Athena is the best service for your project. We will also spend some time peeling back the curtain and discussing how Athena builds upon Presto, an open source SQL engine initially developed at Facebook.

Most of the chapters in this book stand on their own and allow you to skip around as you follow your curiosity. However, we do not recommend skipping this chapter unless you already know Athena well and are using this book to dive deep into specific topics.

In the subsequent sections of this chapter, we will cover the following topics:

- Getting to know Amazon Athena
- What is Presto?
- Understanding scale and latency
- Metering and billing
- Connecting and securing
- Determining when to use Amazon Athena

Technical requirements

This chapter is one of the few, perhaps even the only chapter in this book, that will not have many hands-on activities. As such, there are not any specific technical requirements for this chapter beyond those already covered in *Chapter 1*, *Your First Query*, namely:

- Basic knowledge of SQL is recommended but not required.
- A computer with internet access to GitHub, S3, and the AWS Console; a Chrome, Safari, or Microsoft Edge browser; and the AWS CLI installed.
- An AWS account and IAM user that can run Athena queries.

As always, any code references or samples for this chapter can be found in the book's companion GitHub repository located at `https://github.com/PacktPublishing/Serverless-Analytics-with-Amazon-Athena`.

Getting to know Amazon Athena

In *Chapter 1*, *Your First Query*, we learned that Amazon Athena is a query service that allows you to run standard SQL over data stored in various sources and formats. We also saw that Athena's pricing model is unique in that we are charged by how much data our query reads and not by how many servers or how much time our queries require. In this section, we will go beyond that cursory introduction and discuss the broader set of capabilities that together make Athena a product worth considering for your next analytics project. We do not go into full detail on every item we are preparing to discuss, but later chapters will allow you to get hands-on with the most notable features. For now, our goal is to increase your awareness of what is possible with Athena, so you can perform technical product selection exercises (aka bakeoffs) or steer toward areas of interest.

Understanding the "serverless" trend

The word **serverless** appears dozens, possibly hundreds of times, in this book. At the end of the book, we will run an Athena query over the complete text to find the exact number of times we used the word serverless. So, what is the big deal? Why is serverless such a seemingly important concept? Or is it just the latest buzzword to catch on? Like most things, the truth lies somewhere between the two extremes, and that's why we will spend some time understanding what it means to be serverless.

In the simplest terms, a serverless offering is one where you do not have to manage any servers. **AWS Lambda** is often thought of as the gold standard for serverless technologies since it was the first large-scale offering of this type. With AWS Lambda, you have virtually no boilerplate to slow you down; you literally jump straight into writing your business logic or *function* as follows:

```
def lambda_handler(event, context):
    return {
        "response": "Hello World!"
    }
```

AWS Lambda will handle executing this code in response to several invocation triggers, ranging from SQS messages to HTTP calls. As an AWS Lambda customer, you do not have to worry about setting up Java, a WebService stack, or anything. Right from the beginning, you are writing business logic and not spending time on undifferentiated infrastructure work.

This model has some obvious advantages that customers love. The first of which is that, without servers, your capacity planning responsibilities shrink both in size and complexity. Instead of determining how many servers you need to run that monthly finance report or how much memory your SQL engine will need to handle all the advertising campaigns on Black Friday, you only need to worry about your account limits. To the uninitiated, this might seem easy. You might even say to yourself, *I have great metrics about my peak loads and can do my own capacity planning just fine!* It is true. You will likely have more context about your future needs than a service like Athena can infer. But what happens to all that hardware after the peak has passed? I am not just referring to that seasonal peak that comes once a year but also the peak of each week and each hour. That hardware, which you or your company paid for, will be sitting idle, taking up space in your data center, and consuming capital that could have been deployed elsewhere. *But what about the cloud? I do not need to buy any servers; I can just turn them on and off as needed.* Yes! That is true.

So, let's go down the rabbit hole a bit more. Suppose we used EC2 instances instead of classic servers in our own data centers. We can undoubtedly scale up and down based on demand. We might even be able to use EC2 AutoScaling to add and remove capacity based on a metric such as CPU usage. This is a good start, and AWS encourages customers to take advantage of these capabilities to drive down costs and improve performance. Should you run this infrastructure fully on-demand or use some mix of reserved instances? On-demand capacity has no up-front expenses and grants you the flexibility to turn it on and off whenever you like. Reserved capacity is more expensive up-front, but it is guaranteed to be there, unlike on-demand, which is first-come-first-served. Or perhaps you are advanced and can take advantage of EC2 Spot instances, which are often available at a 90% discount but can be taken from you at a moment's notice if EC2 needs the capacity elsewhere.

The journey does not end here. Suppose you built an autoscaling infrastructure that reacts to changes in demand, like the one we whiteboarded thus far. In that case, you know that generating demand forecasts, capacity forecasts, calculating ROI on CapEx, and then actually starting and stopping servers on the fly is only the beginning. Your application needs to be capable of running on an infrastructure that is continuously changing shape. For classic web services, simple request-reply systems, a single instance receives and responds to each customer request. There may not be much work to adapt such an application to this brave new world. In fact, **AWS Fargate** is an excellent example of how well most containerized workloads can *just work* in the serverless world. For analytics applications, adapting to serverless infrastructure gets trickier. Even a simple query like the following one may enlist the combined computational power from dozens of instances to help read the raw data, filter relevant records, aggregate the results, perform the sort, and finally generate the output:

```
SELECT sum(col1) as mysum FROM my_table WHERE col3 > 10 ORDER
BY mysum
```

If our elastic infrastructure wants to scale down to reduce waste during idle periods, how does it know which instances it can safely turn off? This is not purely an infrastructure problem. In the case of distributed analytics applications such as **Apache Spark** or **Presto**, the application has an inbuilt scheduler that dispatches work to the infrastructure. In this context, work might be reading a file from S3, filtering a batch of rows, or any number of other operations required to complete your query. When assigning this work, the scheduler has multiple choices for choosing which instances the task will run on. For example, the scheduler can choose to place as few concurrent units of work on each host as possible. This is commonly described as *going wide* and can offset adverse effects associated with contention caused by a noisy neighbor process. Alternatively, the scheduler can choose to co-locate units of work to improve utilization or reduce the overhead associated with network communication.

Simple metrics such as CPU or MEMORY usage will not tell the story of how a distributed analytics engine is using (or not using) the underlying compute instances. Solving this problem well is extremely difficult. Even a mediocre solution requires integration between the analytics engine itself and the infrastructure.

Noisy neighbors

When one workload, process, or application negatively affects a neighboring process running on the same shared resource, we refer to the offending process as a noisy neighbor. If the people in the apartment above you or the house across the street played loud music deep into the night, it would disturb your ability to go about your activities. It's the same for workloads in a multi-tenant system. If the system doesn't provide strong isolation between workloads, those workloads may interact in undesirable ways.

Beyond "serverless" with 'fully managed' offerings

By now, you hopefully have a much better understanding of why the industry, cloud providers, and customers alike are rushing to build and use serverless offerings. While the word **serverless** probably seems a bit self-describing at this point, we've yet to discuss what is arguably the more meaningful benefit of many serverless offerings, including Athena. We often refer to Athena as a **Fully Managed** service because it handles far more than the vision of automated infrastructure management we mentioned earlier.
The Athena service is also responsible for the configuration, performance, availability, security, and deployments of the underlying analytics engine. When talking about Athena with prospective customers, I tend to use three scenarios to convey the benefits of using fully managed offerings.

Analytics engines such as Apache Spark, Presto, and traditional RDBMSes frequently implement multiple approaches for executing your query. You may even have heard of these engines producing logical and physical query plans. These plans result from applying a series of rules and statistics to your query before deciding the fastest way to get you a result. For example, suppose your query is joining two tables. In that case, the engine can choose between a broadcast join, which exploits the relative size of the two tables, or a fully distributed join, which can scale to larger sizes but takes longer to complete. The critical optimization in the broadcast join is that if one of the two tables is small enough to *broadcast* to every instance participating in the query, then each instance can operate independently, with less data shuffling and associated communication overhead. Being fully managed, Athena has the responsibility to determine an appropriate memory limit, beyond which broadcast joins are not reliable or underperform due to memory constraints.

Athena could also decide that it should raise the available memory in its fleet by adding more hosts or hosts of a different type to increase the broadcast join limit for a particular query that will significantly benefit from it. Athena's actual approach to join optimization is not publicly documented, but the point we are illustrating is that this is no longer your challenge to solve. The hundreds, or in some cases thousands, of tuning parameters available in these algorithms are squarely in the hands of Athena. In the next chapter, we will touch on Athena's automatic engine upgrades and self-tuning capabilities.

This is an excellent segue into the second differentiator for fully managed offerings. With Athena, you do not have to worry about deploying new versions of the analytics engine. If you run your own Spark or Presto cluster on servers, and even if you run them in AWS Fargate, you'll need to handle deploying updates to get bug fixes, new features, and security updates. On the surface, this might seem straightforward. After all, you did set it up the first time. Deploying updates on an ongoing basis to a live system is more complicated. How do you avoid downtime? How do you handle rollbacks? How do you know the new version is backward compatible or what changes your queries need to succeed on the latest version?

In 2020, Athena publicly announced the self-tuning technology used internally to manage upgrades of its Presto fleet. To ensure seamless upgrades, Athena is continually running your queries on varying versions of its engine with numerous configurations allowing Athena to identify the best settings for each query. It also means Athena knows when a new version of Presto, its underlying engine, is or isn't safe for your workload. As a fast-moving open source project, Presto does not always ensure backward compatibility before cutting a new release. Athena allows you to experiment with new versions before you are auto upgraded or roll back to a previous version with the click of a button. You can even perform targeted upgrades or downgrades of specific queries! You do not need to worry about having a fleet of the old and a new fleet while simultaneously updating apps to point at one or the other.

The third and final scenario centers around availability. If you are running your analytics engine on EC2 or Fargate, you've likely encountered scenarios where the infrastructure was running, but your queries fail in a seemingly random fashion. After the number of initially uncorrelated user complaints mount, you finally register that something strange is happening. Some instances of your engine, executors in Spark parlance, and workers in Presto nomenclature, seem to have a higher error rate than their peers. You are facing a classic gray failure. The root causes can vary from slow resource leaks to noisy neighbors, but identifying them can be challenging because they often masquerade as a user error. If you use long-lived clusters, this problem becomes even more prevalent. You will find yourself rejuvenating instances periodically by restarting or tracking per-instance success metrics to find outliers that need to be removed from service. As a managed service, Athena owns this in addition to the easier availability problems where an instance is entirely unresponsive and requires replacement.

As you can see, there is a non-trivial amount of infrastructure work and capital that are required to ensure your applications have the compute capacity ready when customers click the button. For all the benefits of using a fully managed, serverless offering, there are also drawbacks. Suppose your functional, performance, or other needs diverge from Athena's roadmap. In that case, you may find yourself needing to build significant pieces of infrastructure just to gain enough control to affect the relatively small change you wanted. This is generally only a meaningful point of consideration for large or sophisticated customers who have both the ability to build their own solution or whose use cases are outliers compared to Athena's target audience. The good news is that AWS's customer obsession is world-renowned, so Athena is incentivized to continually add features and improve performance as part of their strategy to remain a great place to run your analytics workloads. These reasons are precisely why so many customers love Athena.

Obsessing over customers

You've probably noticed our tendency to mention AWS as being customer-obsessed. This notion comes from one of *Amazon's leadership principles*, which states: *"Leaders start with the customer and work backward. They work vigorously to earn and keep customer trust. Although leaders pay attention to competitors, they obsess over customers."* This philosophy drives everything AWS does. You can learn more about the Amazon leadership principles by reviewing the links at the end of this chapter.

Key features

Thus far, we have spent a lot of time discussing the unique advantages that come with Athena's promise of serverless analytics. Now we will go through the compelling analytics features that Athena offers. While reading this section, keep in mind that our objective is to build an awareness of these capabilities. As such, the descriptions will be high-level and intentionally simplified so as not to overwhelm you while we build up to the more advanced sections of this book. Later chapters will guide us through getting hands-on with many of the features we are about to review.

Statement types

Athena supports several different statement types, including DDL and DML. **Data Definition Language (DDL)** statements allow you to interact with your Data Lake's metadata by defining tables and updating those tables' schema or properties. You can also use these statements to add or modify the partitions in your tables. Customers commonly use these statements to ingest new data into their Data Lake. **Data Manipulation Language (DML)** statements allow you to interact with your Data Lake's actual data.

SELECT queries are the most used DML statement type in Athena and can be combined with Create-Table-As-Select (CTAS) statements to create new tables. Like CTAS, INSERT INTO statements can be used along with SELECT to add data to an existing table. Both CTAS and INSERT INTO queries can automatically add new partitions to your metastore, eliminating the need for you to manage partitions manually. While not traditionally a statement type, Athena's TABLESAMPLE feature acts as a modifier in your SELECT statements by instructing the query planner to only consider a subset of the data your query would normally scan. This can be helpful when scanning the full dataset would be too costly or take too long. There are two different sampling techniques available. In *Chapter 1, Your First Query*, we used the BERNOULLI technique, which considers each row in the input table individually. The SYSTEM sampling method is a more coarse-grained sampling technique that groups rows into batches and then considers each batch for inclusion in the query. The batches may be one-to-one with an S3 object or, depending on the file format, aligned to a chunk of rows. BERNOULLI can offer less observation bias than SYSTEM sampling but is often much slower.

The SQL dialect

Athena SQL is ANSI SQL-compliant. Notable variances from ANSI SQL include extensions to better support complex types such as MAPs, STRUCTs, and LISTs. This means you can use all your favorite JOIN types and window functions. You can even craft those oh so easy-to-understand, deeply nested queries. In all seriousness, Athena SQL does have a mechanism to improve the readability of such statements. The WITH-CLAUSE syntax allows you to extract and essentially parameterize the nested sub-queries, making the original statement far easier to digest. We will see some first-hand examples of this later, and you can find more details in the Athena documentation referred to in the *Further reading* section at the end of this chapter.

The specific syntax and available functions vary slightly, depending on which **Engine Version** you are using. Thanks to Athena's auto-upgrade functionality, most customers never realize that Athena supports multiple engine versions or dialects. That is because changes are typically additive, and the few breaking changes that do occur can be handled query by query, so you never see a failure. Athena presently supports two engine versions:

- Athena version 1 is based on Presto 0.172
- Athena engine version 2 was released in December 2020 and is based on Presto 0.217

Unless you have a specific reason to use the older version, you should use Athena engine version 2 or later as it runs up to 30% faster than engine version 1 and includes dozens of new functions.

On the DDL front, Athena uses a subset of HiveQL syntax for managing everything from tables to partitions. The complete list of supported DDL operations can be found in the official Athena documentation, but rest assured that it includes everyday operations such as CREATE TABLE, ALTER TABLE, CREATE VIEW, SHOW, and DROP.

Support for Hive-compliant metastores

In addition to the out-of-the-box support for the AWS Glue Data Catalog, Athena allows you to bring your own **Hive-compliant** metastore. This can help you already run your own Hive metastore for use with other applications, or if you do not intend to use AWS Glue Data Catalog. Customers also use this facility for integrating Athena with their home-grown metadata systems. To attach Athena to your metastore, you provide Athena with a Lambda function to call for all metadata operations. For example, when Athena needs to get the columns and types in a given table, it will contact the Lambda function you provide. The Lambda function should be capable of interfacing with your actual metastore and providing an appropriate response to Athena. Athena expects the Lambda function to support Hive's Thrift protocol and the Athena team provides a ready-made Lambda function capable of talking to your Hive metastore. You can find more details on this feature in *Chapter 4, Metastores, Data Sources, and Data Lakes*, as well as in the official Athena documentation linked from the *Further reading* section at the end of this chapter.

When used with Lake Formation's new ACID transaction capabilities, these form powerful building blocks for any analytics application.

Supported file formats

Amazon Athena supports common file formats such as **CSV**, **TSV**, and **AVRO** in addition to more advanced columnar storage formats, including **Apache ORC** and **Apache Parquet**. You can also query unstructured or semi-structured files in **Textfile** and **JSON** format. The preceding formats can be combined with Snappy, Zlib, LZO, or GZIP compression to reduce file size and cost while improving scan performance. This is notable because Athena charges based on compressed data size. This means that if your data is originally 100 GB, but it compresses down to 10 GB, you will only be charged for 10 GB if you read all the data from an Athena query.

ACID transactions

While Amazon S3 is the world's most popular store for building data lakes, the immutability that contributes to its scalability also creates challenges for use cases that have concurrent readers and writers or need to update existing data. Put another way, this means that if you want to modify or delete 1 row that happens to reside in an object that contains 1,000,000 rows, you will need to read all 1,000,000 rows and then overwrite that original S3 object with a new object containing the original 999,999 rows plus your one new row. This write amplification is a significant scaling challenge. You might be thinking, *thanks for telling me. I can simply avoid updating existing rows.* That would have been a reasonable strategy, but new regulations are making that approach less practical. For example, the European Union's new **General Data Protection Regulation (GDPR)** requires companies to purge data about specific customers upon request. This is worth repeating. GDPR likely requires you to delete data pertaining to individual customers no matter where it resides in your data lake. That could mean deleting a single row from every S3 object in your many petabytes of data.

Similarly, customers are increasingly moving to near real-time data ingestion using technologies such as **Kafka** and **Amazon Kinesis**. These applications reduce the time it takes for new data to become available in your data lake (and therefore your analytics queries) but create many small files. These tiny files can quickly degrade performance for analytics systems such as Athena, Spark, and even Redshift Spectrum because of the increased overhead associated with each read operation. To balance the need for data to become available in a few minutes or a few seconds in extreme cases, customers find themselves running periodic compaction jobs that read the small files, merge them into larger files, and then delete the original small files. However, if you attempt such compaction while also running a query, you will likely see incorrect results or fail. This is because your reader might have processed small file #1, and then your compaction job writes a new file containing the contexts of file #1 through file #100. Your reader might then also read that new file, resulting in duplicate data in your query! It is also possible that your reader will decide it needs to read a file, and the compaction job will delete that file between the reader deciding the file needs to be read and reading it. This will result in a query failure for most engines.

This is where ACID transactions can help. Athena supports Lake Formation transactions for snapshot isolation between any number of concurrent readers and writers. This integration also provides automatic background compaction of small files, among other accelerations. We will cover these topics in detail as part of *Chapter 14, Lake Formation – Advanced Features*. In addition to Lake Formation transactions, Athena also offers partial support for HUDI copy-on-write tables and Delta Lake. Hudi was developed by Uber and primarily attempts to address the consistency and performance concerns emerging from update operations.

Delta Lake is produced and maintained by Databricks as part of their Spark offering. Support in Athena comes from **SymlinkTextInputFormat**, as defined in the Delta Lake documentation linked in the *Further reading* section of this chapter. This provides read-only access to Delta Lake tables from engines that do not natively support Delta Lake's format.

Readers may be happy to learn that this is a rapidly evolving area for Amazon Athena, and we have had to update this section of the book *three* times since we started writing. This is notable because, as you choose technology for your project or company, you want to select ones growing along the dimensions you care about most.

Self-tuning and auto-upgrades

When I think about the most frustrating projects in my career, many of them were related to upgrading software or finding the right combination of cryptic settings to achieve the advertised performance we had been sold on. With Athena, you do not have to concern yourself with either of these responsibilities. It is, however, useful to understand Athena's approach to these disciplines. Other offerings require you to pick the version of the software you want to use. With Athena, you can choose whether or not to use specific versions to get early access to new features. At any time, you can also enable auto-upgrade to have Athena continuously monitor your queries for the best combination of settings and software. It is not uncommon for analytics vendors to publish their TPCH and TPCDS performance results in their marketing materials. These industry benchmarks are crafted by TCP and use a mix of query patterns common in data science and other prototypical workloads. The resulting performance numbers can be used as a decision support tool. Unfortunately, many vendors overfit these exact tests, resulting in solutions that do not perform well for use cases that don't closely match the industry benchmark. Since Athena learns from your specific workloads, you can expect it to do well both in cases where your workloads follow well-known industry patterns and when you're running that oddball query for a new idea you had.

Federation and extensibility

One of my favorite Athena features is **Athena Query Federation**, with just a small fraction of my enthusiasm stemming from my personal involvement in its development. Athena Federation allows you to extend Athena with your own custom data sources and functionality. The Athena Federation SDK and many of the data source connectors are 100% open source and are available on GitHub. We've included a link to the GitHub repository in the *Further reading* section at the end of this chapter. A growing community is contributing to its development, with several integration partners joining the Athena team in publishing connectors and UDFs to the **AWS Serverless Application Repository** where you can 1-click deploy them. There are more than 30 available data sources, including 14 open source connectors provided by the Athena team, including:

- **Amazon Timestream**: This connector enables Amazon Athena to communicate with Timestream, making your time series data accessible from Athena. A great use case would be identifying anomalous IoT devices in Timestream and joining those with details of the site that houses the sensor from elsewhere.

- **Amazon Neptune**: This connector enables Amazon Athena to communicate with your Amazon Neptune instance(s), making your graph data accessible from SQL. This connector has a unique way of translating vertices and relationships to tables.

- **Amazon DynamoDB**: This connector enables Amazon Athena to communicate with DynamoDB, making your DDB tables accessible from SQL.

- **Amazon DocumentDB**: This connector enables Amazon Athena to communicate with your DocumentDB instance(s), making your DocumentDB data accessible from SQL. The also works with any MongoDB-compatible endpoint.

- **Elasticsearch**: This connector enables Amazon Athena to communicate with your Elasticsearch instance(s), making your Elasticsearch data accessible from SQL.

- **HBase**: This connector enables Amazon Athena to communicate with your HBase instance(s), making your HBase data accessible from SQL.

- **JDBC**: This connector enables Amazon Athena to access your JDBC-compliant database. At launch, this connector supports MySQL, Postgres, and Redshift. For the latest list, check the connector's README.md.

- **Redis**: This connector enables Amazon Athena to communicate with your Redis instance(s), making your Redis data accessible from SQL.

- **CloudWatch Logs**: This connector enables Amazon Athena to communicate with CloudWatch, making your log data accessible from SQL.

- **CloudWatch Metrics**: This connector enables Amazon Athena to communicate with CloudWatch metrics, making your metrics data accessible from SQL.

- **AWS CMDB**: This connector enables Amazon Athena to communicate with various AWS services (EC2, RDS, EMR, S3, and so on). Using this connector, you could run a query to identify all the EC2 instances in a particular VPC. Yes, you could do this using the EC2 API, but with this connector, you can use one API, Athena SQL, to query many different resource types.

- **TPC-DS**: This connector enables Amazon Athena to communicate with a source of randomly generated TPC-DS data for use in benchmarking and functional testing.

Unstructured and semi-structured data

Athena's support for a wide range of file formats, rich text, and JSON manipulation functions, as well as support for custom UDFs, make it an excellent choice for analyzing unstructured and semi-structured data. Whether you are trying to count the number of Tweets with negative sentiment in the previous hour (spoiler, the answer is all of them) or use the **Levenshtein distance** to correlate log lines, Athena can help you generate that result. We will go through a few examples of using unstructured and semi-structured data with Athena in *Chapter 8, Querying Unstructured and Semi-Structured Data.*

> **The Levenshtein distance**
>
> The Levenshtein distance is a handy technique for performing fuzzy matching between strings, including spelling errors, variations in spacing or punctuation, and other differences that are challenging to classify. It is named after the Soviet mathematician **Vladimir Levenshtein** who first described the algorithm for quantifying the difference or similarity between two strings. The approach counts the minimum number of single-character edits (insertions, deletions, or substitutions) required to change one word into the other. You might be surprised to learn that the Levenshtein distance is part of many systems capable of fuzzy matching to accomplish that feat, including the search mainstay Elasticsearch. You can use this algorithm directly from an Athena query from the built-in `levenshtein_distance(string, string)` function.

Built-in functions

Since Amazon Athena is based on Presto, it shares many of the same functions. These functions range from standard string or timestamp manipulation capabilities common in many databases to more advanced geospatial functions. You can find the full list of functions, grouped by type, in the Athena documentation (`http://amzn.to/2KoHAKE`), and I'm sure you'll find it to be a close match for Presto's documentation (`http://bit.ly/3nKaHFS`).

This is perhaps a great time to shift gears for our next topic, where we will peel back the curtain just a bit and talk about how Athena works under the hood. Much of that conversation will focus on Presto and its architecture.

What is Presto?

As we have mentioned a few times already, Athena is based on a fork of the Presto open source project. By understanding Presto, what it is, and how it works, we can gain greater insight into Athena.

Presto is a distributed SQL engine designed to provide response times in the order of seconds for interactive data analysis. While it may be tempting to do so, it is essential not to confuse Presto with a database or data warehouse as Presto has no storage of its own. Instead, Presto relies on a suite of **connectors** to plug in different storage systems such as HDFS, Amazon S3, RDBMS, and many other sources you may wish to analyze. This simple but inventive approach allows Presto to offer the same consistent SQL interface regardless of where your data lives. It's also why Athena claims that *"there is no need for complex ETL jobs to prepare your data for analysis."*

If you have an existing data lake, you may be familiar with Apache Hive or Hadoop tools. Presto was, in part, intended as a high-performance alternative to the Hadoop ecosystem for queries requiring interactive performance on data ranging in size from gigabytes to many terabytes. The evolutionary pressure exerted on Presto by Hive has its roots at Facebook, where both analytics tools were created and later open sourced. As of 2012, the last time Facebook published these figures, Facebook's Hive data warehouse had reached a staggering 250 petabytes in size. Having architecture limitations and lacking the right code-level abstractions to meaningfully scale Hive and its shared Hive infrastructure beyond the tens of thousands of daily queries it already handled, the engineers at Facebook sought a fresh start in creating Presto. The inertia of the existing 250+ petabytes of HDFS data and the emergence of other, siloed data stores across Facebook influenced the critical architecture decision to separate storage and compute in Presto. Naturally, one of the first and most mature Presto connectors was the Hive connector. This allowed Presto's new distributed SQL engine to access the wealth of existing data without taking on the effort of migrating the data itself. In 2013, roughly a year after the journey started, Facebook ran its first production Presto workloads. The first open source version of Presto was released later that year.

In the ensuing 7 years, a rich community grew around Presto, with Netflix, Uber, and Teradata making significant private and public investments in the engine. AWS did not engage with Presto until 2015 when it added support for Presto in AWS EMR, positioning the distributed SQL engine along with side-related technologies such as Spark and Hive.

It was not until 2016 that Athena sought to make Presto even easier to use and scale by making Presto a core part of the newly minted service. Then, in 2018, the Presto community started to fracture with the original engineering team leaving Facebook over differences in the open source project's stewardship. That original team went on to establish the Presto Software Foundation, forking the original Presto repository in the process. Not wanting to lose face (pun very much intended) over the split, Facebook joined with Uber, Twitter, and Alibaba to form the Presto Foundation under the Linux Foundation's governance. If you are following along at home, we now have a Presto Foundation and a Presto Software Foundation developing divergent forks of Presto. It should then come as no surprise that in late 2020, the Presto Software Foundation, comprised of the original developers who left Facebook, was required to rebrand its fork as Trino. Only time will tell which fork ultimately wins. In the meantime, many sophisticated customers are merging features from both distributions to get the best of both worlds.

Now that we know what Presto is, as well as some of the history that led to its creation, you can take advantage of Facebook's experience in trying to scale suboptimal tools for a job that needed something new. By understanding the motivations for creating Presto, you may even identify similar struggles or requirements in your organization and be better equipped to explain why Presto and, by extension, Athena, is a good fit to meet those needs. Next, we will look at how Presto works in relation to a service like Athena.

Presto architecture

As an engineer who has spent the last decade working on and supporting large-scale, multi-tenant analytics applications, I have experienced joy, frustration, and honestly, the full range of human emotions in those pursuits. Those experiences have shaped how I define architecture. Unlike many other books or white papers that you may read, I'll be describing Presto's architecture as it relates to executing a query, not how you deploy it. After reading this section, you may want to compare and contrast the explanation given in the original Presto white paper that we've provided a link to in the *Further reading* section of this chapter.

Most, perhaps even all, SQL engines start by parsing your query into an **Abstract Syntax Tree (AST)**. Presto uses **ANTLR** to generate parser and lexer code that help Presto's SQL planner turn your SQL string into an AST. In *Figure 2.1*, you can see a simplified AST for the following query:

```
SELECT table_1.col_a, table_2.col_1
FROM table_1 LEFT JOIN table_2 ON col_b = col_2
WHERE col_a > 20 and col_1 = 10
```

The SQL engine's planner operates on a tree representation of your SQL because it perfectly captures the relationship between the different operations needed to generate the result.

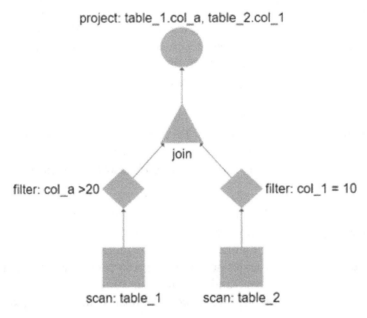

Figure 2.1 – A diagram of a hybrid AST and a logical query plan

As Presto begins planning how to execute your query, it runs several transformations over the AST. One such modification is injecting **Operators** into the tree. Aggregations such as max, min, or sum are examples of an operator. Similarly, reading from a table in S3 would be akin to a **TableScan** operator. Referring back to *Figure 2.1*, we can read the plan for our query from the bottom up. Our engine needs to perform independent `TableScan` operations of `table_1` and `table_2`. These can occur in parallel since they are on different branches of the tree. Each TableScan leads into a filtering operation that applies the relevant portion of the WHERE clause. Data from both `TableScan` operations converge at a `Join` operation before passing through a project operator that trims down the set of columns to only those required by our SELECT clause.

At this point, you might be asking yourself, *what does this have to do with architecture? I thought Presto had a coordinator node that handled all the query planning and one or more worker nodes that did the heavy lifting.* **Coordinators** and **Workers** are the units you deploy when running Presto yourself. Still, the exciting part of Presto's architecture is how it can reshape the relationship between those components on the fly for each query. You'll frequently see this called a physical plan. So far, everything we have discussed happens in the ether because the AST and logical plan don't connect to the physical world of servers and processes.

After the coordinator node generates what it believes to be the best logical plan, it needs to decide which worker nodes to involve in the actual execution. The result is a **physical plan** influenced by the number of available workers, the parallelism the logical plan offers, and even the workers' current workload.

While the Presto coordinator does play a unique role in orchestrating your queries' execution, all nodes in your Presto fleet can run the same software. Upon starting, each node attempts to contact the coordinator. This discovery process allows the coordinator to build an inventory of resources, including what capabilities each node offers. For example, you may have many nodes configured to run the Hive connector because you have lots of data in S3, but you only have two nodes with the JDBC connector installed since you rarely federate queries to your sole MySQL instance. In my experience, there are more advantages to having your fleet be homogenous than taking on the complexity of running different configurations on different nodes. The node discovery mechanism and self-differentiating workers allow multi-tenant services such as Athena to remove the need to manage *clusters*. Instead, Athena custom crafts a serverless resource plan for each of your queries. This is a fancy way of saying the servers come into the picture just in time to execute their share of the work and then rapidly move to the next job or query that needs them.

A lot of Presto's architecture may seem familiar. The broad strokes are similar to that of Hive, Spark, and many other distributed analytics engines. A leader node, homogenous workers, and logical and physical plans are all concepts that pre-date Presto. There is, however, one area where Presto significantly diverges from its peers. Hive, Spark, and Presto all break their query plans down into stages. Stages usually demark a boundary between dependent but discretely different operations. Sometimes, these boundaries are useful for marking changing resource requirements or creating checkpoints to recover from partial failures. Presto's execution engine is deeply pipelined, often executing all query stages simultaneously. Hive and Spark currently wait for a stage to complete before the next stage can start. Deep pipelining gives Presto a structural advantage for queries that don't have blocking operations because later stages can attempt to make partial progress even while early stages are still completing. Spark attempts to approximate this by collapsing pipeline-able operations into the same stage, but that isn't always possible.

Similarly, Presto doesn't always benefit from deep pipelining. Queries having a subquery with paired ORDER BY and LIMIT clauses are one case where pipelining benefits can be limited. In this case, the outer query can't make meaningful progress until the LIMIT clause of the inner query gets results from the preceding ORDER BY clause. Unfortunately, the ORDER BY clause can't generate results until everything before it completes, thereby stalling the pipeline. Exceptions aside, all Presto nodes continuously send intermediate results to the next worker in the physical plan. Like a real tree, water, or data, flows from the bottom of the tree to the top.

The flow of data, or more precisely the location of the data you query, is another notable aspect of Presto's architecture. Earlier in this chapter, we mentioned that Athena supports querying data in over 14 different sources, including S3, Elasticsearch, and MySQL. Querying data across multiple sources is made possible, in part, by Presto separating storage and compute. Presto's creators knew that running traditional data warehouse systems was expensive both operationally and in terms of licensing. Companies frequently hire entire teams to manage the data warehouse and help police use of storage and SLA, thereby impacting job contention. Presto takes a different view and is a semantic layer over your data – a virtual data warehouse. If the separation of storage and compute makes Presto a good choice for querying a data lake, then federation may make it the best option. Suppose you are moving your organization to a data lake or have some awkward data sources to feed into an existing data lake. In that case, Presto's connector suite lets you query across multiple sources as if they were one. There is no need to ETL data from one source to another just to run queries over it. You can run the same ANSI SQL over all connected sources, regardless of their underlying query languages.

Beyond architectural choices, Presto also does a lot of small things well. Each worker makes use of an in-memory parallel processing model that heavily multi-threads query execution to improve CPU utilization. When appropriate, Presto even rewrites its own code to execute your query more quickly. This technique is known as code generation, and it can help improve CPU branch prediction and exploit machine-specific instruction sets. If you've never worked on a code generator, this might seem rather theoretical, so let's look at an example. In the following code, our imaginary SQL engine is copying only the columns selected by our query from a page of intermediate results to `targetPage` representing the query's final output:

```
for(nextColumn in selectedColumn){
    sourcePage.copy(nextColumn, targetPage)
}
```

What's the big deal? I only selected five columns. How could this possibly matter? Well, this code runs for every ROW! So those seemingly meaningless comparisons and small copy operations add up and degrade performance when your query processes millions or billions of rows. Instead, Presto generates very targeted pieces of code with generalization. In our hypothetical example, Presto creates the following code:

```
sourcePage.copy(column1, targetPage)
sourcePage.copy(column2, targetPage)
sourcePage.copy(column3, targetPage)
sourcePage.copy(column4, targetPage)
sourcePage.copy(column5, targetPage)
```

This seemingly contrived example was a real issue Athena patched in Presto. For queries exceeding 6,000 projected columns in any stage, Presto's code generator would fail and revert to using the original `for` loop approach, resulting in a 20% increase in query runtime. Removing one column or fixing the code generator restored the original performance. By making the CPU operations required to complete the query more predictable, we were able to make better use of the deep execution pipeline in modern x86 CPUs. This technique isn't unique to Presto, but it is useful to know how Presto uses it.

In this section, we've tried to highlight the fluidity of Presto's architecture because its creators made a conscious choice to go with this model over more prescriptive but more straightforward approaches. This is just the tip of the iceberg in terms of how Presto works. If you'd like to learn more about this topic, I encourage you to read the Presto white paper. Next, we will learn more about the kind of performance and scale Athena delivers using Presto.

Understanding scale and latency

Ever wonder why companies ambiguously describe their products as *fast* or *highly scalable* without quantifying those superlatives? For a long time, I thought it was because they were hiding something. Maybe they didn't provide hard numbers because they weren't the fastest or had a terrible *gotcha*. As it turns out, performance is personal, with dozens of variables affecting how long a query will run. Even the differences between a successful query and an unsuccessful query can come down to random chance associated with your data's natural ordering. These are some of the reasons why companies do not provide straightforward performance figures for their analytics engines. However, this doesn't mean we can't identify useful dimensions for anticipating a workload's performance.

When evaluating Athena's performance, the first thing to understand is that Athena is not likely to be the fastest option. This may be the most controversial statement in the entire book. Earlier in this very chapter, we discussed the trade-offs in ease of use and added control when using fully managed services. As a managed service, Athena is in charge of deciding most aspects of how your queries execute, including the number of key resources such as CPU and memory. So, it comes as no surprise that Athena doesn't have any setting you can use to influence those resources. As good as Athena's query planning and resource allocation technology can be, it is not likely to guess your SLA needs. This is important because Athena, as part of removing the need for customers to tune cryptic performance settings, closed a standard mechanism for increasing performance. In the future, such settings may get added, but today Athena simply doesn't know that your urgent query needs to finish by the start of that 9 a.m. meeting.

Many other products in this space, including Google BigQuery, allow you to change the price/performance balance by influencing the amount of hardware parallelism the underlying engine will give your query. In BigQuery parlance, you can choose to use more *slots* to try making a query run faster. The added control enables these alternatives to outperform Athena frequently. It also makes them more expensive than Athena. In the case of Google BigQuery, it is relatively easy to create queries that run 50% faster in BigQuery than Athena, but cost more than 10x what Athena charges for the same result.

Beyond individual query performance, we also need to consider how the system behaves when we have concurrent queries. According to the Amazon Athena Service Quota page, customers using the US-EAST-1 region can submit 20 concurrent DDL and 25 concurrent DML queries. The documentation also notes that these default values are soft limits for which you can request an increase from the **Athena Service Quotas** console. These limits consider both running and queued queries. Lack of capacity is the overwhelming reason a query might find itself in the queued state. Such a capacity shortfall can result from Athena itself being low on capacity and maintaining fairness between customers. It can also result from you exceeding your account limits. The specific reasons for queuing aren't important as they are most likely related to internal details of how Athena schedules queries. Instead, we should focus on things we can control. A quick Google search for *Athena Queuing* turns up many Stack Overflow and AWS Forum posts where customers didn't consider their concurrency needs before building on Athena. The point you should remember is to include concurrency testing in your evaluation of Athena. If your anticipated workload needs twice the advertised default concurrency within the next 2 years, engage with the Athena service team early to understand how they can accommodate your workload. Soft limits offer a useful data point about how a service scales, but it isn't surprising to see a serverless offering sensitive to concurrency. As we saw in our Presto architecture overview, all queries get mapped into the physical world of servers and processes at some point.

In the last year, Athena has more than doubled many of its default limits. I expect to see Athena continue that trend and perhaps even offer more controls for customers to manage performance while maintaining the current ease of use. For now, Athena provides a one-size-fits-all balance between price and performance, but that doesn't mean there aren't other levers you do have control over that directly influence performance and cost. We will talk about some of those next.

Price versus performance

We have made this point already, but it is worth reinforcing. When an analytics engine builds a query plan, it often has to balance opposing goals. For example, a broadcast join can require considerably more memory (RAM) than a distributed join. If your system currently has excess memory and limited CPU, a query plan that dedicates surplus memory to the join stage to qualify for a broadcast join can make sense. Conversely, if the environment had extra CPU and network bandwidth, you might opt for the distributed join plan even though it will use more expensive hardware. It would be the only choice you had if you didn't want to fail the query. In each of these cases, we optimized for something different. In the first example, we tried to preserve scarce CPU while the second path reserved limited memory for future needs. Knowing when to make trade-offs can be challenging. You may not even know what trade-offs are available, let alone when to use each. Athena values ease of use and doesn't want you to be bothered with these trade-offs. Earlier, we described Athena's performance as *one size fits all*. Not unlike the clothing items from which we borrowed that classification, there are outliers at the margins who won't be entirely happy with the fit. For the vast majority of people, however, Athena's ability to reshape itself to your needs will be indistinguishable from magic.

TableScan performance

Now, you may recall Presto's query execution pipeline `TableScan` operations to read data from your tables, from *Figure 2.1*. Lucky for us, Athena was built to take full advantage of Amazon S3's scalability as a storage layer for data lakes. By following the best practices covered in *Chapter 4, Metastores, Data Sources, and Data Lakes*, you can expect a typical Athena query to scan, filter, and project data at more than 100 Gbps. If your data is mostly numeric and stored in a columnar format such as Parquet, you can easily see scan performance above 200 Gbps. Things get even more impressive when your queries include a predicate that can be pushed down to the scan operation. You will often see this abbreviated as a `ScanFilterProject` operator since it combines three steps into a single more efficient operation. In such cases, Athena is smart enough to use metadata within your ORC and Parquet files to reduce the actual amount of work it does per row. The net effect is that the perceived scan performance can improve by orders of magnitude.

Your choices of storage and file format play outsized roles in achievable `TableScan` performance. For example, if you store the same data in Amazon S3 and MySQL, and then count the rows in each table, Athena will struggle to achieve 64 Mbps against MySQL, while throughput from the table in S3 will be well above 100 Gbps. That was not a typo. The difference was more than 99.9 Gbps. This isn't a fair comparison since Athena does not yet take full advantage of MySQL's ability to run the count operation itself. However, it illustrates that few data sources can keep up with S3-based data lakes.

If you do anticipate using Athena to federate queries to data stores other than Amazon S3, you should be aware of the current incarnation of Athena Federated Query functions as a `TableScan` operation. This means that as Athena is producing a query plan for your federated source, it is mostly unaware of that source's capabilities. Except for pushing down conjunctive predicates, Athena will ask your source to return all the row data for any subsequent operations, such as aggregations and joins. It is not always possible to push more of the work into the source system, even when that system is as capable as MySQL. Still, many federated queries can benefit from the data transfer reductions offered by aggregate pushdown. In the previous example, MySQL could have completed the count, an aggregation operation, if Athena had pushed that part of the query down below the TableScan operation. Such an optimization would effectively hide the fact that MySQL cannot transfer row data as fast as Amazon S3. To be crystal clear, MySQL was never intended to transfer data externally at high rates. Athena Federated Queries can achieve scan rates above 100 Gpbs, but the actual figures are highly dependent on the source. Athena Federation is covered in full detail, including how to write a connector for a custom data source, in *Chapter 12*, *Athena Query Federation*.

Memory-bound operations

From our walk-through of Presto's architecture, we learned that Presto favors in-memory columnar representations of data for their speed. The flip side to that coin is that Presto, and thus Athena, can be sensitive to memory-intensive operations such as joins and distinct value operations. Until Athena engine version 2, which loosely correlates to Presto's 0.217 release, Athena rarely spilled to disk when physical memory was under pressure. If you are not yet running your queries against Athena Engine Version 2 or tried Athena in the past and had issues with queries failing due to resource exhaustion, you should try them again. Athena still lags Spark in large joins and performing distinct operations on high - cardinality datasets, but it has made significant improvements in this area over the past year. Memory exhaustion remains one of the most common causes of query failures in Athena. This was true in our testing in writing this book and also a commonly asked question online.

Writing results

One of the final performance dimensions to keep in mind when considering Athena centers around how quickly you can write results. If all your queries return a limited number of rows, this section won't be a concern. If, however, your queries generate hundreds of megabytes of results, you should consider which of the three ways Athena can write query results may be best for you. Usually, when you run a DML statement like the one following, Athena will return the results from a single file in S3:

```
SELECT sale_date, product_id, sum(sales)
FROM product_sales
GROUP BY sale_date, product_id
```

Athena also provides a convenient API called `GetQueryResults` to return pages of results to you without your client ever needing to interact with Amazon S3. Based on the S3 access logs, it would seem Athena is reading from S3 for you when you use this API. This is the slowest method of getting results from Athena. It works perfectly well for relatively small result sizes, but when your queries start to generate larger result sets, you'll find yourself bound by the throughput of writing results. For those cases, we recommend you look at Athena's `CTAS`, `INSERT INTO`, and `UNLOAD` queries. These statements tell Athena it is OK to parallelize writing results. You'll end up with multiple files in S3, which you'll be able to consume in parallel, removing the bottlenecks that come with regular `SELECT` statement results.

By now, I hope my earlier statement about performance being very personal is starting to make sense. There is an incredible number of variables at play. Some factors are independent, but many are partially correlated. It would take a degree in advanced physics to approach the problem without apprehension. Don't go rushing to buy the top-rated differential equations book on Amazon just yet. Our next topic is refreshingly straightforward. Athena pricing is as simple as it gets and is one of the dimensions where Athena is in a league of its own.

Metering and billing

Amazon Athena meters the amount of data Athena must read to satisfy your query. The data your query reads is then billed at the rate of $5 per terabyte. This pricing model's simplicity makes it easy to quickly estimate how much the query you are about to run might cost. If your table is 1 terabyte in size, it's a reasonably safe assumption that querying such a table should not cost more than $5. You might think that this is the end of the pricing conversation, and for all practical purposes, it is. However, in classic AWS fashion, the model's simplicity hides the real value of what that $5 is actually buying you.

As of this writing, several alternative offerings are also charging $5/TB scanned for a similarly rich SQL interface. Beyond informing you of how Athena is priced, the goal of this section is to help you understand what that $5/TB is buying.

Let's double-click on the metering aspect first. Amazon Athena charges you for the bytes it reads from S3, or, in the case of federation, the bytes returned by the connector. More precisely, Athena charges you for the raw size before any interpretation of the data. This is significant because it means the bytes are counted before decompression. If you have 10 TB of data in CSV format and compress the CSV files down to 1 TB using gzip before you query it with Athena, you just cut your Athena bill by 90%! Many of the other offerings in this space charge you for the logical size, known as the size after decompression, deserialization, and interpretation. In my time working with Athena, this was easily one of the most overlooked benefits of the service.

Later in this section, we will examine how different file formats and compression techniques compare concerning file size and performance.

Athena Query Federation metering

Athena natively supports querying data stored in Amazon S3. This feature allows Athena to read data from any source that implements a *connector* using the open source Athena Federation SDK. Data from federated sources is metered at the same $5/TB as data originating in S3, but the point at which the bytes are counted is subtly different. If your connector reads 10 TB from a MySQL database, but manages to filter that data down to 1 TB before passing it to Athena, you are charged for 1 TB, not 10 TB.

You may be wondering whether your Athena costs will vary between long-running or short-running queries. Regardless of the runtime of your query, you will be charged the same $5/TB. If your queries are longer because they read more data, they cost more than shorter queries that read less data. There are no surprise bills associated with executing a CPU-intensive sort or memory-hungry join. You should, however, keep in mind that there are few free rides in this world. So, while it might be Athena's problem to execute such queries within the agreed $5/TB pricing structure, that does not mean your queries have access to infinite memory or unlimited query runtimes. By default, Athena DML queries are allowed to run for no longer than 30 minutes. You can request an increase to this soft limit from the service quota console.

On top of the charges that will come directly from Amazon Athena, your queries will incur additional costs associated with other services that Athena interacts with on your behalf in the course of executing your queries. We'll cover those next.

Additional costs

Firstly, don't be alarmed. Nearly all AWS services can incur additional costs from interacting with other AWS services on your behalf. In the case of Amazon Athena, these additional costs rarely add more than 0.1% to your queries' total costs. The services that Athena interacts with most often are listed in the additional costs section of the Athena pricing page. Regardless of the documentation, you can try to self-identify other cost sources by removing Athena from the picture and imagining what you would need to do if you were Athena. The first thing Athena does when you run a query is to get the details of any tables used in the query by talking to AWS Glue Data Catalog. Athena calls Glue's **GetTable** API once per table and the **GetPartitions** API for each batch of 1,000 partitions in your table. AWS Glue's free tier offers one million API calls and just $1 for each one million API calls beyond that. An Athena query against one table that follows the best practices in this book is unlikely to generate more than 11 API calls to AWS Glue. For more information about AWS Glue's pricing, check out the AWS documentation.

Putting ourselves back in Athena's shoes, our next step after gathering metadata from AWS Glue is to start reading data from S3. We would need to list the objects in each partition to enumerate all the objects we need to read. Then we would need to reach each object. If we are using an advanced format such as Parquet or ORC, reading the objects might require *seeking* different offsets within the object. This allows Athena to skip large chunks of the file, saving you costs with respect to the bytes read by Athena, but increasing the number of S3 calls. Considering 1,000 S3 requests cost just $0.005, it is easy to see why seeking within an object in order to skip chunks of data is well worth the effort. More concretely, a well-organized table containing one million objects totaling 128 TB of data would cost $640 to read in Athena fully. That same query would incur less than $0.50 (0.0007%) of additional costs from Amazon S3.

Once Athena has read the data from S3, or in the case of S3 server-side encryption, the data may need to be decrypted before Athena can make sense of what it read. In these cases, Athena will call **AWS Key Management Service** (**KMS**) to get the appropriate data key for the file being read. It is a recommended best practice to use a different data key with each S3 object. Accordingly, Athena or S3 may need to call KMS one or more times per object. AWS KMS charges $0.03 per 10,000 requests. Our query exceeding the preceding hypothetical table would generate $3 (0.004%) of additional KMS charges. You can find full details of AWS KMS pricing on the KMS pricing page.

If these additional costs are indeed so inconsequential, why are we giving it so much attention? The short answer is that you will see these costs, and they won't always look like such small percentages even though they are.

Since Athena charges by the terabyte scanned, customers are incentivized to reduce their data sizes through compression and columnar formats such as Parquet. Let's apply some of these techniques to the hypothetical 128 TB table from the previous examples. After converting to Parquet and changing our query to use a more targeted filter, our Athena charges have been reduced to $6. Parquet allowed Athena to evaluate our query's filter using only statistics from each row-group's header without reading the entire S3 object. The net effect is that Athena could skip 90% of each object's contents, cutting our Athena bill by a proportional 90%! However, skipping 90% of the data required many more calls to S3 and KMS. In this example, we'll assume 10 times more calls. At the end of our query, our KMS and S3 costs are now a combined $35 compared to Athena's $64 line item. Our additional costs have ballooned to more than 50% of our total costs! Yes, that is true, but don't forget that we spent that extra $30 to save $576. We aren't highlighting this because we feel you should gladly accept these additional costs. Instead, we hope you will approach the delicate art of optimization with an informed understanding of the drivers that impact each cost dimension. In this particular example, you might be tempted to cut the additional $30 that comes from KMS by disabling KMS encryption. This might be reasonable, or it might be an unnecessary risk if your data is sensitive. It is likely easier to make an additional $30 of revenue than it will be to rebuild customer or regulator trust if the lack of encryption exposes sensitive data.

Details of your query and the file formats involved can affect your costs in subtle ways. We've used extreme examples to illustrate that point. Additional costs are expected to be an insignificant portion of your overall cost. Knowing what drives them will help you understand which scenarios apply to your workloads. Besides total data size, your choice of file formats and compression techniques is the most significant factor in your queries' cost. We'll cover these in more detail now.

File formats affect cost and performance

Your file format choice affects the raw data's size that Athena will need to read from S3 to answer your query. For example, if your data comprises one field containing the quantity of each item you have in stock at your stores, you can represent that data in multiple ways. The first and perhaps simplest is as a CSV. While easy to get started with, CSVs are a poor choice for storing numeric values. The number *30,000* would occupy 5-10 bytes in CSV format, but just 2 bytes in columnar formats such as Parquet. If you have millions of rows, this 80% size difference can add up quickly. Beyond literal cost implications, it takes more CPU and, by extension, time for Athena to deserialize the text representation of numeric values found in a CSV to the type of appropriate representation required for most operations, including addition and subtraction. This deserialization penalty can slow down your queries.

We can use Athena itself to run a quick experiment with different file formats and compression algorithms. The following CTAS statement reads from the `nyc_taxi` table we created in *Chapter 1, Your First Query*, and then rewrites that table's contents into a new table using Parquet with SNAPPY compression instead of the original GZIP CSV format:

```
CREATE TABLE nyc_taxi_parquet
WITH (
        format = 'Parquet',
        parquet_compression = 'SNAPPY',
        external_location = 's3://packt-serverless-
analytics-888889908458/tables/nyc_taxi_parquet/')
AS SELECT * FROM nyc_taxi;
```

By running this query for various formats and then inspecting the resulting S3 objects from the S3 console, we constructed the table in *Figure 2.2*:

Format	Compression	Size	Performance Ranking
CSV	None	50.7 MB	6
CSV with GZIP	GZIP	9.7 MB	5
Parquet	Run-length encoding	7.9 MB	1
Parquet	Run-length encoding + SNAPPY	10.9 MB	3
ORC	Run-length encoding	10 MB	2
ORC	Run-length enconding + SNAPPY	15.6 MB	4

Figure 2.2 – Table comparing different file formats for the NYC Taxi data

By studying the table in *Figure 2.2*, we can see that columnar formats such as Parquet and ORC can reduce our costs while also improving performance vis-à-vis simpler formats such as CSV. Columnar formats exploit the similarity between rows for a given column to provide a more compact representation of the data without requiring computationally intensive compression techniques such as GZIP. Here we've compared the most common approaches. CSV, while simple and broadly supported, is the least compact. It also has the most deserialization overhead due to its textual representation of everything from strings to small integers.

Even when coupled with an intensive compression algorithm such as GZIP, it still underperforms the size reduction capability of Parquet while using considerably more CPU. Parquet and ORC performed similarly, and given the relatively narrow testing we did here, little can be learned about the two approaches relative to one another.

Interestingly, both Parquet and ORC performed worse when we enabled SNAPPY compression. This is likely because of run-length encoding doing such a good job, leaving SNAPPY to compress data that contained minimal repetition. Most compression algorithms fare poorly against data that is entirely random, though I wouldn't have expected ORC to be more vulnerable to this than Parquet. One of Parquet and ORC's main differences originates in the frequency and size of the metadata they store for the chunks of underlying row data. By default, ORC tends to favor more metadata in anticipation of more significant query-time benefits. This has a side effect of higher overhead, which may have been magnified by our example's meek 10 MB of data.

Much of the Athena documentation strongly recommends the use of Parquet as your format of choice. This book partly takes the same view because of the performance, size reduction, and rich engine support for Parquet. ORC is a close runner-up with many of the same features.

Run-length encoding (RLE)

Run-length encoding is an inventive form for compression that uses relatively little CPU or memory to compute while still offering substantially smaller data sizes. Unlike related techniques used in video processing, RLE is lossless, making it ideal for Parquet and ORC formats. When used in conjunction with sorted data, RLE can reduce data sizes by upwards of 10x. At its core sits an algorithm for exploiting *runs* of data that have a shared or common base. Instead of storing the repeated information in adjustment rows, you merely store a delta from the previous value. For example, the string ABBBBBBBBBBBCAAAAAAAAAAA could be natively run-length encoded to A10BC12A, yielding nearly 10x lossless compression.

Amazon Web Services has reduced prices on one or more services more than 70 times in the last decade. Prices can and do vary between regions, and prices may have changed since this book was written. Even though we could not find a single documented case of pricing going up, please verify the current pricing details before using any services. You can find the latest pricing for Athena in the AWS documentation (`http://amzn.to/3r5pYTD`). Now that we understand what drives our costs, we can look at options for controlling them.

Cost controls

Athena offers several tools for helping you control costs. This includes mechanisms for capping the data scanned by individual queries or by grouping your applications into organizationally relevant buckets with accompanying budgets. On the **Workgroup** settings page shown in *Figure 2.3*, you can set a per-query limit for each workgroup. Once a query reaches that limit, it will be killed. Further down on the same page, you can configure a budget for the entire workgroup. Once the queries that run in the workgroup have cumulatively exceeded the limit, further queries in that workgroup will be killed until enough time has passed that the budget resets.

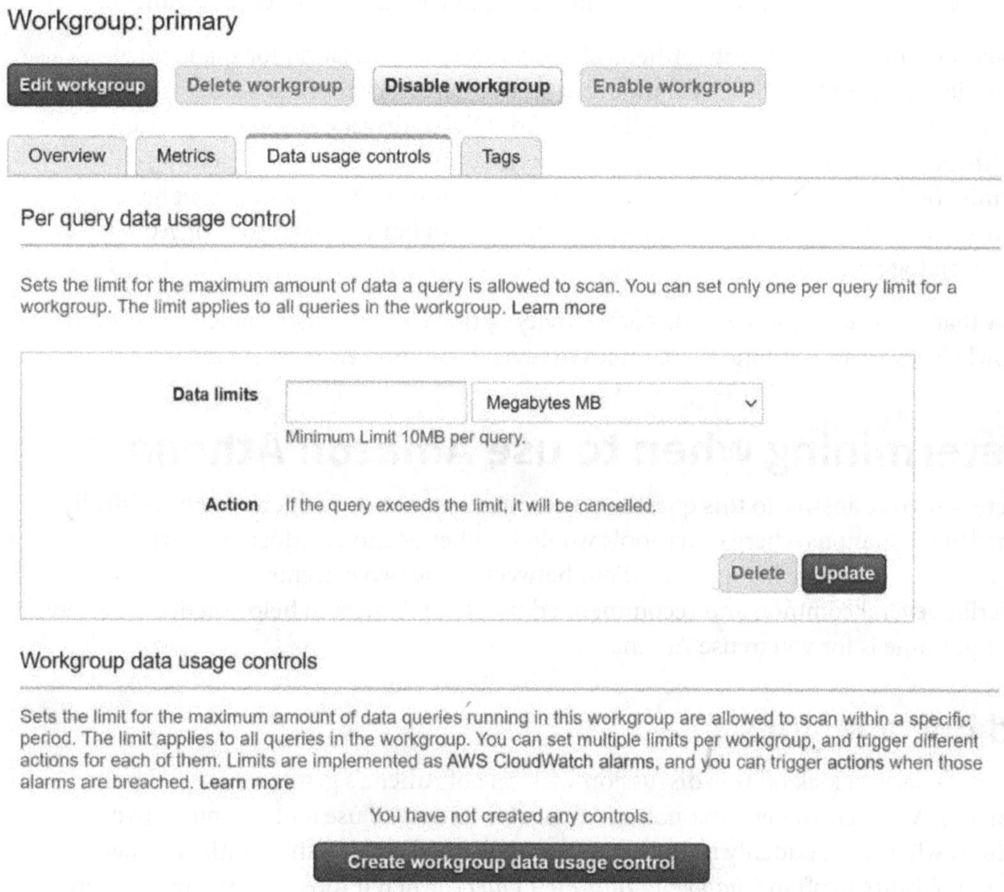

Workgroup: primary

| Edit workgroup | Delete workgroup | Disable workgroup | Enable workgroup |

| Overview | Metrics | Data usage controls | Tags |

Per query data usage control

Sets the limit for the maximum amount of data a query is allowed to scan. You can set only one per query limit for a workgroup. The limit applies to all queries in the workgroup. Learn more

Data limits   [   ]   Megabytes MB   ⌄

Minimum Limit 10MB per query.

Action   If the query exceeds the limit, it will be cancelled.

Delete   Update

Workgroup data usage controls

Sets the limit for the maximum amount of data queries running in this workgroup are allowed to scan within a specific period. The limit applies to all queries in the workgroup. You can set multiple limits per workgroup, and trigger different actions for each of them. Limits are implemented as AWS CloudWatch alarms, and you can trigger actions when those alarms are breached. Learn more

You have not created any controls.

Create workgroup data usage control

Figure 2.3 – Athena Workgroup settings page; Data usage controls tab

In addition, you can enable CloudWatch metrics for your queries. Once active, Athena will send updated metrics about in-flight and completed queries to CloudWatch, where you can monitor them with your own custom rules, reports, or automation.

Connecting and securing

Connectivity and authentication features are often overlooked. Like all AWS services, Athena offers a set of APIs for interacting with the service from your applications or from the command line when using the AWS CLI. These APIs allow you to submit a new query, check the status of an already running query, retrieve pages of results, or kill a query. These same APIs are used from within Athena's JDBC and ODBC drivers. When connecting to Athena, you can use the standard endpoints if you have an internet gateway in your VPC or opt to call Athena from a VPC endpoint and avoid the need to have internet connectivity from your application VPC. This gives you added control over your security posture by pushing the responsibility of securely connecting to your data sources onto Athena.

In addition to VPC endpoints, Athena also offers SAML federation for managing identities outside of AWS. This allows your Active Directory users to seamlessly authenticate to Athena when using the JDBC or ODBC drivers. At the cornerstone of Athena's access control system lies Lake Formation. Lake Formation allows you to permission IAM users or roles for specific tables, columns, and rows without having to write complex IAM policies to coordinate access to millions of S3 objects or AWS Glue Data Catalog resources.

Now that we've added some basic connectivity options to our performance and cost driver knowledge, we can combine these topics to review common Athena use cases.

Determining when to use Amazon Athena

There is no one answer to this question. There are use cases for which Athena is ideally suited and situations where other tools would be a better choice. Most potential applications of Athena lie in the gradient between these two extremes. This section will describe several common and recommended usages of Athena to help you decide when the right time is for you to use Athena.

Ad hoc analytics

We might as well kick off this discussion with one of Athena's greatest hits – ad hoc analytics. Many customers first notice Athena for its ease of use and flexibility, two key features when you suddenly need to have an unplanned conversation with your data. We saw this firsthand in *Chapter 1*, *Your First Query*, when it took us just a few minutes to load up the NYC Taxi trip dataset and start finding relevant business insights. Ad hoc analytics can be used to describe unplanned queries, reports, or research into your data for which a pre-made application, tool, or process does not exist. These use cases often require flexibility, quick iteration times, and ease of use so that a highly specialized skillset is not needed.

For this class of usage, there are relatively few things to consider. The first is where your data is stored. If your data is already in S3 and perhaps already cataloged in AWS Glue, it should be effortless to use Athena as your preferred ad hoc analytics tool. If not, then you will need to think about how you will manage metadata. If your users are savvy, they can create table metadata on the fly using Athena's DDL query language. If not, you may want to consider adding Glue Crawlers to your tool kit. Glue Crawlers automatically scan and catalog data in S3. When complete, the crawlers populate AWS Glue, so you never need to run `table create` statements manually. Many organizations that are not yet considering or are just starting their data lake journey notice the benefits that come with democratizing access when using Athena for ad hoc analytics. Some organizations go a step further and create a business data catalog. This allows employees to discover datasets and learn their business relevance in addition to the technical details of how and where it is stored. In short, a business data catalog often has more documentation than what is currently offered by the AWS Glue Data Catalog.

Related to the cataloging of data, managing access to that data is another facet to consider. Athena offers two mechanisms for controlling who can read and write analytics data. The first is traditional IAM policies, where you grant individuals or IAM roles access to the specific S3 paths that comprise your tables. This can work well if your data is well organized in S3 and your permission needs are limited to a handful of non-overlapping S3 prefixes. If your needs are more complex, necessitating column or even row-level access control, you'll want to use Athena's Lake Formation integration to manage permissions. In this model, you never have to write IAM policies and instead use an analytics-oriented management console (or APIs) to grant and revoke permissions.

Since ad hoc analytics is a frequent Athena use case, the service has worked with several partners to release driver support in popular BI tools. **Tableau** and **Looker**, two popular BI tools, both natively support Amazon Athena. You can also leverage Athena's ODBC and JDBC drivers to query Athena from a slew of other tools, including Microsoft Excel.

The final criteria for using Athena for ad hoc analytics is purely about the kinds of queries you want to run. As we've seen in this chapter, Athena offers limited options for tuning your queries' scale or performance. If your analytics queries often require large amounts of working memory or another extreme scale, you'll want to test how well Athena runs your queries. The good news is that if you eventually encounter a query that Athena struggles with, you can run that outlier with AWS Glue ETL, a serverless form of Spark. That's why it is essential to consider the surrounding ecosystem in addition to Athena's product-specific capabilities. With AWS, the whole is usually greater than the sum of its parts. In *Chapter 7, Ad Hoc Analytics*, we will get hands-on with more examples of using Athena for this popular recipe.

Adding analytics features to your application

Another popular pattern is to use Amazon to add decision-support information to your application. Imagine we are the authors of a digital advertising campaign system. Our customers use the application to set up new campaigns, monitor the budget of existing campaigns, and even understand the available impression inventory. All this is fancy advertising lingo for understanding different elements of who their campaign is reaching and when they'll exhaust their advertising budget. It would be useful to show some historical trends alongside the current budget remaining number. We can easily use Athena's APIs or JDBC driver to run a query that will return both the hourly impressions, conversions, and budget burndown for the last 24 hours, 7 days, or other relevant timeframes. Because we don't need this data to update live, we can avoid building an OLTP data store. Instead, we need only to feed our existing application logs, or possibly simple metrics, to S3 in a location our Athena queries can access. If we want to be clever, we can even write the metrics to S3 paths based on campaign identifiers and reduce query costs while boosting performance. Thanks to Athena Query Federation, you can go a step further and allow embedded dashboards that show the near-real-time campaign performance only for live campaigns or those within 10% of exhausting their budget. One way to do that is by joining the live status of the campaigns from an OLTP store such as **AWS DynamoDB** with your historical data from Amazon S3.

We'll go through one more example for good measure. Suppose we are using a machine learning algorithm, such as **DeepAR** in **Amazon SageMaker**, to predict demand in our inventory ordering system. The system then automatically reorders ingredients or parts that will be used to replenish our supply of the finished product. For the best results, we've found that the prediction accuracy increased substantially when we supply a week of the most recent sales data as context for the prediction API calls. Unfortunately, our inventory system doesn't keep track of historical inventory burn rates or sales. Well, we could call Amazon Athena to query our data lake's historical order table just before calling our SageMaker prediction endpoint for the next. With a relatively minor enhancement to our application and even less new infrastructure, we've just enabled our inventory system to provide the recent inventory data that will improve our forecasting capabilities.

When considering this usage pattern, you should pay close attention to your anticipated concurrency needs and how the new dependency will affect your application's liveliness. Athena is built for high availability. You don't need to worry about having it in the critical path of your application flow. Still, it's always a good idea to limit critical path dependencies to those that are absolutely necessary. In *Chapter 10, Building Applications with Amazon Athena*, you will get a chance to see this pattern up close.

Serverless ETL pipeline

With the advent of serverless infrastructure has come a wave of new serverless use cases. Anywhere you previously had a server or cluster of servers running big data jobs has become fair game for Athena's serverless promises. So, it comes as no surprise that customers use Athena to build serverless ETL pipelines. As we stroll toward our imaginary system design whiteboard, let's pretend we work for a hedge fund. Our team is responsible for calculating the company's short risk in response to substantive changes in the stock market. The software that runs our various trading desks generates a file every hour, containing a summary of our long and short positions. Whenever one of these files comes in, we need to recalculate each of the updated stocks' overall positions. Our goal is to ensure that our hedge fund doesn't unknowingly take on too much risk, as was the case with the great *Reddit GameStop uprising of 2021*.

GameStop won't stop

The saga of **GameStop**, GME, began in August 2020 when an anonymous user on Reddit posted an in-depth analysis and justification for why GameStop would *go to the moon*. In addition to a few solid fundamental theories, this person highlighted the absurd reality that GameStop shares had a short interest greater than 100% of the available shares. This means that for every share of GameStop stock, more than one share had been sold. This happens when people, or companies, bet against the stock by *borrowing* shares from their broker and sell them to someone else. You are said to be *short* with the stock because you now owe someone else a share you borrowed. What began as a way to make money sticking to the *shorts* morphed into a socio-economic movement pitting the underdog retail investor against some of the biggest hedge funds in the world. No matter which side you were on, it was unprecedented. It also generated many amusing memes.

Using Athena and AWS Lambda, a serverless technology for responding to events, we can configure S3 to send an event to Lambda whenever a new trade summary file arrives in our S3 bucket. When the file comes, a Lambda function gets invoked. Within that function, we can run custom code to have Athena query the newly added file and join it with relevant information from a dimension table before writing the results to our data lake in Parquet format. After the initial transform and load, we trigger another Lambda function, which reruns our overall risk analysis Athena query to determine whether we are overexposed to one or more securities. Without touching a single server, we built an entire ETL pipeline, albeit a relatively simple one. Depending on the data sizes involved, it's not unreasonable for this ETL pipeline to cost mere pennies a day.

While simple ETL pipelines can be appropriate for Athena, you should consider the number and size of jobs you expect to run in your ETL system. Like earlier examples, the AWS ecosystem has complementary capabilities in AWS EMR and AWS Glue ETL, which can help if you outgrow or run into requirements that Athena cannot satisfy. AWS Glue ETL is also serverless, though it is based on Apache Spark and charges you for compute time instead of Athena's Presto-based engine, which charges by the terabyte scanned. In *Chapter 9, Serverless ETL Pipelines*, we will go step by step and build out a reactive ETL pipeline.

We will conclude our review of common Athena use cases by discussing a few miscellaneous examples that, while too small to dedicate a full section to, are equally valid.

Other use cases

While less prominent than the use cases described in previous sections, some customers use Athena as an operational tool or a rapid prototyping tool. Athena's filtering performance makes it a rather performant choice for rapidly scaling and filtering log data without the need to keep a costly infrastructure running all the time. This is ideal for operation situations that arise from nowhere. Customers filter and parse everything from VPC flow logs to application logs, looking for root causes or quantifying impact. Athena's flexibility also makes it a great choice to quickly iterate on complex reports or ETL jobs that you'll later implement in a different system. This is not unlike other data preparation use cases from machine learning or data quality checking.

Just because a use case you have in mind wasn't explicitly mentioned in this chapter doesn't mean you should consider Athena. We've only listed examples of good use cases so that you can extrapolate and apply what you've learned to your projects and environment.

Summary

In this chapter, we formalized your introduction to Athena by going over the service's high-level capabilities, including ACID transactions, federation, ETL operations such as CTAS, and open source file formats. We went inside Athena by learning more about Presto, the open source distributed SQL engine that sits at its core. As part of that exercise, we experimented with supporting our own multi-tenant analytics infrastructure. This allowed us to see all the value-added functionality that sets Athena apart from other serverless technologies that fall short of being fully managed. As if that wasn't enough of a reason to hop on the serverless analytics bandwagon, we unpacked the marketing hype to find that Athena's $5/TB price tag is significantly cheaper than many of its competitors who also claim to charge $5/TB, but count the uncompressed bytes!

We also learned that performance is personal and that we'd have to test our access patterns and data models to see how Athena would perform for us. Regardless of the numerous variables that impact performance, we covered how to control common cost and performance drivers by using columnar storage formats such as Apache Parquet. Using these techniques dramatically reduces our costs but subtly increases the additional costs associated with the other services Athena calls on our behalf, including S3, AWS Glue, and AWS KMS. In addition to pre-emptive actions to control costs, Athena also gave us mechanisms to limit the total cost of each query or group of queries through workgroup-level limits on data scans.

Lastly, we combined all these points when reviewing several common usage patterns for Athena. We walked through a real-world example using a hypothetical system design for each of the frequently seen patterns. We'll be revisiting each of these design patterns in later chapters, where we will get hands-on and build one of each.

The next chapter will conclude our introduction to Amazon Athena by exploring built-in features you can use to make your reports or application more powerful. This includes approximate query techniques to speed up analysis of large datasets, **CTAS** (**CREATE TABLE AS SELECT**) statements for running queries that generate significant amounts of result data, and getting hands-on with several of the topics discussed in this chapter.

Further reading

In this section, we've gathered links to additional materials that you may find useful in diving deeper into some of the primary sources regarding the topics mentioned in this chapter. Many of these topics will be covered in more detail later in this book, but it can often be useful to know where to go for authoritative details:

- *Presto SQL Dialect* documentation can be found here: `http://bit.ly/39kMJeW`.

- *Amazon Athena SQL Dialect* documentation can be found here: `http://amzn.to/35tRT7w`.

- *Amazon leadership principles* can be found here: `http://bit.ly/3k79PuB`.

- *Amazon Athena Engine Version 1* specification can be found here: `http://bit.ly/3boEoty`.

- *Amazon Athena Datasource and External Hive Metastore* documentation can be found here: `http://amzn.to/3bvU9y`.

- The official GDPR regulation and associated data retention requirements discussed in this chapter can be found here: `http://bit.ly/38RlolU`.

- Details for connecting Athena to Delta Lake from `SymlinkTextInputFormat` can be found in the *Delta Lake* documentation here: `http://bit.ly/3ozT9gG`.

- More information about the TPC organization and the industry benchmarks it maintains (TPCH/DS) can be found here: `http://bit.ly/39HMXgJ`.

- You can find the **Athena Federation SDK** on GitHub here: `http://bit.ly/38NfRg4`.

- **Trino**, formerly PrestoSQL, documentation can be found here: `http://bit.ly/39JLGFE`.

- The original Presto white paper from Facebook can be found here: `https://bit.ly/38vQgI8`.

3
Key Features, Query Types, and Functions

In *Chapter 1, Your First Query,* we got our first taste of serverless analytics by building and querying a mini-data lake for **New York City** (**NYC**) taxicab data. *Chapter 2, Introduction to Amazon Athena,* continued that introduction by helping us understand and perhaps appreciate what goes into enabling Athena's easy-to-use experience. This chapter will conclude our introduction to **Amazon Athena** by exploring built-in features you can use to make your reports or applications more powerful. Unlike the previous chapter, we will return to a hands-on approach that combines descriptive instruction with step-by-step activities that will help you connect with the material. The exercises should also offer you a basis to experiment with your own ideas, should you choose to do so.

After completing this chapter, you will have enough knowledge to begin using and integrating Athena into **proof-of-concept** (**POC**) applications. *Chapter 4, Metastores, Data Sources, and Data Lakes,* begins *Part Two* of this book, which transitions to broader topics associated with building and connecting your data lake as part of delivering sophisticated analytics strategies and applications at scale.

In the subsequent sections of this chapter, you will learn about the following topics:

- Running **extract-transform-load** (**ETL**) queries

- Running approximate queries

- Organizing workloads with WorkGroups and saved queries

- Using Athena's **application programming interfaces** (**APIs**)

Technical requirements

Wherever possible, we will provide samples or instructions to guide you through the setup. However, to complete the activities in this chapter, you will need to ensure you have the following prerequisites available. Our command-line examples will be executed using **Ubuntu**, but most Linux flavors should work without modification, including Ubuntu on **Windows Subsystem for Linux** (**WSL**).

You will need internet access to GitHub, **Simple Storage Service** (**S3**), and the **Amazon Web Services** (**AWS**) console.

You will also require a computer with the following installed:

- The Chrome, Safari, or Microsoft Edge browsers

- The AWS **Command-Line Interface** (**CLI**)

This chapter also requires you to have an **AWS account** and an accompanying **Identity and Access Management** (**IAM**) user (or role) with sufficient privileges to complete this chapter's activities. Throughout this book, we will provide detailed IAM policies that attempt to honor the age-old best practice of least privilege. For simplicity, you can always run through these exercises with a user that has full access. Still, we recommend using scoped-down IAM policies to avoid making costly mistakes and we advise you to learn more about using IAM to secure your applications and data.

You can find the suggested IAM policy for this chapter in the book's accompanying GitHub repository listed as `chapter_3/iam_policy_chapter_3.json`, here: `http://bit.ly/37zLh8N`. The primary changes from the IAM policy recommended for *Chapter 1*, *Your First Query*, include the following:

- `glue:BatchCreatePartition`—Used to create new partitions as part of `CTAS` or `INSERT INTO` statements.

- Restricted Athena workgroup actions to WorkGroups beginning with `packt-*`.

- Added read/write access for AWS CloudShell, a free Linux command line in the AWS console. You only pay for the other services you interact with, such as Athena.

Running ETL queries

While this book's goal is not to teach **Structured Query Language** (**SQL**), it is beneficial to spend some time reviewing everyday SQL recipes and how they relate to Athena's strengths and quirks. Transforming data from one format to another, producing intermediate datasets, or simply running a query that outputs many **megabytes** (**MB**) or **gigabytes** (**GB**) of output necessitates some understanding of Athena's best practices to achieve peak price/performance. As we did in *Chapter 1, Your First Query*, let's start by preparing a larger dataset for our exercises.

We will continue using the `NYC Yellow Taxi` dataset, but we will prepare 2.5 years of this data this time. Preparing this expanded dataset will entail downloading, compressing, and then uploading dozens of files to S3. To expedite that process, you can use the following script to automate the steps. To do so, add all the files from `yellow_tripdata_2018-01.csv` through `yellow_tripdata_2020-06.csv`. Each file represents 1 month of data. The NYC Taxi and Limousine Commission has not updated the data since June 2020 due to the impact the pandemic has had on their day-to-day operations. If you have the option, we recommend downloading a copy of the pre-made script with added error checking from the book's companion GitHub repository in the `chapter_3/taxi_data_prep.sh` file or by using this link: `http://bit.ly/3k4bMYU`. The following script has been edited for brevity, but the one in GitHub is ready to go without modification. Regardless of which script you use, you can run it on any Linux system with `wget` and the **AWS CLI** installed and configured by executing it and passing the name of the S3 bucket where you'd like the data uploaded. You can even reuse the S3 bucket we created in *Chapter 1, Your First Query*, to save time.

> **AWS CLI**
>
> The `taxi_data_prep.sh` script will use your system's AWS CLI to upload the compressed taxicab data to the S3 bucket you specify. The script expects you to have configured the AWS CLI ahead of time with appropriate credentials and a default region that corresponds to where you are running the exercises in this book. To review or update your default AWS CLI configuration, you can run `aws configure` at the command line.

The configuration is shown here:

```bash
#!/bin/bash

BUCKET=$1
array=( yellow_tripdata_2018-01.csv
        yellow_tripdata_2018-02.csv
        # some entries omitted for brevity
        yellow_tripdata_2020-06.csv
      )
for i in "${array[@]}"
do
      FILE=$i
      ZIP_FILE="${FILE}.gz"
      wget https://s3.amazonaws.com/nyc-tlc/trip+data/${FILE}
      gzip ${FILE}
      aws s3 cp ./${ZIP_FILE} s3://$BUCKET/chapter_3/tables/
nyc_taxi_csv/
      rm $ZIP_FILE
done
```

Code 3.1 – NYC taxi data preparation script

Speeding things up!

Depending on the speed of your internet connection and the type of **central processing unit (CPU)** you have, this script may take over an hour to prepare the 4.5 GB of data for the recommended 2.5 years of historical data. We recommend using AWS CloudShell (https://aws.amazon.com/cloudshell/) to run this script natively within the AWS ecosystem. AWS CloudShell provides a Linux command line with AWS CLI and other common tools preinstalled at no extra charge. You are only charged for the other services you interact with, not for your usage of CloudShell itself. In our testing, AWS CloudShell took roughly 23 minutes to prepare our test data, thanks in part to its high-speed connectivity to Amazon S3. Alternatively, you can reduce the amount of historical data you use in the exercise by reducing the number of monthly files used in the script.

Once the script completes execution, you can verify your data is now in the proper location by listing S3 from the command line using the following command or navigating to /chapter_3/tables/nyc_taxi_csv/ from the S3 console in your browser. If all went well, you'd see 30 files in this path:

```
aws s3 ls s3://YOUR_BUCKET_NAME_HERE/chapter_3/tables/nyc_taxi_
csv/
```

Our final data preparation step is to use Athena to define a table rooted at the path we uploaded the data to. To do this, we'll apply our final *Chapter 1, Your First Query,* refresher in the form of a CREATE TABLE query. If you have the option, we recommend downloading a copy of the following CREATE TABLE query from the book's companion GitHub repository in the chapter_3/create_taxi_table.sql file or by going to https://bit.ly/2TOinOs:

```
CREATE EXTERNAL TABLE 'packt_serverless_analytics'.'chapter_3_
nyc_taxi_csv'(
  'vendorid' bigint,
  'tpep_pickup_datetime' string,
  'tpep_dropoff_datetime' string,
  'passenger_count' bigint,
  'trip_distance' double,
  'ratecodeid' bigint,
  'store_and_fwd_flag' string,
  'pulocationid' bigint,
  'dolocationid' bigint,
  'payment_type' bigint,
  'fare_amount' double,
  'extra' double,
  'mta_tax' double,
  'tip_amount' double,
  'tolls_amount' double,
  'improvement_surcharge' double,
  'total_amount' double,
  'congestion_surcharge' double)
ROW FORMAT DELIMITED
  FIELDS TERMINATED BY ','
STORED AS INPUTFORMAT
  'org.apache.hadoop.mapred.TextInputFormat'
```

```
OUTPUTFORMAT
  'org.apache.hadoop.hive.ql.io.HiveIgnoreKeyTextOutputFormat'
LOCATION
  's3://<YOUR_BUCKET_NAME>/chapter_3/tables/nyc_taxi_csv/'
TBLPROPERTIES (
  'areColumnsQuoted'='false',
  'columnsOrdered'='true',
  'compressionType'='gzip',
  'delimiter'=',',
  'skip.header.line.count'='1',
  'typeOfData'='file')
```

Code 3.2 – CREATE TABLE SQL query

You can execute this query right from the Athena console, but be sure to update the S3 bucket in the LOCATION portion of the CREATE TABLE statement. The table creation should complete almost instantaneously. The most common errors at this stage are related to insufficient permissions, using an incorrect database or table name, or already having a table with that name in your catalog. In the event you do encounter an issue, retrace your steps, and double-check these items. It's always a good practice to run at least one query to ensure our table is properly set up since the CREATE TABLE operation is purely a metadata operation. That means it didn't actually list or read any of the data we prepared in S3. A simple COUNT(*) query, as illustrated in the following code snippet, will suffice to ensure our table is ready to be used in more ambitious queries:

```
select count(*) from chapter_3_nyc_taxi_csv
```

After running the preceding query from the Athena console, you should get a result of 204,051,059. The query should have scanned around 3.4 GB of data and completed after roughly 8 seconds. We have just completed one of the most common activities you'll encounter in Athena or any data lake analytics tool. The table we just created is commonly described as a landing zone. It is a place where newly arrived source data lands before being cleaned up and made available to applications in your data lake. Data ingestion is always where it begins, but the table we just created is sub-optimal for a number of reasons, and we may not want to let applications or analysts use it directly. Instead, we will reorganize this table for peak performance as a way to demonstrate some of Athena's advanced query types, such as CTAS, INSERT INTO, and TABLESAMPLE.

Using CREATE-TABLE-AS-SELECT

Athena's **CREATE-TABLE-AS-SELECT (CTAS)** statement allows us to create new tables by applying a SELECT statement to an existing table. As part of doing that, Athena will shun the SELECT portion of the statement to generate the data to be stored as part of the new table. Both CTAS and VIEW statements can be thought of as a SELECT statement that forms a new table as a derivative of one or more existing tables but with one key difference in how the underlying data is handled. CTAS is like a materialized view since it runs the SELECT portion of the statement one time and stores the resulting data into a new table for later use. On the other hand, a VIEW statement requires the underlying SELECT statement to be rerun every time the VIEW statement is queried.

Suppose we want to use our NYC taxi data to run reports for daily and weekly periods as well as rate codes such as *Standard and JFK (Airport)*. We could use the current chapter_3_nyc_taxi_csv table we just created, but running even basic queries against that table requires Athena to read all 204,051,059 rows and all 3.4 GB of data. Even for such a small table, this is rather wasteful if we only care about data from a specific week. On larger datasets, it is even more important to model our table along dimensions to give the best performance and cost. *Chapter 4, Metastores, Data Sources, and Data Lakes,* will go deeper into how your table's structure affects performance. For now, we will focus on using CTAS to create a new copy of our table that converts our compressed **comma-separated values (CSV)** files into columnar Parquet and partitions for efficient time filtering and rate code aggregation.

In *Code 3.1*, we have prepared a CTAS statement that reads all columns and rows from the chapter_3_nyc_taxi_csv table created earlier. Once Athena has read all the data, we ask that the resulting table be stored in Parquet format using Snappy compression. Changing formats from CSV to Parquet should result in more compact and faster data to query, especially for simple operations such as COUNT, MAX, and MIN. Using Parquet also has the side-effect of making our queries cheaper since there is less physical data for Athena to read. Our CTAS statement also reorganizes our data by creating two new columns that correspond to the year and month when the taxi ride began. These columns are used to physically partition the data so that Athena can use AWS Glue Data Catalog for *partition pruning* and significantly reduce the data scanned when our queries contain filters along these dimensions.

> **Data bucketing**
>
> We could also have bucketed the rows by the `ratecodeid` column. Bucketing data can help reduce the amount of computation required to generate aggregates when grouping by bucket column. Bucketing by GROUP BY columns helps ensure rows with the same `ratecodeid` column are processed together, reducing the number of partial aggregations Athena's engine will have to calculate. Bucketing has a similar effect to partitioning, without adding additional overhead that can arise from increasing metadata sizes that accompany high numbers of partitions. We'll discuss this more in later chapters, but if you find yourself creating tables with more than 10,000 partitions, you'll want to understand why you have so many partitions and if the benefits outweigh the drawbacks. We excluded bucketing from this example because a later part of this chapter will use INSERT INTO for this table, and Athena doesn't presently support INSERT INTO for bucketed tables.

Now that we understand our CTAS statement, let's go ahead and execute the query in *Code 3.3*. If you have the option, we recommend downloading a copy of the CTAS query from the book's companion GitHub repository in the `chapter_3/ctas_nyc_taxi. sql` file or by using this link: `http://bit.ly/3s6HCXM`. This query shown here should take around 14 minutes to complete and will scan all 3.4 GB of our NYC taxi ride dataset:

```
CREATE TABLE packt_serverless_analytics.chapter_3_nyc_taxi_
parquet
WITH ( external_location = 's3://YOUR_BUCKET_HERE/chapter_3/
tables/nyc_taxi_parquet/',
      format = 'Parquet',
      parquet_compression = 'SNAPPY',
      partitioned_by = ARRAY['year', 'month']
)
AS SELECT
      vendorid, tpep_pickup_datetime, tpep_dropoff_datetime,
      passenger_count, trip_distance, ratecodeid, store_and_fwd_
flag,
      pulocationid, dolocationid, payment_type, fare_amount,
extra,
      mta_tax, tip_amount, tolls_amount, improvement_surcharge,
      total_amount, congestion_surcharge,
      year(date_parse(tpep_pickup_datetime,'%Y-%m-%d %H:%i:%s'))
as year,
```

```
        month(date_parse(tpep_pickup_datetime,'%Y-%m-%d
%H:%i:%s')) as month
FROM packt_serverless_analytics.chapter_3_nyc_taxi_csv
```

Code 3.3 – CREATE TABLE AS SELECT query for partitioned and bucketed NYC taxi data

As you watch Athena crunch away at the CTAS query, you might be wondering why it will take 14 minutes to run this query but only took 8 seconds to read all the CSV data in the earlier test query. The CTAS statement takes considerably longer for two reasons. Firstly, the Parquet format is more computationally intensive to create than CSV. Secondly, we asked Athena to arrange the new table by year and month. Organizing the data in this way requires Athena's engine to shuffle data much in the same way as a GROUP BY query would. Once your query finishes, you should see a new folder in S3 with many subfolders that correspond to the year and month of the data. Now that our new table is ready, let's rerun our simple COUNT query to test it out, as follows:

```
select count(*) from chapter_3_nyc_taxi_parquet
```

After running the preceding query from the Athena console, you should get a result of 204,051,059. The query should have scanned around 0 **kilobytes (KB)** of data and completed after roughly 1 second. While the COUNT query matches the result of 204,051,059 we found in our CSV formatted table from before our CTAS operation, this query's results are far different. The COUNT query against our new Parquet table was 8 times faster than the CSV table and was 340 times cheaper thanks to having read 0 KB of data. You might be asking yourself how this query generated a result if it read no data. This is another happy side-effect of using the Parquet format. Each Parquet file is broken into groups of rows, typically 16-64 MB in size. While generating the Parquet file, the Parquet writer library keeps track of statistically significant information about each row group, including the number of rows and minimum/maximum values for each column. All this metadata is then written as part of the file footer that engines such as Athena can later use to how and if they read each row group. COUNT is one example of an operation that can be fully answered by reading only the row group metadata, not the contents of the files themselves. This leads to the significantly better performance we saw. It also happens to be the case that Athena does not presently consider file metadata to be part of the bytes scanned by the query. So, this query was charged Athena's minimum of 10 MB or **US Dollars (USD)** 0.00005 compared to our earlier COUNT query, which cost USD 0.017.

Reorganizing data for cost or performance reasons is just one of the many reasons you'll find yourself running CTAS queries. Sometimes, you'll want to use CTAS to fix erroneous records or produce aggregate datasets. A less common use case is to speed up result writing. With regular SELECT statements, Athena writes results to a single output object. Using a single output file makes consuming the result easier since ordering and other semantics are inherently preserved. Still, it also limits how much parallelism Athena can apply when generating output. If your query returns many GB of data, you will likely see faster performance simply by converting that query from a SELECT query to a CTAS query. That's because CTAS queries give Athena more opportunity to parallelize the write operations.

For all its benefits, CTAS also has drawbacks, the most prominent of which relates to limited control over the number and size of the created files. Even in our NYC taxi ride example, you can find plenty of files under the recommended 16 MB minimum for Parquet. If our query has to read too many small files, we'll see the overall performance suffer as Athena spends more time waiting on responses from S3 than processing the actual data. Bucketing is one way to help limit the number of files CTAS operations create, but it comes at the expense of making the CTAS operation itself take longer due to increased data shuffling. Without bucketing, we could easily have had three times the number of small files. The final thing to keep in mind with CTAS is that there is a limit to the number of new partitions Athena can create in a single query. This example would have been even better with daily partitions instead of the year and month partitions we included. However, Athena presently limits the number of new partitions in a CTAS query to 100. Since our exercise used 2.5 years of data, we'd have exceeded this limit when using daily partitions. This limit is unique to CTAS and INSERT INTO queries, which create new partitions. SELECT statements can interact with millions of partitions since they are a read-only operation with respect to partitions.

As we've seen, CTAS makes it easy to create new tables by applying one or more transforms to existing tables and storing the result as an independent copy that can be queried without the need to repeat the initial transform effort. INSERT INTO is a related concept that allows you to add new data to an existing table by applying a transform over data from another table. We'll get hands-on with INSERT INTO, sometimes called SELECT INTO, in the next section.

Using INSERT-INTO

Our new and optimized table was a hit with the team. They are now asking if we can add even more history to the dataset and keep it up to date as new data arrives in the landing zone. We could rebuild the entire table using CTAS every time new data arrives, but it would be great if we could run a more targeted query to process and optimize only the newly landed data. That is precisely what INSERT INTO will allow us to do. As we did in the earlier example, our first step will be to download the new data from the NYC Taxi and Limousine Commission. For this exercise, let's add the 2017 trip data to our landing zone by modifying the script from *Code 3.1* to include only our new desired dates. In *Code 3.4*, we've shown how to get started with changing the script. Be sure to run the following script in a directory that has sufficient storage space. If you are using CloudShell, consider running the script in /tmp/, which has more space than your home directory:

```bash
#!/bin/bash


BUCKET=$1
array=( yellow_tripdata_2017-01.csv
        yellow_tripdata_2017-02.csv
        # some entries omitted for brevity
        yellow_tripdata_2017-03.csv
        )
for i in "${array[@]}"
do
     FILE=$i
     ZIP_FILE="${FILE}.gz"
     wget https://s3.amazonaws.com/nyc-tlc/trip+data/${FILE}
     gzip ${FILE}
     aws s3 cp ./${ZIP_FILE} s3://$BUCKET/chapter_3/tables/
nyc_taxi_csv/
     rm $ZIP_FILE
done
```

Code 3.4 – Additional NYC taxi data preparation script

After you run the script, you'll have added 12 new gzipped CSV files to the landing zone table in S3. The way we've created the landing zone means we won't need to run any extra commands once we upload the files—they are immediately available to query once the upload completes. Now, we can add the data to our Parquet optimized table using an `INSERT INTO` query, as illustrated here:

```
INSERT INTO packt_serverless_analytics.chapter_3_nyc_taxi_
parquet
SELECT
    vendorid,
    tpep_pickup_datetime,
    tpep_dropoff_datetime,
    passenger_count,
    trip_distance,
    ratecodeid,
    store_and_fwd_flag,
    pulocationid,
    dolocationid,
    payment_type,
    fare_amount,
    extra,
    mta_tax,
    tip_amount,
    tolls_amount,
    improvement_surcharge,
    total_amount,
    congestion_surcharge,
    year(date_parse(tpep_pickup_datetime,'%Y-%m-%d %H:%i:%s'))
as year,
    month(date_parse(tpep_pickup_datetime,'%Y-%m-%d
%H:%i:%s')) as month
FROM packt_serverless_analytics.chapter_3_nyc_taxi_csv
WHERE
    year(date_parse(tpep_pickup_datetime,'%Y-%m-%d %H:%i:%s'))
= 2017
```

Code 3.5 – INSERT INTO query for adding 2017 data to our Parquet optimized table

The INSERT INTO query should take a bit over 2 minutes to complete, and it will automatically add any newly created partitions to our metastore. You may also have noticed that our INSERT INTO query read more data than you'd expect. We uploaded roughly 1.8 GB of new data, but the INSERT INTO query reports to have read 5.2 GB. Let's dig into why that is by running some ad hoc analytics over our tables. We'll run a query to count distinct tpep_pickup_datetime values in both our landing zone table and our optimized Parquet table for rides that started in 2017. *Code 3.6* contains the query to run against our landing zone tables, and *Code 3.7* has the query you can use against the optimized Parquet table. When you run these queries, you'll notice a couple of interesting differences in how they perform, the amount of data they read, and also that the queries themselves have some differences despite accomplishing the same thing.

You can see the first query here:

```
SELECT
        COUNT(DISTINCT(tpep_pickup_datetime))
FROM packt_serverless_analytics.chapter_3_nyc_taxi_csv
WHERE
  year(date_parse(tpep_pickup_datetime,'%Y-%m-%d %H:%i:%s')) =
2017
```

Code 3.6 – Landing zone distinct vendorid value query

The alternate query is shown here:

```
SELECT
        COUNT(DISTINCT(tpep_pickup_datetime))
FROM packt_serverless_analytics.chapter_3_nyc_taxi_parquet
WHERE year = 2017
```

Code 3.7 – Landing zone distinct vendorid value query

The first difference to understand is that the query in *Code 3.6* must first parse and transform the `tpep_pickup_datetime` column before it can be used to filter out records that aren't from 2017. This is significant because it indicates that our dataset may not be partitioned on the filtering dimension. A closer look at the landing zone table's definition from *Code 3.2* confirms there are no partition columns defined in the table creation query. Applying a function, transform, or arithmetic to a column as part of the `WHERE` clause is not a guarantee that you aren't querying along a partition boundary. However, Athena achieves peak filtering performance when partition conditionals use literal values. This is because Athena can *push* the filtering clauses deeper within its engine, or possibly down to the metastore itself. In this case, we are using the `date_parse` function because the landing zone table isn't partitioned on `year`; it's not partitioning on anything at all. That's why any query we run against the landing zone table may be forced to scan the entire table.

Contrast this with the query in *Code 3.7*, which has an explicit `year` column and can use a simple, literal filter of `year=2017`. The second query runs much faster than the first and scans only a subset of the data (396 MB) that is in the 2017 partition. This is much closer to what we'd expect because it seems natural that filters reduce the data scanned. You might also be wondering why we chose to use `COUNT( DISTINCT vendorid)` instead of something more straightforward such as `COUNT(*)`. The reason is simple. Our optimized Parquet table can answer `COUNT(*)` operations without actually reading the data because it stores basic statistics in every row group's header. Using `DISTINCT` is one way to bypass many Parquet optimizations that apply only to special-case queries such as `COUNT(*)`, `MIN()`, and `MAX()`. Had those optimizations kicked in for our investigation, we'd have formed the wrong impression about how much data was in our Parquet table or how long it might take to query. In practice, these optimizations are precisely why Parquet is increasingly becoming a go-to format.

At this point, it might be obvious why we'd normally want to upload new data into a dedicated folder within the landing zone. Partitioning the landing zone allows us to run targeted queries against only the latest data. This can be done by treating that folder as a new partition or temporary table. For simplicity, we omitted that step from this example.

In the next section, we will learn about the final type of advanced query covered in this chapter. You'll learn how the `TABLESAMPLE` decorator allows you to reduce the cost and runtime of exploratory queries while bounding the impacts of sampling bias.

Running approximate queries

In *Chapter 1*, *Your First Query*, we used TABLESAMPLE to run a query that allowed us to get familiar with our data by viewing an evenly distributed sampling of rows from across the entire table. TABLESAMPLE enables you to approximate the results of any query by sampling the underlying data. Athena also supports more targeted forms of approximation that offer bounded error. For example, the approx_distinct function should produce results with a standard error of 2.3% but completes its execution 97% faster while also using less peak memory than its completely accurate counterpart, COUNT(DISTINCT x). We'll learn more about these and several other approximate query tools by exploring our NYC taxi ride tables.

TABLESAMPLE is a somewhat generic technique for running approximate queries. Unlike the other methods we discuss in this section, TABLESAMPLE works by sampling the input data. This allows you to use it in conjunction with any other SQL features supported by Athena. The trade-off is that you'll need to take care to ensure you understand the error you may be introducing to your queries. This error most commonly manifests as observation bias since your query is now only "observing" a subset of the data. If the underlying sampling is not uniform, you may draw conclusions that are only relevant to the subset of data your query read but not the overall dataset.

To demonstrate, let's try running a query to find the most popular hours of the day for riding in a taxi. We'll run the query in three different ways, first using the following query, which scans the entire table and produces a result with 100% accuracy:

```
SELECT
    hour(date_parse(tpep_pickup_datetime,'%Y-%m-%d %H:%i:%s'))
as hour,
    count(*)
FROM
    packt_serverless_analytics.chapter_3_nyc_taxi_parquet
GROUP BY hour(date_parse(tpep_pickup_datetime,'%Y-%m-%d
%H:%i:%s'))
ORDER BY hour DESC
```

Code 3.8 – Hourly ride counts query

Our second query, shown here, adds the `TABLESAMPLE` modifier. This query uses the `BERNOULLI` sampling technique to read only 10% of the table's underlying data:

```
SELECT
    hour(date_parse(tpep_pickup_datetime,'%Y-%m-%d %H:%i:%s'))
as hour,
    count(*) * 10
FROM
    packt_serverless_analytics.chapter_3_nyc_taxi_parquet
    TABLESAMPLE BERNOULLI (10)
GROUP BY hour(date_parse(tpep_pickup_datetime,'%Y-%m-%d
%H:%i:%s'))
ORDER BY hour DESC
```

Code 3.9 – Hourly ride counts query with 10% BERNOULLI sampling

The following code block contains our third and final query. It again uses the `TABLESAMPLE` modifier but swaps `BERNOULLI` sampling for the `SYSTEM` sample technique to read only 10% of the table's underlying data:

```
SELECT
    hour(date_parse(tpep_pickup_datetime,'%Y-%m-%d %H:%i:%s'))
as hour,
    count(*) * 10
FROM
    packt_serverless_analytics.chapter_3_nyc_taxi_parquet
    TABLESAMPLE SYSTEM (10)
GROUP BY hour(date_parse(tpep_pickup_datetime,'%Y-%m-%d
%H:%i:%s'))
ORDER BY hour DESC
```

Code 3.10 – Hourly ride counts query with 10% SYSTEM sampling

After you run all three queries, you'll see a pattern form. We've collated the data for the most popular hours as a table in *Code 3.11*. The original query took 6 seconds and scanned 1.42 GB of data but produced results that are 100% accurate. The second query used the BERNOULLI sampling technique to uniformly select 1 out of every 10 rows for inclusion in the result. That query took 3.4 seconds to complete and still scanned 1.42 GB of data but incurred an error of just 0.006%. That's a nearly 50% speedup while sacrificing minimal accuracy. Our last query used SYSTEM sampling to include 1 out of every 10 files in the dataset. This final query scanned 92% less data (116 MB) and ran 30% faster (4.3 seconds) than the original query but was 9% less accurate on average.

You can see the results here:

Hour	Actual Count	BERNOULLI	SYSTEM
17	18206794	18223390	15679710
18	20352240	20378540	19000690
19	19580912	19553590	17711660
20	17871911	17903240	16079030
21	17712882	17720660	16195500

Table 3.1 – Ride count by hour using different sampling techniques

Our NYC taxi ride data is mostly uniformly distributed, so both sampling techniques did reasonably well. If our data had not been uniformly distributed with respect to the dimensions we queried on, then SYSTEM sampling would be more vulnerable to sampling bias. BERNOULLI sampling is more resistant to skew in the data's physical layout but isn't completely immune from sampling bias. In general, both sampling techniques speed up the query by reducing how much data is considered, but they do it differently. SYSTEM sampling discards entire files, which is why it scanned less total data from a billing perspective. BERNOULLI sampling applies the same determination at a row level, which means reading all the data before discarding it.

That wraps up the generic approximation facilities. Next, we'll use more targeted functions that can speed up specific analytical operations. A common exercise is to understand how a value in your data compares to the rest of the data; for example, is this distance traveled in a given taxi ride an outlier or relatively common? One way to answer this question is to understand what percentile the given ride presents. Put another way, what percentage of rides were less than or equal to the length of the ride we are inspecting? Percentiles are a great way to accomplish that. Unfortunately, calculating the percentiles for a large dataset can be resource-intensive and require scanning the entire dataset. We can do better than the generic sampling techniques offered by TABLESAMPLE. The following query calculates five different percentiles for our dataset while scanning only 462 MB of the total 1.4 GB in our table yet still manages to achieve a standard error of 2.3%. The approx_percentile function we are leveraging also supports supplying your own accuracy parameter:

```
SELECT approx_percentile(trip_distance, 0.1) as tp10,
  approx_percentile(trip_distance, 0.5) as tp50,
  approx_percentile(trip_distance, 0.8) as tp80,
  approx_percentile(trip_distance, 0.9) as tp90,
  approx_percentile(trip_distance, 0.95) as tp95
  FROM  packt_serverless_analytics.chapter_3_nyc_taxi_parquet
```

Code 3.11 – Approximating ride duration percentiles with approx_percentile(…)

After running the query, you'll see that 90% of rides traveled at least 6.9 miles and 10% of rides traveled just .6 miles. You can see how basic outlier detection can be implemented using approx_percentile to compare any given value to the broader population of values. In addition to approx_percentile, Athena also supports approx_distinct and numeric_histogram functions of other memory-intensive calculations that typically require scanning the entire dataset.

Quantile Digest (Q-Digest): Using trees for order statistics

As with many other engines, Athena uses a special data structure to facilitate the time and memory-saving capabilities offered by approx_percentile. Q-Digest is a novel usage of binary trees whose leaf nodes represent values in the population dataset. By propagating infrequently seen values—and their frequency—up to higher layers of the tree, you can bound the memory required to generate percentiles. The memory allocated to the construction of these trees directly influences the rate of error in the resulting statistics.

We've run quite a few queries so far in this chapter. You might be wondering how to find that fascinating query we ran at the start of the chapter or where you can see the error associated with a particular query once you've closed your browser. In the next section, we'll review options for organizing workloads and reviewing our query history.

Organizing workloads with WorkGroups and saved queries

Athena WorkGroups allow you to separate different use cases, applications, or users into independent collections. Each workgroup can have its own settings, including results location, query engine version, and query history, to name a few. In *Figure 3.1*, you can see the various WorkGroups we have created while authoring this book. This view lets you see the status of each workgroup at a glance. More in-depth settings or the creation of new WorkGroups are just a click away. Every Athena query runs in a workgroup. So far, we haven't set any specific workgroup for our queries, so they've been running the "primary" workgroup. The primary workgroup is special and is automatically created for you the first time you use Athena.

You can see an overview of the Athena WorkGroups screen here:

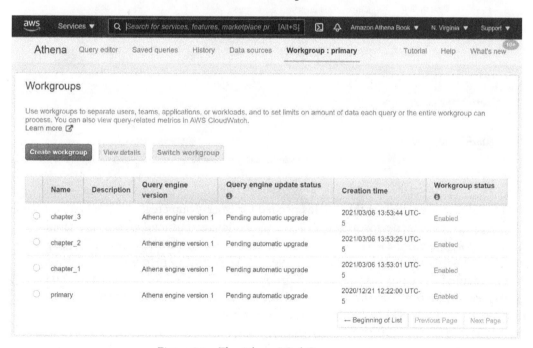

Figure 3.1 – The Athena WorkGroups screen

Athena customers often choose to use different WorkGroups for different kinds of queries. You can start getting into the habit of doing this right now by creating a new workgroup that you can use to run the remainder of the exercises in this book. To begin, click the **Create workgroup** button on the **Workgroups** page shown in *Figure 3.1*. You can get to that page by clicking on the **Workgroup: primary** tab at the top of the Athena console. If you are using the IAM policy recommended for this chapter, clicking the **Create workgroup** button will take you to a new page with the **Create workgroup** form, as shown in the following screenshot, *Figure 3.2*, and *Figure 3.3*:

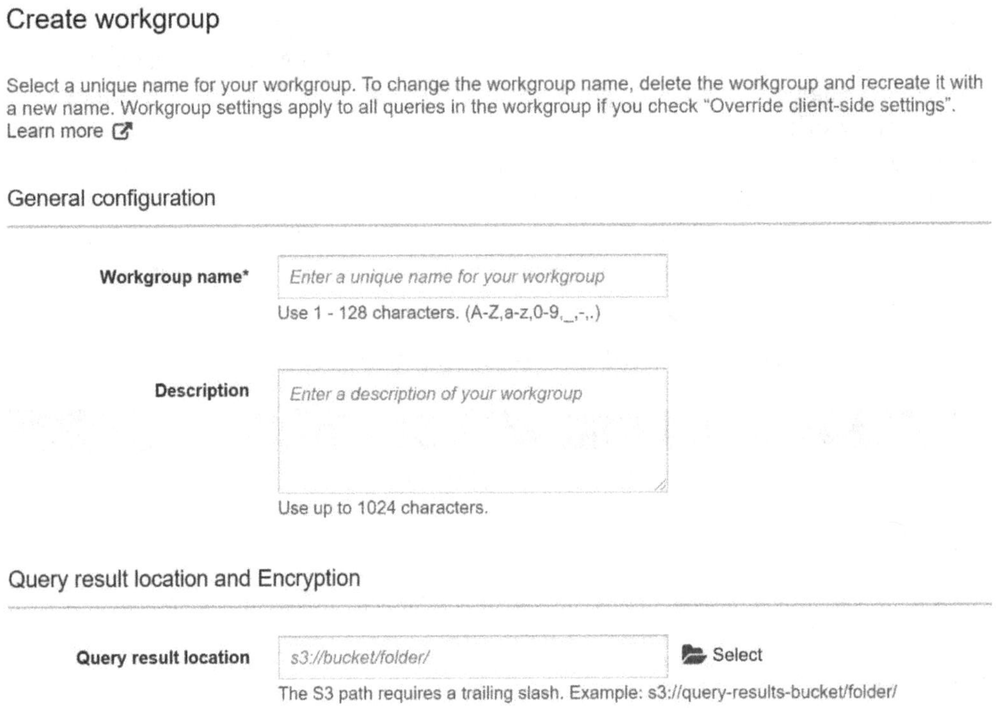

Create workgroup

Select a unique name for your workgroup. To change the workgroup name, delete the workgroup and recreate it with a new name. Workgroup settings apply to all queries in the workgroup if you check "Override client-side settings". Learn more ⤢

General configuration

Workgroup name*

Enter a unique name for your workgroup

Use 1 - 128 characters. (A-Z,a-z,0-9,_,-,.)

Description

Enter a description of your workgroup

Use up to 1024 characters.

Query result location and Encryption

Query result location

s3://bucket/folder/   📂 Select

The S3 path requires a trailing slash. Example: s3://query-results-bucket/folder/

Figure 3.2 – Creating an Athena workgroup form Part 1

In *Figure 3.2*, you see the first three fields needed for workgroup creation. The first is simply the name of the new workgroup. The IAM policy recommended for this chapter will allow you to create new WorkGroups as long as they begin with packt-. You can try packt-athena-analytics as an example. The **Description** field is optional, purely used to document the purpose of the workgroup. Lastly, we need to set the default query results location for this workgroup. You may recall from previous chapters that Athena stores query results in S3 before making them available to your client or the Athena console. This allows you to reread the results as many times as you like, without needing to pay or wait for the query itself to run again. Naturally, we need to tell Athena where we'd like to store the results of queries run in this workgroup.

Aside from any organization naming conventions you may need to follow, there are two important factors to keep in mind when configuring these settings. The first is that Athena won't clean up this data after it's no longer needed. In fact, Athena has no idea if you are done using this data. You'll minimally want to set up an S3 Lifecycle policy to automatically delete data from this location that is older than a threshold you deem appropriate. If you need the results to be available longer than that, you should explicitly move them to a different location for long-term retention or consider running such queries in their own workgroup. Lastly, you'll want to consider who else has access to this S3 location. Imagine you have two personas in your organization: an Administrator who can read from any table and an intern who only has access to non-sensitive datasets. If the Administrator is running queries in a workgroup with a result location that is readable to the intern, you may be inadvertently providing a path for privilege escalation. The intern may accidentally stumble across the results of highly sensitive queries run by an Administrator. The same is true for a malicious actor. They no longer need to attack your permissioning system. You've unintentionally poked a hole in the armor by picking an overly permissive or shared query results location.

In the following screenshot, we are presented with four more settings to create our new workgroup:

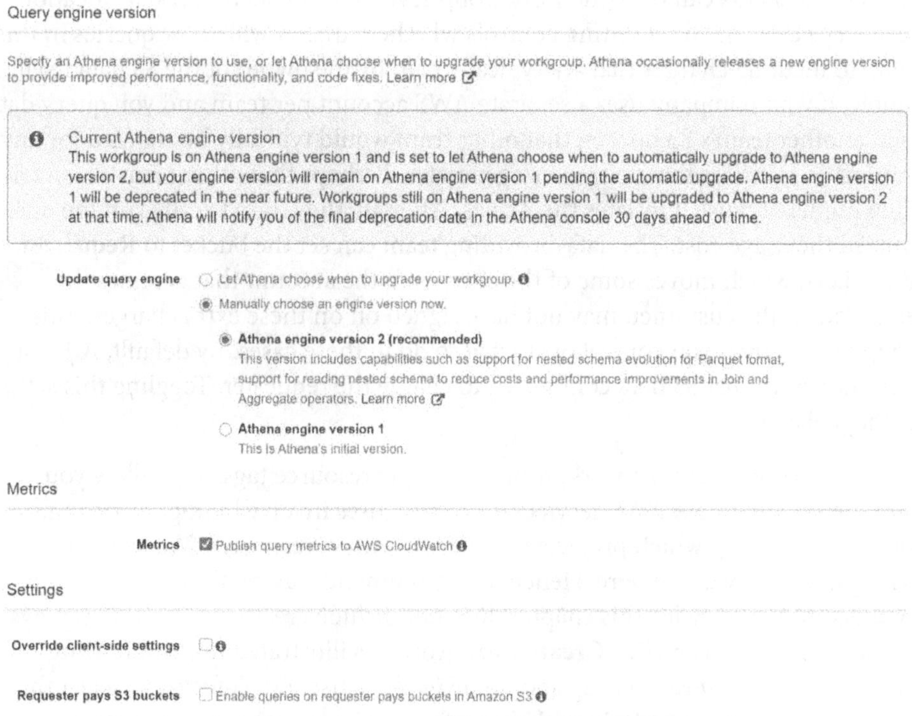

Figure 3.3 – Creating an Athena workgroup form Part 2

Athena's underlying engine, a hybrid of Presto and Trino, is rapidly evolving. As such, Athena has built-in facilities to handle upgrades. We'll talk more about Athena's automatic testing and upgrade functionality later in this chapter. For now, all you need to know is that Athena offers you full control over which engine version you use per workgroup. This allows you to isolate sensitive workloads to prevent auto-upgrades and enables you to take a sneak peek at upcoming versions so that you can prioritize upgrades that have an outsized benefit for you. It is highly recommended to set this to **Let Athena choose when to upgrade your workgroup** unless you've been advised otherwise by the Athena service team or are attempting to run a test against a specific version. This book's exercises will include new features that are only available in Athena engine version 2 or later, so be sure to pick **Manually choose an engine version now** and pick **Athena engine version 2** or later. Failing to set the appropriate engine version on your workgroup may result in failures later, as Athena may or may not have auto-upgraded you when running through the exercises in this book.

The next setting determines if Athena will emit query metrics to AWS CloudWatch for all activities in the workgroup. We recommend leaving **Metrics** enabled as this will make troubleshooting, reporting, and auditing much easier. The last two settings are uncommon but enable interesting applications and integrations. As the Administrator of a workgroup, you can decide if clients can override workgroup-level settings such as results location on a per-query basis. The final setting controls whether Athena will allow queries in this workgroup to incur S3 charges that are typically paid by the owner of the S3 data itself. For example, if your company uses a separate AWS account per team and you query data that sits in another team's S3 bucket, that other team would typically be charged for any S3 operations or transfers that your query generates. Perhaps that other team doesn't like this billing model because it inflates their costs. After all, they didn't really run the query that incurred the usage cost. The data-providing team can set the bucket to **Requester pays S3 buckets**, which moves some of the charges to the account that accesses the S3 objects. You, as the customer, may not have signed off on these extra charges. This workgroup setting gives you control over what to do in these cases. By default, Athena will abort queries against S3 data configured to charge the requester. Toggling this setting changes that behavior.

The final option we can set on a workgroup is to apply resource tags. Tags allow you to organize resources across AWS services. Common uses involve billing, reporting, or simply understanding which projects make use of which resources. We won't be covering tagging in any depth here. Hence, we recommend leaving these blank as the recommended IAM policy for this chapter does not include creating or modifying tags. Once you are ready, you can click **Create workgroup**, as illustrated in the following screenshot, and your new workgroup should be ready to use. Don't forget to select that new workgroup by clicking **Switch workgroup** from Athena's **workgroup** page:

Tags

A tag is a label that you assign to an Athena workgroup resource. It consists of a key and a value. Use tags to categorize workgroups by purpose, owner, or environment. You can also use tags in IAM policies to allow or deny access to workgroup actions based on a tag key/value pair, or on specific values for a tag key. Use best practices and create a consistent set of tags. Do not use duplicate tag keys the same workgroup. Learn more ☑

Key		Value (Optional)	
	Enter key		*Enter value*

Use 1 - 128 characters. (A-Z,a-z,0-9,   Use up to 256 characters. (A-Z,a-z,0-9,
._.:::/,=,+,-,@)   ._.:::/,=,+,-,@)

Cancel   Create workgroup

Figure 3.4 – Creating an Athena workgroup form Part 3

Now that we have our new `packt-athena-analytics` workgroup, let's see how we can save our most frequently used queries as **named queries** in our workgroup. **Named Queries**, also called **saved queries** in some parts of the Athena console, allow you to quickly load and run a query without re-entering the entire text of the query. To begin creating a named query, start typing a new query into the Athena query editor, just as you did for the previous queries we've run. For simplicity, you can use a COUNT(*) query by year over our taxi ride data, as illustrated in the following code snippet:

```
SELECT year, COUNT(*)
FROM packt_serverless_analytics.chapter_3_nyc_taxi_parquet
GROUP BY year
```

Go ahead and run the query so that we know it works and we didn't mistype anything. Once the query completes, click **Save as** below the query editor, as shown in the following screenshot:

Figure 3.5 – Creating a named query

After you click **Save as**, you'll be prompted to give the query a name and a description. The saved query will only be visible to users of the workgroup you saved it to. Athena will remember this query and allow you to run or edit the query as many times as you like until you delete it. You can access the current set of saved queries by click on the **saved queries** tab at the top of the Athena console. This feature is good for bookmarking frequently used queries as part of an operational runbook or ad hoc analysis.

So far, all our Athena usage has been via the AWS console. As we begin to conclude *Part 1* of this book, *Fundamentals Of Amazon Athena*, we'll introduce you to Athena's rich APIs. Virtually everything we've done with Athena's console can be done via the AWS **software development kit** (**SDK**) or AWS CLI. If you plan to build applications or automate analytics pipelines using Athena, you'll find using these APIs an easier route. If you aren't a developer or rarely use the command line, don't be intimidated. We will go step by step through each command, its arguments, and common reasons for failure.

Using Athena's APIs

As an introduction to Athena's APIs, we will demonstrate how to run basic geospatial queries with Athena using the AWS CLI. The AWS CLI provides a simple wrapper over each of the APIs supported by Athena. This allows us to get familiar with the APIs without having to make any choices about programming language. The APIs we use in this section are available in all supported languages such as Java, Golang, and Rust. Now that we've got a better understanding of the basic Athena concepts, we'll also use a slightly more advanced example dataset that will give us a chance to experiment with Athena's geospatial capabilities.

Use Athena engine version 2 or later

In case you skipped the instructions in the previous section pertaining to the creation of a new workgroup with Athena engine version 2, please take a moment to either switch to that workgroup now or change your current workgroup to explicitly use Athena engine version 2 or later to avoid errors in this exercise. Athena's geospatial functions have dramatically improved since Athena engine version1, so we'll be targeting features from Athena engine version 2 or later.

First, we will need to download two geospatial datasets from the **Environmental Systems Research Institute (Esri)**, an industry leader in geospatial solutions, and upload that data to S3. The first dataset contains earthquake data for the state of California. The second dataset includes information on borders between all the different counties in California. California is an extremely seismically active area of the **United States (US)**. Next, we will use Athena's APIs to run two **Data Definition Language (DDL)** queries to create tables for each of the datasets we downloaded. These datasets are less than 5 MB each. The book's GitHub repository contains a script that fully automates these steps to make this process easier. You can run the following commands in your AWS CloudShell environment, right from your browser. Alternatively, you can run these commands in most Linux-compatible environments with `wget` and the AWS CLI installed. After you run these commands, we'll quickly walk through what the `geospatial_api_example.sh` script does. Remember to supply the script with the S3 bucket you've been using to store data related to our experiments and the name of an Athena workgroup in which the script's queries will run:

```
wget -O geospatial_api_example.sh https://bit.ly/3sZZRia
chmod +x geospatial_api_example.sh
./geospatial_api_example.sh <S3_BUCKET> <ATHENA_WORKGROUP_NAME>
```

If successful, the script will have created two new tables in the `packt_serverless_analytics` database and printed the details of the accompanying Athena DDL queries to the Terminal. Let's go section by section through the script you downloaded earlier. We'll skip the uninteresting bits such as documentation or boilerplate error handling. Here we go:

```
#!/bin/bash
BUCKET=$1
WORKGROUP=$2
```

Bash scripts always start with a special sequence of characters, `#!`, called a *shebang*. This tells the system that what follows is a series of commands for a particular shell. In this case, we are using the **Bash** shell located at `/bin/bash`. This is mostly unrelated to Athena and its APIs, so don't worry if it is new or confusing. The only interesting bit in this first section is that the script treats the first argument as an S3 bucket and the second argument as an Athena workgroup. We'll see how these arguments get used later in the Athena APIs that get called.

The script then downloads the first dataset from the Esri GitHub repository using wget and then uploads it to S3 using the S3 bucket provided by the first argument to the script, as illustrated in the following code snippet. This process is repeated for the second dataset:

```
wget https://github.com/Esri/gis-tools-for-hadoop/blob/master/
samples/data/earthquake-data/earthquakes.csv

aws s3 cp ./earthquakes.csv s3://$BUCKET/chapter_3/tables/
earthquakes/
```

So far, the script hasn't interacted with Athena at all. This section prepares a CREATE TABLE query that it will send to Athena via the start-query-execution API. The script again uses some Bash magic in the form of read -d" VARIABLE << END_ TOKEN to make the multiline CREATE TABLE query more human-readable. The code is illustrated here:

```
read -d '' create_earthquakes_table << EndOfMessage

CREATE external TABLE IF NOT EXISTS packt_serverless_analytics.
chapter_3_earthquakes

(/* columns omitted for brevity */)

ROW FORMAT DELIMITED FIELDS TERMINATED BY ','

STORED AS TEXTFILE LOCATION 's3://${BUCKET}/chapter_3/tables/
earthquakes/'

EndOfMessage
```

The CREATE TABLE query preparation is repeated for the second dataset before we finally get to our first Athena API calls. Here, we are using the start-query-execution API to run a DDL statement to create an earthquakes table. A nearly identical API call also gets made for the California counties dataset. The API takes two parameters, the query to run and the workgroup in which to run the query, as illustrated in the following code snippet:

```
aws athena start-query-execution \
--query-string "${create_earthquakes_table}" \
    --work-group "${WORKGROUP}"
```

The vast majority of Athena's APIs are asynchronous. This means that the API calls complete relatively quickly, but the API's work isn't necessarily done when the API call completes. The start-query-execution API is a perfect example of this asynchronous pattern. When you run the script or this API call directly, you'll see that it returns almost immediately, even for queries that may take many minutes or hours to run. That's because completion of this API means Athena has accepted the query by doing some basic validations, authorization, and limit enforcement before giving us an **identifier** (**ID**) that we can later use to check the status of the query. This ID is called an Athena query execution ID and will also be used to retrieve our query's results programmatically.

Let's use the output of the script's two `start-query-execution` calls to check the status of our `CREATE TABLE` queries. Replace `QueryExecutionId` in the following command with one of your `QueryExecutionId` instances:

```
aws athena get-query-execution --query-execution-id
<QueryExecutionId>
```

When you run this API call, you'll get output similar to the following. If your query ran into any issues, including permissions-related problems, you'd see root cause details too:

```
"QueryExecution": {
        "Query": "<QUERY TEXT OMMITED FOR BREVITY>",
        "StatementType": "DDL",
        "ResultConfiguration": { "OutputLocation": "s3://… "},
        "Status": {
            "State": "SUCCEEDED",
            "SubmissionDateTime": "2021-03-
07T18:25:14.736000+00:00",
            "CompletionDateTime": "2021-03-
07T18:25:15.902000+00:00"
        },
        "Statistics": {
            "DataScannedInBytes": 0,
            "TotalExecutionTimeInMillis": 1166
        },
        "workgroup": "packt-athena-analytics"
}
```

In addition to giving us information about if the query succeeded or failed, the `get-query-execution` API also returns information about the type of query, how much data it scanned, how long it ran, and where its results were written. Using this API, you can embed lifecycle tracking and query scheduling functionality in your own applications. Now that we have a basic understanding of how to use Athena's APIs via the AWS CLI, let's try running queries that leverage Athena's geospatial functions. For this final exercise, let's imagine we work for an insurance company trying to build an automated claim-handling website. Our customers will go to this website and fill out forms to make insurance claims against their homeowners' insurance. We'd like to automatically approve or reject obvious claims before they get to a human. This saves time by giving customers rapid responses and helps ensure we prioritize essential claims. We've been asked to ensure that claims pertaining to natural disasters get escalated quickly. All our customers are in California, so we decided to start by automating earthquake claims. Whenever a customer selects `earthquake` as the cause for a claim, we need to run a query to determine if there were any recent earthquakes in their area. Luckily, Athena's geospatial function suite offers several ways to do this. A straightforward way is to understand if the county that the homeowner lives in has had any recent earthquakes. Here, we are using the county as a bounding box and then searching for any earthquakes in that vicinity. The `ST_CONTAINS` and `ST_POINT` functions allow us to treat the county as the search area and the earthquake's epicenter as a point; then, we can count how many earthquakes originated in each given county. In practice, a better method would also be to treat the earthquake as an area of impact and then the homeowner's house as a point, but that would be a much more challenging sample dataset to create.

The following Athena API call will run a query that uses our new `earthquake` and `counties` tables in conjunction with the `ST_CONTAINS` and `ST_POINT` functions to count how many earthquakes happened in each county:

```
aws athena start-query-execution \
--query-string "SELECT counties.name, COUNT(*) cnt \
FROM packt_serverless_analytics.chapter_3_counties as counties \
CROSS JOIN packt_serverless_analytics.chapter_3_earthquakes as earthquakes \
WHERE ST_CONTAINS (counties.boundaryshape, ST_POINT(earthquakes.longitude, earthquakes.latitude)) \
GROUP BY  counties.name \
ORDER BY  cnt DESC" \
--work-group "packt-athena-analytics"
```

If you repeat our `get-query-execution` API call for this query, you may see that it is in a RUNNING state since it takes much longer than our CREATE TABLE queries. You can keep running the `get-query-execution` API call shown here until the query transitions to a SUCCEEDED or FAILED state:

```
aws athena get-query-execution --query-execution-id
<QueryExecutionId>
```

Assuming your query succeeds, you can then use Athena's `get-query-results` API to fetch pages of rows containing your query results. The command is shown in the following code block. Remember to substitute in a quoted version of your query's execution ID:

```
aws athena get-query-results --query-execution-id
<QueryExecutionId>
```

The `get-query-results` API returns data as rows of **JavaScript Object Notation (JSON)** maps. The first row contains the column headers, while subsequent rows have the values associated with each row. The output can be very verbose, so many applications choose to access results directly from the S3 location. The code is illustrated in the following snippet:

```
{"ResultSet": {   "Rows": [
  {"Data": [{"VarCharValue": "name"},{"VarCharValue": "cnt"}]},
  {"Data": [{"VarCharValue": "Kern"},{"VarCharValue": "36"}]},
  {"Data": [{"VarCharValue": "San
Bernardino"},{"VarCharValue":"35"}]}
                ...Remainder Omitted for Brevity...
```

When integrating with Athena, `start-query-execution`, `get-query-execution`, and `get-query-results` are the most frequently called APIs. Still, there are many others for managing WorkGroups and saving queries and data sources. Hopefully, if you've never used APIs before, this exercise has removed some of the mystery surrounding them. If you're a seasoned developer, you're likely starting to form a view of how you can connect your applications to Athena. *Chapter 9, Serverless ETL Pipelines,* and *Chapter 10, Building Applications with Amazon Athena,* will use more sophisticated examples to demonstrate the power of integrating your application with Athena.

Summary

In this chapter, you concluded your introduction to Athena by getting hands-on with the key features that will allow you to use Athena for many everyday analytics tasks. We practiced queries and techniques that add new data, either in bulk via CTAS or incrementally through INSERT INTO, to our data lake. Our exercises also included experiments with approximate query techniques that improve our ability to find insights in our data. Features such as TABLESAMPLE or approx_percentile allow us to trade query accuracy for reduced cost or shorter runtimes. Cheaper and faster exploration queries enable us to consult the data more often. This leads to better decision-making and less reluctance to run long or expensive queries because you proved their worth with a shorter, approximate query. This may be hard to imagine given that all the queries in this chapter took less than a minute to run and, in aggregate, cost less than USD 1. In practice, many fascinating queries can take hours or days to complete and cost hundreds of dollars. These are the cases where approximate query techniques can show their merit.

Next, we saw how to organize our workloads into WorkGroups so that our queries can use different settings such as Athena engine. Then, we closed out with an excursion into using Athena's APIs, instead of the AWS console, to run queries. This example was simple but demonstrated how a fictional insurance company could use these APIs to enhance their application by running geospatial workloads on Athena.

While your introduction to Athena is now complete, the next part of this book will begin an introduction to building data lakes at scale. Understanding how data modeling affects your Athena applications' performance and security will enable you to ensure you have the right data in place for your application or analytics needs. Tools such as **AWS Lake Formation** will help you automate many of the activities you'll need to have in place before *Part 3* of this book, *Using Amazon Athena*, brings us full circle to write our applications on top of Athena.

Section 2: Building and Connecting to Your Data Lake

In this section, you will learn how to build, secure, and connect to a data lake with Athena and Lake Formation.

This section consists of the following chapters:

4
Metastores, Data Sources, and Data Lakes

One of the best features of **Athena** is that it allows you to query data where it lives. That data can be sitting on S3, in a relational database, your EC2 environment, or any other source from which business value can be derived. However, the vast majority of Athena's usage is to query data on S3. Before Athena can query this data, it needs to know where the data is and how to read it, as data on S3 can be in many different file formats. Athena needs to translate the databases and tables referenced in SQL statements into physical S3 locations, and then choose the right libraries to interpret the data that's been read from that location. The place where Athena goes to look up these translations is called the **metastore**.

This chapter will dive into the metastore and the information stored there. We will cover what information is required to register tables in a metastore. The metastore is just one of three key pieces that make up a data source; the other two components are the data that we want to query and a connector that lets Athena access the metastore and data. We will break down the data source by looking at the two different S3 data sources that Athena natively provides in depth. We will then compare the two to help you decide which one is appropriate for your use case.

Metastores need to be populated to be useful. We will go over some common ways to register tables. Manually entering datasets into our catalog can be a painful and error-prone process, so we will look at how **AWS Glue Crawlers** can help. Crawlers can automatically discover and register datasets in the metastore. We will also go through the process of creating one and see it in action.

Lastly, we will look into the **data lake architecture** and appreciate the value that it can bring to an organization. Building data lakes requires a central catalog (that is, a metastore) that can be used by an organization to discover datasets and query data that was not possible with traditional on-premises storage.

In this chapter, we will cover the following topics:

- What is a metastore?
- What is a data source?
- Registering S3 datasets in your metastore
- Discovering your datasets on S3 using AWS Glue Crawlers
- Designing a data lake architecture

Technical requirements

For this chapter, you will require the following:

- Internet access to GitHub, S3, and the AWS console.
- A computer with Chrome, Safari, or Microsoft Edge and the AWS CLI version 2 installed (`https://amzn.to/3sYabba`).

- An AWS account and accompanying IAM user (or role) with sufficient privileges to complete this chapter's activities. For simplicity, you can always run through these exercises with a user that has full access. However, we recommend using scoped-down IAM policies to avoid making costly mistakes and learn how to best use IAM to secure your applications and data. You can find a minimally scoped IAM policy for this chapter in this book's accompanying GitHub repository, which is listed as `chapter_4/iam_policy_chapter_4.json` (https://bit. ly/3qAcNtU). This policy includes the following:

  - Permissions to create and list IAM roles and policies. We will be creating a service role for an AWS Glue Crawler to assume.

  - Permissions to read, list, and write access to an S3 bucket.

  - Permissions to read and write access to Glue Data Catalog databases, tables, and partitions. You will be creating databases, tables, and partitions manually and with Glue Crawlers.

  - The ability to create and run permissions for Glue Crawlers.

  - The ability to gain access to run Athena queries.

- An S3 bucket that is readable and writeable. If you have not created an S3 bucket yet, you can do so from the CLI by running the following command:

```
aws s3api create-bucket --bucket <YOUR_BUCKET_NAME>
--region us-east-1
```

Ensure that the *NYC Taxi* dataset has been copied into your bucket. If you have not done so, you can run the commands located at `https://bit.ly/2XW1LCA`.

What is a metastore?

Metastores are a critical component for Athena. Metastores tell Athena which datasets are available for it to query and how to process the underlying data. When a user submits a SQL statement to Athena for execution, Athena parses the query's text, identifies the tables and columns needed, and looks up a description of them from the metastore. Once it knows where the data lives, how it is stored, and the format, Athena requests the data, interprets it, and executes the query.

The metastore also serves as a directory of available datasets that can be queried. Datasets are represented by tables stored in databases, although in this context, the terms tables and databases do not refer to physical databases or tables. We refer to tables and databases as **metadata**, data that describes other data, and metastores store metadata. In the big data space, analytics engines usually store metadata and data separately. Athena's most common metastore is AWS Glue Data Catalog, and its most common data store is S3. The following diagram shows the separation of the metastore and data:

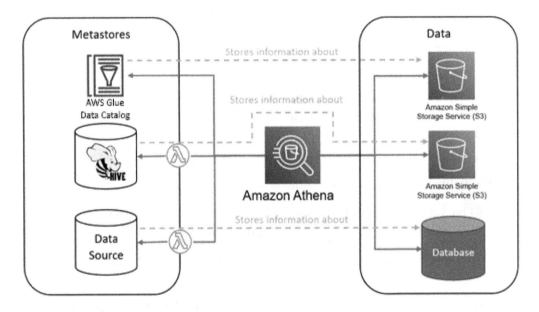

Figure 4.1 – Metastores are stored separately from the underlying data

The data in a metastore is organized into a hierarchy of databases and tables. The following diagram shows the relationship between the various objects that are stored in a metastore:

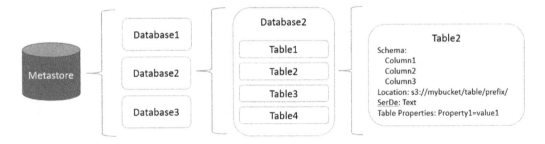

Figure 4.2 – The essential components of a typical metastore

Adding, updating, and removing metadata from Athena's metastore can be done using Apache Hive's **Data Definition Language (DDL)**. Under the hood, Athena executes DDL statements using Apache Hive.

In this section, we will go through the information stored in a metastore so that we can learn how to start populating our metastore with databases and tables. But before we do, let's look at what a data source and its components are.

Data sources, connectors, and catalogs

Metastores, connectors, and catalogs can sometimes be seen being used interchangeably, although there are subtle differences between the terms. A metastore contains a catalog of available datasets and their metadata. A connector allows Athena to read the metastore and data. Lastly, a data source includes all three components: the metastore, data, and connector. There is almost always a one-to-one relationship between a metastore and data source and a connector:

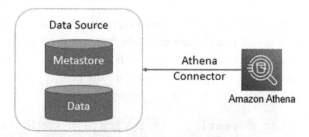

Figure 4.3 – Data source versus metastore versus connector

The preceding diagram shows the relationship between these components. Athena can register many different data sources. We will look at this in more detail in the *What is a data source?* section.

Databases and schemas

Athena and Apache Hive use the term database to refer to a collection of tables. In MySQL and other relational databases, the term schema is usually used instead. Still, many people, including myself, use these terms interchangeably. We will use the term database exclusively for the remainder of this book.

Using databases can help organize tables into categories based on usages, such as keeping tables owned by the same group within an organization. You can also group tables based on user roles, for example, having a database for finance users and one for analysts. Ultimately, it is suggested that you keep the rules that are used to group tables together consistent.

To create a database in Athena, you can run a DDL query like the following:

```
CREATE DATABASE packt_serverless_analytics LOCATION 's3://<S3_
BUCKET>/tables/'
```

This query has two parts: creating the database called `packt_serverless_analytics` and the location where that database should point to.

> **Important Note**
>
> The database's location is not very important to Athena because it doesn't use it. However, it will be essential to use the same metastore with Apache Hive or PrestoDB on Amazon EMR or elsewhere. The location represents where managed tables will be stored and can cause a lot of headaches. It is strongly recommended to set the location of the database to a path in S3. We will look at this in more detail in the *Tables/datasets* section when we look at managed tables.

When running an Athena query, any references to tables that are not prefixed with a database will be assumed to run in a default database that you specify. To refer to tables in other databases, you can prefix the table name with the database. For example, the following statement refers to the `nyc_taxi` table from the `packt_serverless_analytics` database:

```
SELECT * FROM packt_serverless_analytics.nyc_taxi where column1
> 10
```

This would be required if the default database was configured as `elb_logs`.

Tables/datasets

A table represents a dataset that users can query. A table has many properties that need to be specified when created, such as the table's location, the table's schema, the file format of the data stored, and more. This data is then stored in the metastore. The following is a sample `CREATE TABLE` statement that can be used to register a table. We will look at each part to illustrate the information needed for a table to be queried:

```
CREATE EXTERNAL TABLE 'packt_serverless_analytics'.'nyc_taxi_
partitioned'(
  'vendorid' BIGINT,
  'tpep_pickup_datetime' STRING,
  'tpep_dropoff_datetime' STRING,
  'passenger_count' BIGINT,
```

```
   'trip_distance' DOUBLE,
   … see rest of the columns on GitHub https://bit.ly/3odawDa
   'congestion_surcharge' DOUBLE)
PARTITIONED BY ('year' INT, 'month' INT)
ROW FORMAT DELIMITED
   FIELDS TERMINATED BY ','
STORED AS INPUTFORMAT
   'org.apache.hadoop.mapred.TextInputFormat'
OUTPUTFORMAT
   'org.apache.hadoop.hive.ql.io.HiveIgnoreKeyTextOutputFormat'
LOCATION
   's3://<S3_BUCKET>/tables/nyc_taxi_partitioned/'
TBLPROPERTIES (
   'areColumnsQuoted'='false',
   'columnsOrdered'='true',
   'compressionType'='gzip',
   'delimiter'=',',
   'skip.header.line.count'='1'
)
```

You can download and run the preceding statement in its complete form by going to `https://bit.ly/3odawDa`. Let's break this code down into each of its components.

External tables, managed tables, and governed tables

The first part of the statement, `CREATE EXTERNAL TABLE`, tells Athena to create an external table. An external table is a table where the user must manage the underlying data, and Athena will not perform any actions when the table is dropped.

If the external table specification is not provided, then the table is a managed table. A managed table differs from an external table in two main ways. First, when creating a managed table, a location is not needed. The table's location will be placed under the database's location property. For example, if we created a managed table called `my_table` in a database whose location property was `s3://packt-serverless-analytics-1234567890/tables/`, then the table's data would be stored at `packt-serverless-analytics-1234567890/tables/my_table`. When a managed table is dropped, the execution engine should delete the table's data and remove the table from the metastore. Managed tables are not supported in Athena.

Another type of table, called a governed table, is specific to AWS. It is a table that AWS Lake Formation manages. It provides additional features, such as supporting **atomic, consistent, isolated**, and **durable (ACID)** transactions and automatic consolidation of data. Small file sizes can have an enormous impact on query performance. We have dedicated an entire chapter to governed tables; that is, *Chapter 14, Lake Formation – Advanced Features*.

Table schema

The schema for a table is the list of columns that can be queried and the column's data type. Athena takes the schema while reading data files and maps the data it finds to the columns with their names. In the previous CREATE TABLE statement, the schema is the columns that are specified.

Note that the data types specified in the table schemas must match or be compatible with the data type stored in the data files. If they do not match, then Athena may fail the query with an error, stating that it cannot convert the data type in the data file into the requested data type. Similarly, some data formats require that the column order specified in the table's schema matches the ordering of columns in the data files. If the file format requires specific ordering, and the order does not match the table schema, then you may get columns with data from other columns, null values for a column, or a failed query.

Partitions

Partitioning a table allows for huge tables to be broken down into smaller slices of data based on one or more virtual columns. This has many advantages, including reducing data scanning, thus reducing Athena's costs and having faster query times. When there is a filter on one or more partition columns, Athena will read only the partitions' data files. Each partition in a table has a directory where the data for that slice is stored. The following diagram illustrates how the nyc_taxi_partitioned table's data is laid out for datasets on S3:

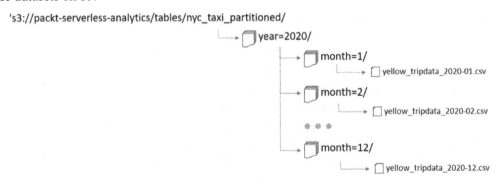

Figure 4.4 – File layout of the nyc_taxi_partitioned partitioned table

The table's base location is s3://packt-serverless-analytics/tables/nyc_taxi_partitioned. Each directory and subdirectory beneath it is a partition with the partition column name and the column's value.

There are two ways to add partitions from Athena. The first is to run the following DDL statement:

```
MSCK REPAIR TABLE nyc_taxi_partitioned
```

This statement instructs Athena to scan the directory under the table's specified location for any partitions. If the partition has not been registered yet, it will add it. This can be an easy and convenient way to discover new partitions, but there is a drawback. First, the directory structure must follow the <partition column name>=<partition value> format. Secondly, the statement can take more than 10 minutes to run if the table has hundreds of thousands of partitions or more.

The second way to add a partition is to add the partitions manually, one by one, if you know ahead of time what the partition values are. The following DDL statement adds the 2020-06-01 partition to the table:

```
ALTER TABLE nyc_taxi_partitioned
    ADD PARTITION (year='2020', month='1')
    LOCATION 's3:// <S3_BUCKET>/tables/nyc_taxi_partitioned/
year=2020/month=1/'
```

This statement tells Athena to add a new partition with a column value of 2020 for the year and 1 for the month and a location of s3://packt-serverless-analytics/tables/nyc_taxi_partitioned/year=2020/month=1/. There are two main advantages to adding partitions this way. First, it is usually faster to run this command to add a single or small number of partitions to a highly partitioned table than to run MSCK REPAIR TABLE. Second, you can specify any location; it does not need to conform to the Hive partition format of partition column=value.

> **Note**
>
> When a partitioned table is created, the table will have 0 partitions registered. Any queries against the table will always return 0 rows. After the creation of a partitioned table, you should perform a partition-adding operation such as using the MSCK REPAIR or ADD PARTITION commands.

Serialization and deserialization and file formats

In the next part of the `CREATE TABLE` statement, we can see the `ROW FORMAT DELIMITED`, `STORED AS INPUTFORMAT`, and `OUTPUTFORMAT` sections. Athena uses this information to select the **serialization and deserialization (SerDe)** library to read the data. Different file formats, such as **Comma-Separated Values (CSV)**, **Apache Parquet**, **Apache ORC**, and others, have their own libraries that are used to read and write the data. In our example, we are telling Athena that the data is stored in CSV format and that we're using the `org.apache.hadoop.mapred.TextInputFormat` library to read the data and the `org.apache.hadoop.hive.ql.io.HiveIgnoreKeyTextOutputFormat` library for any writes.

Athena supports a wide variety of open source data formats. To see the up-to-date list of supported data formats, see Athena's documentation link in the *Further reading* section.

Table properties

The next section is the table properties. These properties are specified as key/value pairs and can be used for a variety of uses. In this case, we are setting properties to configure `SerDe`: we are telling the `SerDe` text input that the fields are not quoted (`areColumnsQuoted=false`), the columns are ordered based on the column order in the table's schema (`columnsOrdered=true`), and that the first row of data is the column header and should be ignored (`skip.header.line.count`). For different `SerDe` instances, these properties can be different.

Another use for table properties is to store information that can be useful for other purposes. For example, you can add an owner, contact information if a user has a question or found a data quality issue, or keywords about what is contained in the table, helping users discover the data they are looking for.

Table statistics

Although it is not specified in the `CREATE TABLE` statement, metastores can also store table-level and column-level statistics. Statistics help execution engines perform **Cost-Based Optimizations (CBOs)** when they're coming up with an execution plan. For example, the join ordering of tables can be optimized if the row counts of each table are known. At the time of writing, Athena does some optimizations, but these statistics will help you perform complex optimizations in the future.

Now that we have a good understanding of what a metastore is, we can discuss data sources.

What is a data source?

If Athena has access to data and the associated metadata, it can read that data. This is one of Athena's greatest strengths, as it can join data from anywhere to enrich and derive business value. For example, suppose an online store has its sales data in a MySQL database, has customer website traffic data in S3, and has product pricing information in **DynamoDB**. In that case, these datasets can be joined together to determine which pricing changes caused the most traffic to the website's stores and drove the most sales. You can look at the available data sources or add new data sources from Athena's console, as shown in the following screenshot:

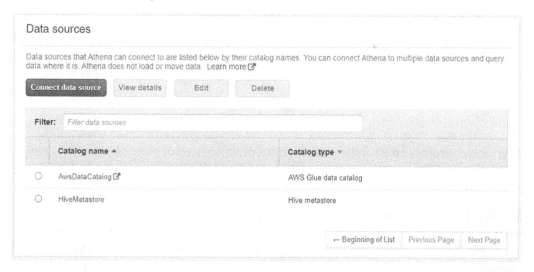

Figure 4.5 – The Data sources tab in the Athena console

For this section, we will mostly focus on querying data on S3. To query data in S3, using the AWS Glue Data Catalog or the Apache Hive metastore are the fastest and easiest ways to store your databases and tables. We will dive into how they are different and when to use one over the other. We will then briefly talk about non-S3 data sources, as an entire chapter, *Chapter 12, Athena Query Federation,* is dedicated to Athena Query Federation, which can read from almost any data source.

S3 data sources

S3 is the most common data source for Athena. It was the first and only data source available when Athena launched. With one DDL statement, Athena can start querying data in S3 in seconds. Athena needed a metastore implementation to store database and table information. AWS Glue Data Catalog was created to be the central catalog for other AWS services while maintaining compatibility with existing execution engines such as Apache Hive, Apache Spark, and PrestoDB. For these engines, AWS Glue created connectors and implementations that adhered to the Apache Hive metastore interface. Where Hive's metastore is used, it can be swapped out for Glue Data Catalog's implementation.

For this reason, Athena's default metastore is Glue Data Catalog. However, some customers want to use Athena without migrating their metastores to Glue or keeping Glue and Hive metastores in sync. Athena introduced support to connect to Hive metastores directly.

When should you use Glue Data Catalog, and when should you use a Hive metastore? What are the advantages of one over the other? We will go over both Glue Data Catalog and Hive metastores and when it would be appropriate to use one over the other.

AWS Glue Data Catalog

AWS Glue Data Catalog is a serverless offering by AWS. There is no infrastructure to manage, and the cost is very reasonable. At the time of writing, the cost of using Glue Data Catalog is free for the first million objects stored and $1.00 for every 100,000 objects thereafter. Also, the first million object requests are free, with each additional million requests costing $1.00. Glue Data Catalog is natively integrated with Redshift, Glue ETL, Lake Formation, and EMR, making sharing the catalog very easy. Glue Data Catalog integrates with open source engines via connectors available for Apache Spark, Apache Hive, PrestoDB, and Trino. Glue Data Catalog also supports versioning of tables. As changes are made to tables, they are saved, and older versions of the tables can be referred to or rolled back when unintentional or breaking changes are made. Lastly, with integration with Lake Formation, data access controls can be applied to Glue tables and columns and applied to AWS services that integrate with Lake Formation. More on that in the next chapter.

There are a few disadvantages of using Glue Data Catalog to be aware of. As with all AWS services, the first main disadvantage is that API calls are subject to service throttling limits and service quotas. If Glue Data Catalog's load is high, requests may be throttled, causing queries to slow down or even fail. These throttling limits are soft and can be raised if an AWS support ticket is entered to increase the limits. The other limits to consider are service quotas, limiting the number of objects stored in Glue. These limits are very high, and it should be challenging to reach these limits.

The second main disadvantage is that it does not support all the features of the Hive metastore. For example, Glue Data Catalog does not support Hive ACID transactional tables, which are not supported in Athena.

There are many other limitations of Glue Data Catalog, but if you are not using Apache Hive, these limitations will not impact you. If you are planning to use Hive, then the documentation (`http://amzn.to/3o0REqS`) provides an exhaustive list.

> **Note**
>
> At the time of writing, AWS accounts have a default limit of 10,000 databases per account, 200,000 tables per database, and 10,000,000 partitions per table. If you breach these limits, it is likely that you may be doing something wrong and should revisit your architecture. These limits are mostly soft limits and they can be increased by entering a support ticket.

Now, let's take a look at the Hive metastore.

Apache Hive metastore

Apache Hive was initially released in 2010 by Facebook to provide SQL-like access to data stored on Hadoop clusters. One of the main components was the metastore. As time progressed, it was used by other Hadoop projects such as Spark to store dataset metadata. Hive metastores are a service that is typically backed by a relational database such as MySQL.

The advantages of using a Hive metastore with Athena over Glue Data Catalog are few but significant. The main advantage is that companies that already have a Hive installation may not wish to migrate their metadata for various reasons, such as using third-party tools that are not compatible with Glue Data Catalog. The other advantage is that it is open sourced, which provides two benefits. First, if a bug or enhancement needs to be made, it can be done quickly and deployed. Second, it is portable and does not lock you into a single cloud infrastructure provider.

The disadvantages are plenty. The main disadvantage that I have seen organizations struggle with is that upgrading the version of the metastore can cause substantial operational problems. Some version upgrades break existing installations because they were not designed to be backward and/or forward compatible. A lot of planning and coordination needs to happen when upgrading versions, increasing the operational burden. I worked with a customer who upgraded their metastore, which was shared between beta and production environments, which caused their production environment to break, causing a several-hour production outage.

The other disadvantage is its performance. When Athena is using Glue Data Catalog, it makes direct API calls. When a Hive metastore is used, Athena invokes a Lambda function to call a Hive metastore process. This has a higher cost for all metadata calls. This is even more apparent with tables that have a large number of partitions. When querying a table with 1 million partitions, we found that Glue Data Catalog-backed queries ran at least half the time than when using a Hive metastore.

The last disadvantage to call out is that Athena does not support writing to external metastores at the time of writing this book. Being able to create tables, alter tables, or perform other operations is currently not supported and the only way to update the metadata is through another application.

> **Tip**
> Our recommendation is to use Glue Data Catalog whenever possible. It has a lower operational burden due to it being serverless, it doesn't need to perform version upgrades, has better auditing capabilities, the ability to update metadata, and provides native integration with other AWS services.

Here is the comparison of AWS Glue Data Catalog and Apache Hive metastores with Athena:

Feature	AWS Glue Data Catalog	Hive Metastore
Stores databases, tables, and columns	Yes.	Yes.
Stores table statistics	Yes.	Yes.
Supported operations	SELECT, CTAS, CREATE TABLE, ALTER TABLE, SHOW COLUMNS, SHOW TABLES, SHOW SCHEMAS, SHOW CREATE TABLE, SHOW TBLPROPERTIES, SHOW PARTITIONS.	SHOW COLUMNS, SHOW TABLES, SHOW SCHEMAS, SHOW CREATE TABLE, SHOW TBLPROPERTIES, SHOW PARTITIONS.

Feature	AWS Glue Data Catalog	Hive Metastore
Cost	First 1 million objects free, $1.00 per 100,000 objects. First 1 million object requests free, $1.00 per 1 million requests afterward.	Cost of hardware to host metastore processes, plus database costs. Athena usage will incur AWS Lambda usage costs.
Operational load	Low.	Medium to high.
Availability	High.	Low to high.
Performance	High.	Low to medium.
Open source engine compatibility	Apache Hive, Apache Spark, PrestoDB, Trino.	Apache Hive, Apache Spark, PrestoDB, Trino, Apache Pig.
AWS service compatibility	Amazon Athena, Amazon EMR, AWS Glue, Amazon Redshift.	Amazon Athena, Amazon EMR.
Table versioning	Yes.	No.
Auditing of changes	Yes, through AWS CloudTrail.	Yes, through log files.
Data access controls	Yes, IAM provides authentication, and authorization can be implemented using Lake Formation or IAM resource policies. See *Chapter 5, Securing Your Data*, for details.	Yes, Hive metastore processes can employ different authentication and authorization methods. See the Further reading section for links to the available mechanisms.

Figure 4.6 – Comparison of AWS Glue Data Catalog versus Apache Hive metastore

Of course, Hive and Glue are not the only possible data sources you can use. Let's quickly look at some other alternatives.

Other data sources

Athena supports a variety of data sources out of the box. At the time of writing, the supported data sources are S3 with AWS Glue Data Catalog or Apache Hive metastores, **Amazon CloudWatch Logs**, **Amazon CloudWatch Metrics**, **Amazon DocumentDB**, **Amazon DynamoDB**, **Amazon Redshift**, **Apache HBase**, **MySQL**, **PostgreSQL**, and **Redis**.

You can register a new data source from the Athena console. The following screenshot shows the Athena console:

Figure 4.7 – Connecting a new data source to Athena in the Athena console

By following a few steps, you can connect to a new data source within seconds and query any of the aforementioned data sources. We will look at adding new data sources and creating custom data sources in more depth in *Chapter 12*, *Athena Query Federation*.

Next, we'll look at how to register S3 datasets so that they can be used with your metastore.

Registering S3 datasets in your metastore

Before you can query your data with Athena, the data must be registered in a data catalog. This section will review the different ways an S3 dataset can be registered in your metastore.

Using Athena CREATE TABLE statements

Athena's console allows you to create databases and tables in your metastore through two methods. The most often used method is generating and executing DDL statements. We have already seen a few examples of CREATE TABLE SQL statements to create tables. If you need a refresher, refer to the *Tables* section earlier in this chapter. Alternatively, you can click on the **CREATE TABLE** template within the Athena console, as shown in the following screenshot:

Figure 4.8 – Using a SQL template in the Athena console for CREATE TABLE

Using Athena's Create Table wizard

This method uses the **Create Table** wizard from the Athena console. It takes the necessary information, generates a CREATE TABLE statement, and submits it to Athena. To run the wizard, click on the **Create table** link in the Athena query editor tab, next to **Tables**, as shown in the following screenshot, and follow the steps:

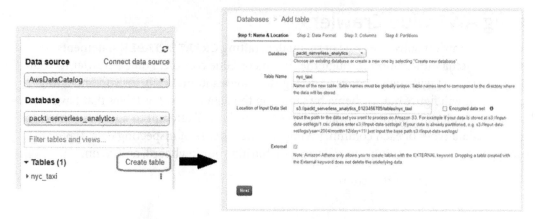

Figure 4.9 – Creating a table using Athena's console

This method of creating tables is very rarely used because it takes a lot of manual effort, especially if your table has many columns.

Using the AWS Glue console

The AWS Glue console supports the creation of databases and tables using wizards, very similar to the Athena console. Creating tables, especially those with many columns, is time-consuming and error-prone. To get to the wizard, click on the **Tables** link on the left-hand side, click on **Add tables**, and select **Add table manually**, as shown in the following screenshot. Follow the steps, and voilà – your table will be created:

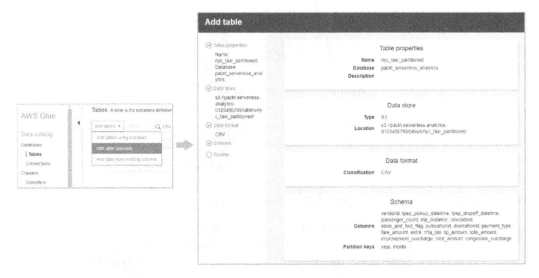

Figure 4.10 – Creating a table using AWS Glue's console

Using AWS Glue Crawlers

AWS Glue Crawlers solve the issue of manually crafting CREATE TABLE statements or entering schema information through the Athena or Glue console. Point a crawler to your S3 location and it will scan, discover, and register tables automatically for you, including partitions. It will figure out a schema by sampling the data it sees. For tables that have hundreds of columns, point a crawler to the table, and it will attempt to figure out the name and data types of each column, saving countless hours of inspecting data, typing in data types, trying and having your queries fail, hitting your keyboard, and so on.

We'll create one in the next section.

Discovering your datasets on S3 using AWS Glue Crawlers

Let's say that you have a lot of data that you are outputting to S3, and you want to query it. Before you can, you need to register that data. However, the data sitting in S3 is in many different formats and schemas. Going through each dataset, inspecting files, and determining the file format, partitions, and columns is a very time-consuming task. If a table contains incorrect column names, incorrect ordering of columns, or any other form of error, then the table may not be queryable until it is corrected. AWS Glue Crawlers solve these issues. Glue Crawlers can scan data on S3, inspect the S3 directory structure and data within it, and automatically populate the data catalog. This section will look at how they work and set up a Glue crawler to discover a sample dataset.

How do AWS Glue Crawlers work?

There are three actions that a Glue crawler takes when scanning S3:

1. It scans S3 directories for data files. File formats such as Parquet, ORC, and Avro are self-describing, meaning they include the data file's schema. If the data format is not self-describing, it will sample the file's data to guess the columns and their data types.

2. As the crawler traverses the directory and sees multiple directories containing data files with similar schemas, it may consider it a partitioned table. If the schema is sufficiently different, then it will consider each of the directories as separate tables.

3. Finally, once it has traversed all the directories, it will register the tables and table partitions to the catalog. If the tables already exist, they will update the schemas and add any undiscovered partitions.

Running your first AWS Glue Crawler

In this section, we will create our first Glue Crawler, which will traverse our bucket and register the tables that are found. We will create a new database to store the crawled tables, and then we will query them in Athena.

If you have already set up your S3 bucket with the example datasets, you can skip this section. If not, please follow the instructions in this book's GitHub repository, which is located at `https://bit.ly/2XW1LCA`. Remember to replace `<YOUR_S3_BUCKET>` with the bucket that you are using.

Getting to the Glue Crawler wizard

There are two ways to get to the Glue Crawler wizard. Let's take a look:

4. The first option is through the Athena console, by clicking on the **Create Table** link and then **from AWS Glue Crawler**. The second option is from the Glue console by clicking on **Crawlers** on the left-hand side menu and then clicking on the **Add crawler** button. Both methods are shown in the following figure:

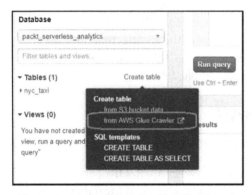 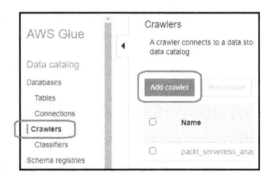

Figure 4.11 – Getting to the Glue Crawler wizard (left is the Athena console, right is the Glue console)

5. Let's call the crawler `packt_serverless_analytics_chapter_4`, as shown in the following screenshot:

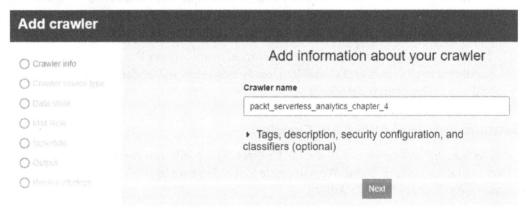

Figure 4.12 – Creating a new crawler info page

6. We will forgo giving the crawler any tags, security configuration, or registering custom classifiers. For the subsequent screens, input the following values. Descriptions of some of the fields are also included:

    - **Crawler name**: This is the name to assign to your crawler. Value to set: `packt_serverless_analytics_chapter_4`.

- **Tag key and value**: Key/value pair that provides metadata on this crawler. Value to set: `None`.

- **Description**: A description of the crawler. Value to set: `None`.

- **Security configuration**: A configuration that specifies the encryption key to use for logs. Value to set: `None`.

- **Crawler source Type**: This indicates whether you want to specify an S3 location (data stores) or crawl existing table locations. Value to set: `Data stores`.

- **Repeat Crawls of S3 data sources**: When Crawlers run multiple times, should they inspect new directories only or all directories? Value to set: `Crawl all folders`.

- **Choose a data store**: The source that should be crawled. Value to set: `S3`.

- **Connection**: A connection in Glue is a named **elastic network interface** (**ENI**), which is created in a VPC with security groups attached. Connections allow Glue Crawlers and other functionality to talk to data sources from within your network and not over the internet. Value to set: Leave blank.

- **Crawl data in**: This is used to specify whether the S3 bucket we are crawling is owned by the account running the crawler or in a different account. Value to set: `Specified path in my account`.

- **Include Path**: S3 path for the crawler to start in. Value to set: `s3://<MY BUCKET NAME>/tables/`.

- **Add another data store**: Crawlers can crawl multiple data sources in a single run. This can be useful if you want the same crawler to focus on a subset of directories in a single bucket or multiple buckets. Value to set: `No`.

- **Choose an IAM role**: The IAM role that is used by the crawler to access the S3 bucket and other resources. Value to set: If you have not created an IAM role before for a crawler, choose **Create an IAM role** and add `packt-serverless-analytics` as the role name suffix. Otherwise, choose **Update a policy in an IAM role** to ensure that the S3 path is added to an existing IAM role.

- **Frequency**: Crawlers can be scheduled to run at regular intervals or set to run on demand. Value to set: `Run on-demand`.

- **Configure the crawler's output**: The database to add/update tables in, and optionally a prefix for table names. Value to set: For the database, enter `packt-serverless-analytics-chapter-4` or click on **Add database** if the database doesn't exist.

- **Grouping behavior for S3 data**: This is optional. It tells the crawler to create a table for each unique S3 directory rather than group multiple directories into a partitioned table. Value to set: `Leave unchecked`.

- **Configuration Options**: For existing tables, these options determine how the crawler will update existing tables. Value to set: Check the **Update all new and existing partitions with metadata from the table** setting. See *AWS Glue Crawler best practices for Athena* for more information.

7. On the last page of the wizard, you will see a summary of the crawler. Your summary should look like this:

Figure 4.13 – Summary of the crawler

8. Click on **Finish**; the crawler will be created.

9. To run the crawler, select the crawler and hit the **Run crawler** button. Your crawler should run, discover new tables, and add them to your catalog, as shown in the following screenshot:

	Name	Schedule	Status	Logs	Last runtime	Median runtime	Tables updated	Tables added
☐	packt_serverless_analytics_chapter_4		Ready	Logs	1 min	1 min	0	2

Figure 4.14 – Output of running the crawler

Congratulations! You have registered two new tables in the `packt_serverless_analytics_chapter_4` database. You can browse these tables in the **Tables** section of the Glue console:

	Name	Database	Location	Classificatio	Last updated	Deprecated
☐	nyc_taxi	packt_serverless_analyt...	s3://packt-serverless-an...	csv	19 January 2021 2:36 A...	
☐	nyc_taxi_partitioned	packt_serverless_analyt...	s3://packt-serverless-an...	csv	19 January 2021 2:36 A...	

Figure 4.15 – Tables created by the crawler

AWS Glue Crawler best practices for Athena

We could easily write half a book dedicated to Glue Crawlers and their best practices, which we will not do. Glue Crawlers have extensive public documentation, so anything that we don't cover here can be found here: `http://amzn.to/3nWcSWZ`. However, when working with several customers, the issues they tend to have trouble with are the **HIVE_PARTITION_SCHEMA_MISMATCH** error, which is related to issues with partition schemas not matching the table's schema, and issues with CSV/TSV files, as there are two different SerDes that can process these file types.

Designing a data lake architecture

Before cloud platforms, organizations had clusters sitting in their data centers with massive amounts of storage that applications would push their data to for analytics. When storage was running low, the organization would either remove that data on the cluster or increase its storage. Ordering new hardware was costly and was often met with long lead times. As cloud platforms have exploded in popularity, businesses and organizations have leveraged unlimited storage and compute to develop new ways of storing and processing data. One of the most common architectures for data analysis was the data lake. The data lake architecture leverages the unlimited storage that cloud platforms provide and can scale storage and compute independently. It can store an organization's data in a single location, where it can be queried by any user using the best application for the particular use case. Any data that was too large or expensive to store on an on-premises cluster can now be stored much more cheaply. The following diagram shows a simplified data lake architecture where data producers (mobile clients, databases, application clients, and application servers) push their data into a central data store (S3). That data is then consumed and analyzed, and the results are written back to S3:

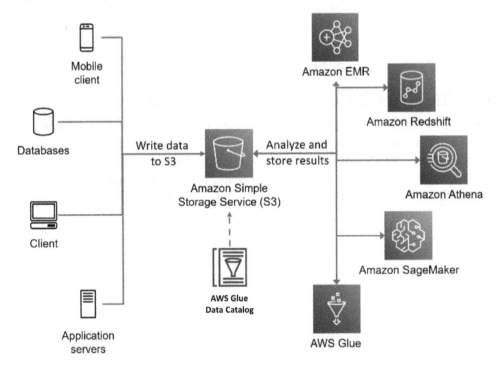

Figure 4.16 – Typical simplified data lake architecture

Because the data is separate from the compute engine, the application that does the analysis can be chosen based on the analytics that is being conducted. For example, for ETL processing, Amazon Athena, Amazon EMR, or AWS Glue ETL can be used. Amazon SageMaker or Amazon EMR can be used for machine learning modeling, or Amazon Athena for ad hoc querying, exploration, or delivering data to applications.

One component that binds all the AWS services together is the central catalog. Without it, each service would need to track its metadata on the datasets on S3, making it challenging to keep in sync. AWS Glue Data Catalog is the only metastore that's leveraged by the suite of AWS analytics services. The catalog can also enforce data access controls and provide auditing, which we will cover in *Chapter 5, Securing Your Data*.

Now that we have an overview of a data lake's basic structure, a few concepts will help you get the most out of Athena and other AWS services. We will look at the three stages or forms of data that typical data lakes utilize and each stage's characteristics before learning how to transform data using Athena.

Stages of data

When data is ingested from applications, databases, and other sources into a data lake, it tends to be in a raw state. They can be in text format, uncompressed, and structured in a suboptimal way. The data can be transformed more optimally to improve the performance and cost of consuming raw data. Let's break down these stages:

- **Raw data**: This is the stage where producers of data push the data that they produce. This data tends to be in suboptimal data formats and structures. The applications generally don't want to spend their resources converting the data. The raw data is seen as the source of truth: therefore, it is recommended that this data is never deleted but archived in S3 Glacier once the data is no longer needed. If the data is required again or there is a data quality issue in any data derived from this data, it can be restored and reused. Also, the data can be difficult to reproduce from the original producer. Raw data usually exists in a separate S3 bucket from other data. The data is highly secure as raw data may contain sensitive data. Querying this data using Athena is great for data exploration and testing the data, but it is not ideal for repetitive querying as it would be slow and expensive.

- **Processed/transformed data**: This stage is the first level of transformation from the raw data. Its purpose is to provide a faster and cheaper way of querying data for general use cases. The data is converted into an optimized file format such as Apache ORC or Apache Parquet. Transforming your data into ORC or Parquet can reduce the data size by 50-90%, which provides a 50-90% reduction in Athena querying costs. The data can also be partitioned based on a frequently used filter in queries, such as the transaction date or region. The transaction helps reduce querying costs as only the selected partitions are queried. This stage also provides an opportunity to filter columns or rows for raw data that is not needed, or that may contain confidential data. For example, suppose customers' **personally identifiable information (PII)** is in a dataset; it would be too sensitive to provide to general users. In that case, it can be removed or encrypted. This stage is great for ad hoc querying from users and general use from other applications such as Amazon Redshift or Amazon SageMaker.

- **Application-specific format**: This stage is most often used for applications that serve data to customers and generally need to run as fast as possible to provide the best experience for end users. The data in this stage is transformed into a specific structure and not meant for general-purpose querying. Datasets can be partitioned based on columns that would not generally be used and other data format optimizations to optimize the application's specific queries or access patterns. We will look at this in more detail in *Chapter 11, Operational Excellence – Maintenance, Optimization, and Troubleshooting*.

Now that we've looked at the stages of data, let's learn how to transform it.

Transforming data using Athena

Transforming data has many advantages, and Athena can perform these transformations using SQL. Two statements can be used with Athena: CREATE TABLE AS SELECT (CTAS) and INSERT INTO. The CTAS statement creates a new table and populates the table from the results of a SELECT statement. The structure of the CTAS query is CREATE TABLE <TABLE NAME> WITH PROPERTIES (...) AS < SELECT QUERY>.

Let's take our NYC Taxi ride dataset in CSV format and convert it into Parquet. We can do this by running the following query (https://bit.ly/3xdzXJb):

```
CREATE TABLE packt_serverless_analytics.nyc_taxi_partitioned_
parquet
WITH (format='PARQUET',
    parquet_compression='SNAPPY',
    partitioned_by=array['year','month'],
    external_location = 's3://<S3_BUCKET>/tables/nyc_taxi_
partitioned_parquet/')
AS
SELECT
    vendorid,
    tpep_pickup_datetime,
    tpep_dropoff_datetime,
    passenger_count,
    trip_distance,
    ratecodeid,
    store_and_fwd_flag,
    pulocationid,
    dolocationid,
    payment_type,
    fare_amount,
    extra,
    mta_tax,
    tip_amount,
    tolls_amount,
    improvement_surcharge,
    total_amount,
    congestion_surcharge,
    --Below are partition columns and are always
    --specified at the end of the SELECT statement
    substr(tpep_pickup_datetime, 1, 4) AS year,
    substr(tpep_pickup_datetime, 6, 2) AS month
FROM packt_serverless_analytics.nyc_taxi;
```

Athena will execute the SELECT statement and create a new table based on the provided file format, file compression, columns, partition columns, and location. The table's schema is defined by the columns specified in the same order using the column's name. If an expression is used, like we did with the year and month columns, a column name cannot be inferred, so it must be specified. You will also notice that the partition columns are defined at the end of the SELECT statement.

The INSERT INTO statement inserts data into an existing table. The data files that are produced will be in the same format and contain the same configuration that was specified in the table properties. The structure of the INSERT INTO statement is INSERT INTO <TABLE> SELECT <QUERY>. Suppose a new month of data arrives and is placed in the nyc_taxi table for July in 2020. The following example query can be used to insert new data:

```
INSERT INTO packt_serverless_analytics.nyc_taxi_partitioned_
parquet
SELECT
    vendorid,
    tpep_pickup_datetime,
    tpep_dropoff_datetime,
    passenger_count,
    trip_distance,
    ratecodeid,
    store_and_fwd_flag,
    pulocationid,
    dolocationid,
    payment_type,
    fare_amount,
    extra,
    mta_tax,
    tip_amount,
    tolls_amount,
    improvement_surcharge,
    total_amount,
    congestion_surcharge,
    --Below are partition columns and are always
```

```
    --specified at the end of the SELECT statement
    substr(tpep_pickup_datetime, 1, 4) AS year,
    substr(tpep_pickup_datetime, 6, 2) AS month
FROM packt_serverless_analytics.nyc_taxi
WHERE tpep_pickup_datetime = '2020-07';
```

Go ahead and give it a try.

> **Important Note**
>
> If the CTAS or INSERT INTO queries fail, you will need to clean up any
> data files created by the process before starting again. Otherwise, incomplete,
> or duplicate data will exist in the destination. If a failure does occur with
> these statements, a manifest file is created in the query results directory as
> QueryID-manifest.csv. The file list can be used to perform the
> necessary clean-up.

Summary

In this chapter, we learned about Athena's data sources and their different components: the metastore, data, and connector. The metastore contains metadata that Athena uses to translate tables and databases into their physical locations and process them. We delved into the information stored within a table and its key components: schema, partition columns, location, serializer/deserializer and associated properties, and table statistics.

We compared the AWS Glue Data Catalog and Apache Hive metastores when data is stored on S3 and looked at other non-S3 data sources. We went through the different ways of registering datasets into a metastore and how AWS Glue Crawlers can make it quick and easy to discover data on S3. Lastly, we looked at the data lake architecture, the different stages of data that are typical in one, and how to transform data using Athena.

Now that we have looked at our metastores and how they relate to our data in S3, we'll look at how we can secure them in the next chapter.

Further reading

For more information regarding what was covered in this chapter, take a look at the following resources:

- Athena's documentation contains the complete list of supported DDL statements: `https://docs.aws.amazon.com/athena/latest/ug/language-reference.html`.

- A complete list of supported SerDes can be found at `https://docs.aws.amazon.com/athena/latest/ug/supported-serdes.html`.

- AWS Glue Data Catalog service quotas can be found at `https://docs.aws.amazon.com/general/latest/gr/glue.html#limits_glue`.

- *Best Practices When Using Athena with AWS Glue*: `https://docs.aws.amazon.com/athena/latest/ug/glue-best-practices.html`.

- *Best Practices When Using Athena with AWS Glue – Using AWS Glue Crawlers*: `https://docs.aws.amazon.com/athena/latest/ug/glue-best-practices.html#schema-crawlers`.

5
Securing Your Data

Data within an organization can be one of its most valuable assets. Data can drive business decisions for an organization, such as to whom and how to advertise, what the behavior of users on a website is, and how they react to sales or help businesses identify inefficient processes. An organization can also package and sell that data to customers or other organizations, getting direct revenue for the information it collects. Regardless, all organizations should protect the data they have from both internal and external entities.

We have all heard stories where a data breach has occurred in a large institution. It is a harrowing and traumatic event for the organization. There could be monetary penalties by governments for breaking laws. Still, for most companies, breaking customers' or the public's trust can be much more damaging. This is why large companies invest large amounts of resources into having dedicated security teams that provide rules of how data should be protected and handled.

Regardless of an organization's size, it is always a good idea to think about security at the beginning of any project. I always tell customers that it is much easier and cheaper to incorporate basic security early and often than later, and most are thankful that they did. By employing security measures later in the process, it becomes much more intrusive to add it. More applications may need to be changed to deal with new rules, or more users consuming data may need to change their processes. You may need to encrypt data in place, which may require system downtime. With some simple guidelines and features, we can avoid many of these headaches later on.

In this chapter, we will cover the following topics:

- General best practices to protect your data on AWS

- Encrypting your data and metadata in Glue Data Catalog

- Enabling coarse-grained access controls with IAM resource policies for data on S3

- Enabling fine-grained access controls with Lake Formation for data on S3

- Managing access through workgroups and tagging

- Auditing with CloudTrail and S3 access logs

Technical requirements

For this chapter, you will require the following:

- Internet access to GitHub, S3, and the AWS Console.

- A computer with Chrome, Safari, or Microsoft Edge and the AWS CLI version 2 installed.

- An AWS account and accompanying IAM user (or role) with sufficient privileges to complete this chapter's activities. For simplicity, you can always run through these exercises with a user that has full access. However, we recommend using scoped-down IAM policies to avoid making costly mistakes and learn how to best use IAM to secure your applications and data. You can find a minimally scoped IAM policy for this chapter in this book's accompanying GitHub repository, which is listed as `chapter_5/iam_policy_chapter_5.json` (https://bit. ly/3qAcNtU). This policy includes the following:

  - Permissions to create and list IAM roles and policies. We will be creating a service role for an AWS Glue Crawler to assume.

- Permissions to read, list, and write access to an S3 bucket.

- Permissions to read and write access to Glue Data Catalog databases, tables, and partitions. You will be creating databases, tables, and partitions manually and with Glue Crawlers.

- The ability to create and run permissions for Glue Crawlers.

- The ability to gain access to run Athena queries.

- An S3 bucket that is readable and writeable. If you have not created an S3 bucket yet, you can do so from the CLI by running the following command:

```
aws s3api create-bucket --bucket <YOUR_BUCKET_NAME> --region
us-east-1
```

General best practices to protect your data on AWS

In this section, we will go over some general best practices. However, before we do, we should understand some security basics. Let's start with what I call the five general pillars of security. They are as follows:

- **Authentication**: Can the user or principal prove who they are? Access to AWS resources depends on IAM authentication through AWS credentials, which are like logins and passwords. These credentials can be long-lived, such as IAM user credentials, or short-lived, such as the AWS credentials that are provided when an IAM role is assumed. Throughout this chapter, we will assume that AWS IAM is the only authentication mechanism that users can use. However, we will also look at other ways to authenticate in *Chapter 7, Ad Hoc Analytics*.

- **Authorization**: Is the user or principal provided permission to access a resource? When an action is requested against an AWS resource, the IAM credentials that are used are checked to see whether those credentials can access the resource.

- **Data protection**: Is the data secure while it is in transit or at rest? Data encryption is the most common way to protect data while transferring it between two parties and storing it.

- **Auditing**: Do you know who is accessing the data, and are they supposed to be accessing it? Auditing is usually the aspect of security that is most forgotten, but it is critical. Auditing serves two purposes: making sure that current access to data is what we expect it to be, and if it is not, then make changes to resource access policies, and assessing the severity of a breach and what was leaked. Severity can be measured by how long a breach occurred, who the actors were, and the sensitivity and amount of data that was accessed.

- **Administration**: How are the policies that grant permission to resources managed? Ideally, there would be a single place where permissions are granted.

Now that we understand the five pillars of security, there is one last point that I would like to make before getting into the best practices: *No system can be 100% secure*. When there is an incident, security policies aim to reduce the **attack surface** and **blast radius**. Attack surface means the different ways a bad actor can try to infiltrate a system. The larger the attack surface, the more ways that a system can be compromised. A blast radius is the amount of potential damage an actor can cause when a system is compromised.

Suppose there were two sets of AWS credentials. One set provides administrative access to an entire AWS account. The other gives read-only access to an S3 bucket that contains cat pictures. If the first set of credentials was obtained by an attacker, they would have access to all the data and be able to perform any action within the account. If the second set of credentials was obtained, they would be able to download cat pictures. The first event would be much more damaging and have a bigger blast radius. To reduce the likeliness of credentials being exposed, or to reduce the attack surface, these AWS credentials can be encrypted and access to them can be limited to only authorized users.

Now that we have a basic understanding of security, let's look at the best practices for securing your data.

Separating permissions based on IAM users, roles, or even accounts

I have seen too many companies use the same IAM credentials across several systems that access different data or services. This increases the *blast radius* if those credentials become compromised as the credentials likely would have been allowed to access all the resources all these systems need. If you need to disable the credentials because of an incident, then it would impact many services. A general rule is that an application should have its own IAM user or preferably an IAM role, and each user should get their own set of credentials.

It may make sense to provision different AWS accounts for each group or application if complexity dictates for larger organizations. This provides isolation for each group or application from others, without it impacting anything outside the account. Using AWS Organizations can help you manage and control accounts. One other frequent use of using separate accounts is in different stages of an application. For example, development, beta, and production environments run within their own AWS accounts, and then changes to policies within the development stage can be propagated to beta and then to production in an automated fashion.

Least privilege for IAM users, roles, and accounts

Within each chapter of this book, we have suggested that you use the IAM policies that we provide when completing the exercises. We do this so that you can use IAM credentials with the least privilege so that you can get into the habit of doing so. Using IAM principals with the least privilege aims to reduce the blast radius if those credentials are ever compromised; for example, if an intern accidentally puts them on GitHub; I am speaking from experience here.

Rotating IAM user credentials frequently

IAM user credentials are long-lived, which means that they can be used until they are rotated or the IAM user is deleted. Rotating credentials means that the old credentials are marked as expired, and a new set is created. This process reduces the attack surface because if credentials leak, they will only be used for a limited time. By the time someone finds them, they may no longer be used, or more importantly, this will limit the amount of time a bad actor can perform their actions for. One common scenario where this helps is if an employee leaves the company and takes credentials with them or, as in the previous section, if an intern accidentally publishes their credentials to GitHub.

Blocking public access on S3 buckets

Many companies recently made news in an embarrassing way. They had set their S3 buckets to be publicly accessible, and their data was available to the world. This scenario can easily be avoided by setting newly created and existing S3 buckets to block all public access. If you are an administrator, you can set this at the account level so that new buckets are not allowed to be made public.

The following screenshot shows the options that are available when setting this setting:

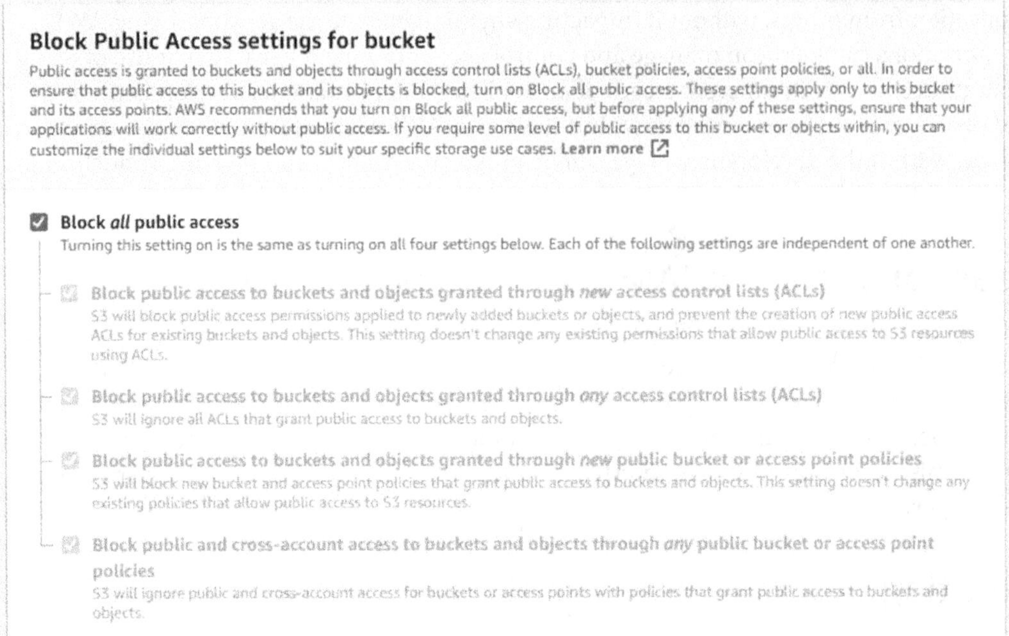

Block Public Access settings for bucket

Public access is granted to buckets and objects through access control lists (ACLs), bucket policies, access point policies, or all. In order to ensure that public access to this bucket and its objects is blocked, turn on Block all public access. These settings apply only to this bucket and its access points. AWS recommends that you turn on Block all public access, but before applying any of these settings, ensure that your applications will work correctly without public access. If you require some level of public access to this bucket or objects within, you can customize the individual settings below to suit your specific storage use cases. **Learn more** [↗]

☑ **Block *all* public access**
Turning this setting on is the same as turning on all four settings below. Each of the following settings are independent of one another.

☑ Block public access to buckets and objects granted through *new* access control lists (ACLs)
S3 will block public access permissions applied to newly added buckets or objects, and prevent the creation of new public access ACLs for existing buckets and objects. This setting doesn't change any existing permissions that allow public access to S3 resources using ACLs.

☑ Block public access to buckets and objects granted through *any* access control lists (ACLs)
S3 will ignore all ACLs that grant public access to buckets and objects.

☑ Block public access to buckets and objects granted through *new* public bucket or access point policies
S3 will block new bucket and access point policies that grant public access to buckets and objects. This setting doesn't change any existing policies that allow public access to S3 resources.

☑ Block public and cross-account access to buckets and objects through *any* public bucket or access point policies
S3 will ignore public and cross-account access for buckets or access points with policies that grant public access to buckets and objects.

Figure 5.1 – Block Public Access settings for bucket page

It is very rare for an organization to want to allow data to be publicly available. If there is an excellent reason to do so, it is recommended that you put safeguards that prevent accidental data from going into the bucket. One approach is to have a separate AWS account and allow only a few trusted people to access it. An even better system would be to set up a process that copies data to the public bucket that a second person approves.

Enabling data and metadata encryption and enforcing it

Enabling data and metadata encryption early on can save a lot of time in the future and be considered before any project. If requirements change and encryption becomes required after data is stored unencrypted, some effort will need to be made to encrypt that data. In addition, any downstream consumers may also need to be changed to decrypt the data. This process can be avoided if the data is encrypted early in the process. To learn how to encrypt your data on S3, see the *Encrypting your data and metadata in Glue Data Catalog* section, later in this chapter.

Ensuring that auditing is enabled

Enabling auditing on AWS is relatively easy and cost-effective. However, the headache that results from not having auditing capabilities can be more costly and cannot be enabled after the fact. For more details on how to enable auditing using CloudTrail logs or S3 server access logs, please see the *Auditing with CloudTrail and S3 access logs* section, later in this chapter.

Good intentions cannot replace good mechanisms

Jeff Bezos was quoted to have said, *"good intentions never work; you need good mechanisms."* A mechanism is a process that enforces that something is done, regardless of if people have the best intentions. For instance, having the intention to wake up at 6 a.m. is not as effective as setting an alarm. When it comes to security, it is always best to have mechanisms by putting in enforcement where possible and auditing to ensure that the mechanisms are working. An example of enforcement would be to put an S3 bucket policy that rejects uploads unless the objects are encrypted.

Encrypting your data and metadata in Glue Data Catalog

There are many ways a malicious person may be able to get access to your data. They may be able to listen on a network for traffic between two applications. They may be able to pull a hard drive from a machine, server, or dumpster. They may be able to gain access to an account that has access to the data they need. Regardless of how the bad actor obtains your data, you do not want them to read the data, and **data encryption** is how that is done. Data encryption takes your data, encodes it using an encryption key, and makes it impossible to read without the decryption key.

Encryption algorithms where the **encryption key** and **decryption keys** are the same are called **symmetric encryption**. Algorithms in which the keys are different are called **asymmetric encryption**.

Let's look at how we can encrypt data on S3.

Encrypting your data

When your data is persisted somewhere, it should be encrypted. All the data that Athena temporarily stores on any disks on their clusters is encrypted and then wiped after each query. However, you will need to choose how to encrypt data that is stored on S3. With AWS services, typically, there are four different ways encryption can be done. The differences between the four relate to where the encryption key is stored and where the encryption/decryption occurs. Encryption can be done server-side or client-side. With server-side encryption, S3 performs encryption and decryption. The client will never see the encryption keys or encrypted data. With client-side encryption, the requester performs encryption and decryption and S3 will never see unencrypted data. With the encryption key, S3's encryption key can't be used, nor is an encryption key stored from a customer's AWS **Key Management Service** (**KMS**), nor is a key provided by the customer. Each of these options has performance, cost, and security considerations, which we will briefly discuss.

Enabling server-side encryption using S3 keys (SSE-S3)

This is the easiest and cheapest way to encrypt your data; that is, by leveraging S3's encryption keys. S3 will encrypt each object with a unique key and encrypt the key with S3's master key. The encrypted key is then stored as metadata for the object, which S3 can use later when reading. If someone did manage to access the raw, unencrypted data, they would still need S3's master key to decrypt the key that was used to encrypt the data. Using this encryption method does not have a financial cost, and its performance penalty should be negligible.

You can enable default encryption within S3. You can set SSE-S3 as the default encryption by configuring your bucket so that any time an object is written, it will automatically be encrypted using S3's encryption keys:

aws-cloudtrail-logs-888889908458-ed59bff1

Objects	Properties	Permissions	Metrics	Management	Access Points

Default encryption

Automatically encrypt new objects stored in this bucket. Learn more 🔗

[Edit]

Default encryption

Disabled

Edit default encryption

Default encryption

Automatically encrypt new objects stored in this bucket. Learn more 🔗

Server-side encryption

○ Disable

● Enable

Encryption key type

To upload an object with a customer-provided encryption key (SSE-C), use the AWS CLI, AWS SDK, or Amazon S3 REST API.

● Amazon S3 key (SSE-S3)

An encryption key that Amazon S3 creates, manages, and uses for you. Learn more 🔗

○ AWS Key Management Service key (SSE-KMS)

An encryption key protected by AWS Key Management Service (AWS KMS). Learn more 🔗

Cancel   Save changes

Figure 5.2 – Enabling default encryption using S3's keys

Next, we'll look at KMS keys.

Enabling server-side encryption using customers' KMS keys (SSE-KMS)

Rather than using S3's master encryption key, you can specify S3 to obtain encryption and decryption keys from your AWS account's KMS. The credentials that are used to read the data must have permissions to access the KMS key, and S3 will use the keys. Using this method is more secure because you can control who can access the keys. If needed, the KMS key can be deleted if you don't want the encrypted data to be readable by anyone, essentially making it useless.

> **Note**
>
> The cost of SSE-KMS is higher than SSE-S3 because there is a cost associated with making API calls to KMS. When S3 is encrypting or decrypting keys using KMS, it will call the service on your behalf. If you are making significant calls, this cost can quickly add up. It is recommended that a single bucket key allows S3 to cache the key to reduce the number of calls to KMS.

You can also mix and match master keys as the key's **Amazon Resource Name (ARN)** is stored in the object's metadata. However, you can enforce a KMS key to encrypt the data if one is not provided:

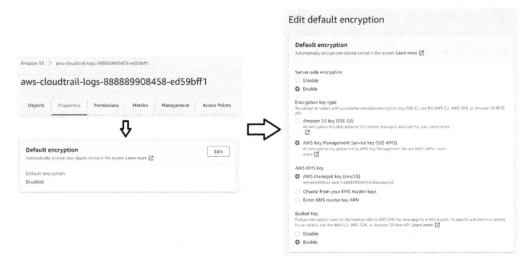

Figure 5.3 – Enabling SSE-KMS

When enabling SSE-KMS, by default, you can specify if the KMS key is managed by S3 or if you are going to maintain it. If S3 manages the key in your account, then the key is free, and it enforces a key rotation that is currently set for every 3 years. If you wish to manage the key, then you will have to pay for the cost of the key. You will also be responsible for rotating the keys, which is a best practice that limits the blast radius.

Enabling client-side encryption using customers' KMS keys (CSE-KMS)

CSE-KMS differs from SSE-KMS and SSE-S3 in that encryption and decryption are done within the client making the request. This method could have a noticeable effect on performance as encrypting and decrypting is a CPU-heavy operation and is not done on S3's fleet of servers. This method can also be much more expensive than SSE-KMS unless KMS key caching is implemented in the caller, which does not exist in Athena today. However, this method can be more secure with its increased cost. First, if there is a middleman attack, they can read your data while transferring it to you. A middleman attack involves a bad actor that has tricked your client into thinking it is talking to S3. At the same time, it proxies messages between S3 and your client. If the data is decrypted on the client side, the middleman does not have access to the decryption keys and won't be able to use the data. This scenario is unlikely to occur because of other mechanisms that AWS uses to prevent such attacks. Secondly, if S3 becomes compromised, the data cannot be decrypted because S3 cannot access the keys. Again, this is an improbable scenario.

When uploading to S3 using the AWS SDK, you need to use the `AmazonS3EncryptionV2` API and provide a KMS ARN. If you're not using AWS SDK, then the `x-amz-meta-x-amz-key` HTTP header must be provided with the encrypted data key. To enable this option for reading within Athena, when specifying your `CREATE TABLE` statement in Athena, set the `has_encrypted_data = true` option in `TBLPROPERTIES`.

> **Reading CSE-KMS Files in Athena Using EMRFS with EMR**
>
> Athena has difficulty reading CSE-KMS encrypted files when using EMRFS with EMR and multipart uploads enabled for Parquet files. If you are writing Parquet files using EMR, ensure that multipart uploads are disabled.

Now, let's compare the different encryption methods.

Comparing encryption methods

The following table compares some important factors regarding the various encryption methods we have just discussed:

Encryption type	Performance	Cost	Level of security
SSE-S3	High	Free	Medium
SSE-KMS	Medium	Cost of KMS key + cost of API costs to KMS. KMS costs can be minimized if the same master key is used.	High
CSE-KMS	Low. Encryption and decryption may not be as parallelly done as if S3 does it.	Cost of KMS key + cost of API costs to KMS. KMS costs can be higher because KMS key caching cannot be done.	Highest

Figure 5.4 – Differences between different encryption methods on S3

Now, let's learn how we can enforce encryption on data in S3.

Mandating encryption at rest with S3

We can create a mechanism by mandating that any data stored in an S3 bucket uses encryption. This can be done by setting a bucket policy that allows only a specific encryption method. See the following example S3 bucket policy, which mandates that all the objects put into this bucket must use SSE-KMS. You can view and download this policy by going to `https://bit.ly/3u4tGiD`.

This policy has two statements. The first statement ensures that the `x-amz-server-side-encryption` header is present on any `s3:PutObject` operation. The second statement contains a condition that prevents any object from being put into the S3 bucket without `x-amz-server-side-encryption` being set to `aws:kms`.

Athena query results can also be encrypted. When an Athena query completes, it stores the results in an S3 bucket that you own. Administrators can set a workgroup to encrypt query results. In the workgroup settings, set the query results to be encrypted using SSE-KMS, CSE-KMS, or SSE-S3 and check the **Override client-side settings** box. The following screenshot shows how to set this up:

Query result location and Encryption

Query result location   s3://my-bucket/athena-query-results/   📁 Select

The S3 path requires a trailing slash. Example: s3://query-results-bucket/folder/

Encrypt query results   ☑ Encrypt results stored in S3

Encryption type   SSE-KMS   ⌄ ❶

Encryption key   athena-kms   ⌄ ❶ ☐ Create KMS key

Settings

Override client-side settings   ☑ ❶

Figure 5.5 – Enforcing encryption on query results

Now that we have learned how to encrypt data, let's look at how we can encrypt our Glue Data Catalog.

Encrypting your metadata in Glue Data Catalog

Some users may want to encrypt their metadata in addition to their data. Metadata may contain sensitive information that you may not want to leave unprotected, such as partition values, table schemas, the location of your sensitive data, and so on. When encryption is enabled in Glue Data Catalog, the non-exhaustive list of information that is encrypted includes databases, tables, partitions, and table versions. Enabling encryption for Glue Data Catalog is relatively simple. The following screenshot shows how to enable encryption with a few clicks:

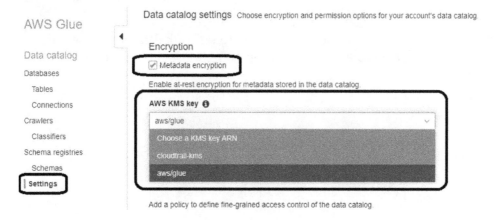

Figure 5.6 – Enabling Glue Data Catalog encryption in the Glue console

Like S3 data encryption, you can specify a Glue service managed KMS key (*aws/glue*) or provide a customer-managed KMS key. If having full control over the key is essential, select your customer-managed KMS key. Otherwise, you can allow Glue to manage the key at no cost.

> **Note**
>
> Glue only supports symmetric keys and will not work if an asymmetric key is provided.

Now that we know how to encrypt our data at rest, let's touch on data in transit.

Encrypting your data in transit

All the data that's read within Athena and between clients and AWS services such as S3 is encrypted using TLS. There is nothing you need to do on your part to enable this.

Now that we know how to encrypt our data, let's look at how we can enable coarse-grained access controls.

Enabling coarse-grained access controls with IAM resource policies for data on S3

Coarse-grained access control (**CGAC**) is a term that does not have an industry-standard definition. Generally, in this book, when we refer to CGAC in the context of data lakes, we are referring to object-level permissions such as individual files on S3. If a user has access to an object, they can access all the data within that file. **Fine-grained access control** (**FGAC**) provides authorization on data within the files, such as columns and rows. We will discuss FGAC in more detail in the next section.

Within AWS, there is one popular way to achieve CGAC with data on S3. That is through bucket policies that limit access to IAM principals. We will look at how to enable this in this section.

CGAC through S3 bucket policies

By default, access to S3 buckets is denied unless there are policies that grant access to it. Regarding a new IAM principal, either an IAM user or role, permissions must be provided to allow them to access S3 resources. There are several ways to grant permissions, but we will focus on two general ways to provide permissions in this section. The first way is to manage permissions to IAM principals within the same AWS account. The permissions that are granted to the IAM principal will be used by Athena to access the underlying data. The second way is to attach S3 bucket policies. Bucket policies allow more flexibility in that they can grant cross-AWS account access. They also have additional conditionals that can fine-tune access and enforce how users interact with that bucket.

If any IAM or S3 bucket policies grant access and there are no policies that deny access to the request, the principal will be able to perform the action on the S3 resource. Otherwise, the request will be rejected. The following diagram illustrates this:

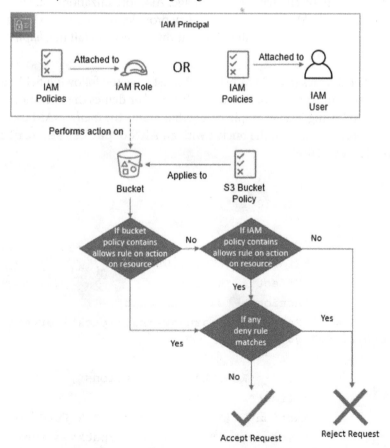

Figure 5.7 – IAM permissions on S3 buckets and objects

Let's look at how we can attach IAM policies to IAM users or roles to control data access. There are two common ways organizations can control access through IAM. First, they can create an IAM user for each of the end users and provide them with AWS credentials and/or console access. Second, they can interact directly with S3 or indirectly using an AWS service such as Athena. An IAM group can be created with specific permissions to S3, and IAM users can be placed within that group. An IAM user that belongs to multiple groups will get a union of all the groups' policies. Using groups is preferred for managing permissions rather than manually setting permissions for individual IAM users because it is a lot less manual work. The other method is to provide permissions to IAM roles and allow your users to assume those roles. Either way, when the IAM principal submits a query to Athena, their permissions will be applied.

For small organizations, providing users with IAM credentials can be a convenient and quick way to control access to AWS resources. However, as organizations grow larger, managing IAM users can be challenging to manage. Also, organizations may want their users to federate into an IAM role using an identity provider to use their existing company login and password credentials. We will talk about this in more detail in *Chapter 7, Ad Hoc Analytics*.

An IAM policy that attaches to an IAM principal must have the following fields: a list of actions and resources and whether the rule grants or denies the operation on the resource. The following is an example policy that grants the *analyst* IAM user permission to perform actions on the bucket with an ARN of packt-serverless-analytics-01234567890:

```
{
    "Version": "2012-10-17",
    "Statement": [
        {
            "Sid": "ListBucketOnBucket",
            "Effect": "Allow",
            "Action": "s3:ListBucket",
            "Resource": "arn:aws:s3:::packt-serverless-
analytics-01234567890"
        }, {
            "Sid": "ReadObjectPermissions",
            "Effect": "Allow",
            "Action": ["s3:GetObject", "s3:PutObject"],
            "Resource": "arn:aws:s3:::packt-serverless-
analytics-01234567890/*"
```

```
        }
    ]
}
```

This policy can be attached to an IAM group, an IAM user, or an IAM role. It will allow the principal to list all the objects within the `packt-serverless-analytics-01234567890` bucket and read and write objects within that bucket.

This is done with two statements. The first statement allows the user to perform the `s3:ListBucket` operation on the bucket. The second statement allows the user to perform `s3:GetObject` and `s3:PutObject` in the same bucket. You may notice that the resource contains `/*` at the end of the second statement and not the first. The reason for this is that the first statement's operation is a bucket-level operation. The operations in the second statement are at the object level.

The previous policy can also be attached to an S3 bucket with one difference. Each of the statements must provide a list of principals. These principals can be applied to entire AWS accounts, IAM principals in the accounts, AWS services, federated users, and anonymous users (public access). There are some benefits to attaching bucket policies rather than attaching them to IAM principals. First, bucket policies have more conditional attributes it can check for. For example, the `x-amz-server-side-encryption` header can be matched to enforce encryption.

Another example is limiting access to the bucket from a VPC or IP address range, although queries that run on a bucket with this condition are not supported with Athena. Instead, you can use the `aws:CalledVia` condition to prevent access to an S3 bucket, except when it's called from Athena. Secondly, you can provide IAM principles in other AWS accounts access to the bucket. For example, an AWS account for a beta environment can be granted access to read-only data in a production account's S3 bucket. The following S3 bucket policy is an example of limiting read access to the `packt-serverless-analytics-0123456789` bucket to only a few IAM users that can only be called from Athena:

```
{
        "Version": "2012-10-17",
        "Statement": [
            {
                "Sid": "ReadObjectPermissions",
                "Effect": "Allow",
                "Principal": {
                    "AWS": [
```

```
                    "arn:aws:iam::9876543210:user/luke",
                    "arn:aws:iam::9876543210:user/leia"
            ] },
            "Action": ["s3:ListBucket", "s3:GetObject",
    "s3:PutObject"],
            "Resource": ["arn:aws:s3:::packt-serverless-
    analytics-01234567890",
                             "arn:aws:s3:::packt-
    serverless-analytics-01234567890/*"],
            "Condition":{
                "ForAnyValue:StringEquals":{
                    "aws:CalledVia":[
                        "athena.amazonaws.com"
                    ]
                }
            }
        }
    ]
}
```

If you decide to go with attaching policies to IAM principals and/or S3 buckets, be aware that there are **service quotas**. There are limits on how large policies can be or the number of policies that can be attached. For S3 bucket policies, you can only have a single policy and it can only be up to 20 KB in size. There are limits to the number of policies you can attach to an IAM user or role, the number of groups an IAM user can belong to, and so on. For a full list, see https://docs.aws.amazon.com/IAM/latest/UserGuide/reference_iam-quotas.html.

Although many use cases can be satisfied using IAM to provide CGAC, FGACs may be needed for other use cases. Let's look at how we can achieve that.

Enabling FGACs with Lake Formation for data on S3

FGAC differs from coarse-grained data access control by providing access control finer than at a file or directory level. For example, FGAC may provide **column filtering** (setting permissions on individual columns), **data masking** (running the value of a column through some function that disambiguates its value), and **row filtering** (allowing users to see rows in a dataset that only pertain to them).

There are many open source and third-party applications that provide this access control level within the big data world. Examples of open sourced software include **Apache Ranger** and **Apache Sentry**. An example of a third-party application is **Privacera**. First-party integration is also available through **AWS Lake Formation**.

One of AWS Lake Formation's major components is providing FGACs to data within the data lake. Administrators can determine which users have access to which objects within Glue Data Catalog, such as tables, columns, and rows. We will discuss setting up and managing Lake Formation access control in depth in *Chapter 6, AWS Glue and AWS Lake Formation*.

Auditing with CloudTrail and S3 access logs

Auditing is an essential part of designing a secure system. Auditing provides validation that existing access policies are working and when there is a security incident, the impact of the incident and hopefully the bad actors. AWS has two native auditing mechanisms for data access that we will look at in detail: AWS CloudTrail and Amazon S3 access logs.

Auditing with AWS CloudTrail

AWS CloudTrail is a service that provides auditing capabilities for API calls that are made to all AWS services that support CloudTrail. When an AWS account is created, CloudTrail logging is enabled by default to help manage APIs. These APIs perform actions on AWS resources such as creating or describing EC2 instances, creating S3 buckets, or submitting Athena queries. The other class of events is data events. These are AWS APIs that are called on a resource itself. At the time of writing, S3 calls to list, get, put, or delete operations and Lambda invocations are considered data events.

Management events are created when an API is called that manage resources, such as starting an EC2 instance or configuring an S3 bucket. The first copy of management events is free, and any additional copies are charged at $2.00 per 100,000 events. The initial events are pushed to CloudTrail's system, which retains the events for up to 90 days, and can be downloaded in JSON or CSV format. If there are requirements to keep this data for longer than 90 days, you will need to create a new trail that stores events in S3, and you will incur a cost for this. You can then use Athena to query the exported audit records. The following screenshot shows what CloudTrail's **Event history** page looks like:

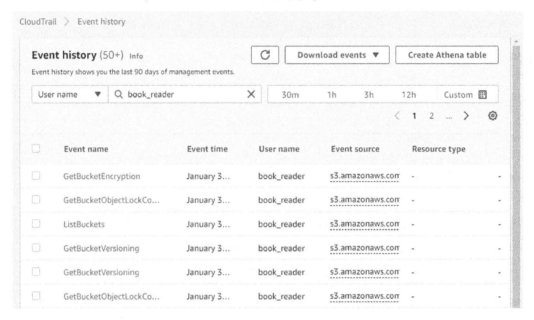

Figure 5.8 – Event history in AWS CloudTrail

The following screenshot shows the type of information stored in the event:

CloudTrail > Event history > GetTables

GetTables Info

Details Info

Event time	AWS access key	AWS region
January 30, 2021, 09:15:26 (UTC-05:00)	ASIA455PXTTVBVV7RXDO	us-east-1
	Source IP address	Error code
User name	173.63.139.155	-
book_reader		
	Event ID	Read-only
Event name	b54cd12d-e64a-462f-abd6-37d9ae81eb2d	true
GetTables		
Event source	Request ID	
glue.amazonaws.com	c7426b98-5835-4758-8662-00268f88cb6e	

Figure 5.9 – CloudTrail event details for a GetTables event

Management events can be useful when tracking the usage of AWS services. For Athena, the StartQueryExecution and GetQueryExecution calls can be tracked, and information about who submitted the query and the query string is logged.

What management events do not provide is data events. For analytics, this means events that retrieve data from S3. To get data events, you will need to enable the data events and incur a cost of $0.10 per 100,000 events, plus any S3 storage the log files may take up. You can set up which buckets and prefixes you want to enable logging on or provide more advanced filters. S3 can generate a massive amount of events, which could lead to high costs. Using filters to capture events from only the buckets containing sensitive information may balance cost and auditability.

If you create new trails that export CloudTrail events to S3, you can use Athena to query the audit logs. Click **Create Athena table** in the top right-hand corner, as shown in the following screenshot:

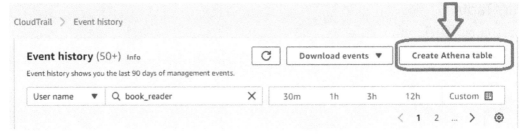

Figure 5.10 – The Create Athena table button in CloudTrail's Event history

This will create a new Athena table.

Auditing with S3 server access logs

S3 access logs differ from CloudTrail logs in a few ways. First, they provide more detailed information about a particular event. Second, it is free to enable, and the only cost that's incurred is the S3 storage costs of the logs. Lastly, the logs' delivery is done with the best effort, meaning that the logs' delivery is not guaranteed. However, from experience, this is rare.

To enable S3 access logs, you will need to enable it on a per-bucket basis and provide a bucket and an optional prefix for where logs are written. You can do this through the console by going to the bucket's **Properties** tab and enabling **Server access logging**, as shown in the following screenshot:

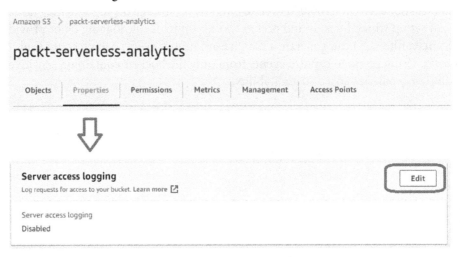

Figure 5.11 – Enabling Server access logging

Now that we have covered the general security aspects of data access on AWS, let's summarize what we learned in this chapter.

Summary

In this chapter, we have gone through some ways that we can protect data from malicious users. We know that no system can ever be 100% secure, but we can take some simple steps to avoid headaches in the future.

We looked at how encrypting your data early in projects can help save time and resources and how to encrypt data at rest and in transit. We looked at the difference between coarse-grained access versus FGACs to implement authorization. Authorization on S3 can be done through S3 bucket policies and/or IAM users, and role policies provide CGACs. Lastly, we looked at how auditing can be enabled and compared these approaches based on their cost and the information they can deliver.

We will dive into Lake Formation, an AWS service that creates and administrates a data lake easier and faster, in the next chapter.

Further reading

For more information regarding what was covered in this chapter, take a look at the following resources:

- *Creating tables based on encrypted datasets in S3*: https://docs.aws.amazon. com/athena/latest/ug/creating-tables-based-on-encrypted-datasets-in-s3.html

- *Encrypt Glue Data Catalog*: https://docs.aws.amazon.com/glue/ latest/dg/encrypt-glue-data-catalog.html

- *Example walkthroughs managing access to S3*: https://docs.aws.amazon. com/AmazonS3/latest/userguide/example-walkthroughs-managing-access.html

- *IAM Best Practices*: https://docs.aws.amazon.com/IAM/latest/ UserGuide/best-practices.html

- *Example S3 Bucket Policies*: https://docs.aws.amazon.com/AmazonS3/ latest/userguide/example-bucket-policies.html

- *Amazon S3 Policy Keys*: https://docs.aws.amazon.com/AmazonS3/ latest/userguide/amazon-s3-policy-keys.html

6
AWS Glue and AWS Lake Formation

Although this book focuses on **Athena** and its rich functionality, you should be aware of **AWS Glue** and **AWS Lake Formation**. These services can be used with Athena to implement use cases that Athena cannot alone. AWS Lake Formation was created to help customers simplify creating data lakes by providing tools to help ingest data, secure data, and reduce the time it takes to get a functional data lake. Lake Formation is a layer that exists on top of AWS Glue and uses Glue's components as building blocks.

One of the main features that Lake Formation brings is fine-grained access controls and auditing to several AWS services, including Athena. Lake Formation augments **AWS IAM** to help secure the data lake. IAM provides authentication of the user, while Lake Formation provides authorization based on the principle that is requesting data. Every authorization request that goes through Lake Formation generates audit events in **CloudTrail** that are reported in the Lake Formation console, providing a single central place to administer and monitor the data lake.

AWS Lake Formation also provides a new table type called the governed table, which provides four key benefits. First, it provides **atomic, consistent, isolated, and durable (ACID)** transactions for metadata and data updates. Second, it provides automatic compaction of data, combining small data files to produce fewer and larger files to optimize query performance. Third, you can run queries on datasets as if they were run at a different point of time to see what the data looked like in the past before certain transactions have been applied. This feature is usually called **time traveling**. Fourth, governed tables provide row and cell-level filtering to enforce user permissions.

Fine-grained access control and governed tables directly integrate with Athena to provide security and enhanced functionality. Lake Formation and Glue can also provide functionality that aids in creating and maintaining a data lake. We will look at some of the functionality that Lake Formation and Glue provide that could solve some of the challenges that Athena cannot solve on its own.

In this chapter, we will cover the following topics:

- What AWS Glue and AWS Lake Formation can do for you
- Securing your data lake with Lake Formation
- What AWS Lake Formation governed tables can do for you

Technical requirements

For this chapter, if you wish to follow some of the walkthroughs, you will need the following:

- Internet access to GitHub, S3, and the AWS Console.
- A computer with either Chrome, Safari, or Microsoft Edge installed on it.
- An AWS account and accompanying IAM user (or role) with sufficient privileges to complete this chapter's activities. For simplicity, you can always run through these exercises with a user that has full access. However, we recommend using scoped-down IAM policies to avoid making costly mistakes and learn how to best use IAM to secure your applications and data. You can find a minimally scoped IAM policy for this chapter in this book's accompanying GitHub repository, which is listed as `chapter_6/iam_policy_chapter_6.json`. This policy includes the following:

  - Permissions to create and list IAM roles and policies:
    - We will be creating a service role for an AWS Glue Crawler to assume.
  - Permissions to read, list, and write access to an S3 bucket.

- Permissions to read and write access to Glue Data Catalog databases, tables, and partitions:

  - You will be creating databases, tables, and partitions manually and with Glue Crawlers.

- Access to run Athena queries.

What AWS Glue and AWS Lake Formation can do for you

Lake Formation and Glue provide tools that aid in creating data lakes and extending functionality to your new or existing data lakes. There is a wide variety of functionality that it provides. In this section, we will go through a non-exhaustive list of features. An entire book could be written on Lake Formation and another on Glue, so we will not go through all of their features in detail in this chapter.

Except for fine-grained access control and governed tables, all features do not directly change how Athena works. If you start by not adopting any of the Lake Formation or AWS Glue features, you can adopt them in the future.

Let's take a look at some of the AWS Glue and Lake Formation features and how they can supplement Athena.

Using AWS Glue to cleanse, normalize, and transform data

Amazon Athena's performance and cost are highly dependent on the data format and layout of the data. In many scenarios, it may be cost-effective and improve performance to provide faster response times to users and applications to transform the data. We will dive into the details of the scenarios and decisions regarding when to perform this in *Chapter 9, Serverless ETL Pipelines*, so it may be a good idea to skip ahead if you are not familiar with this process.

This is where AWS Glue ETL can be really helpful for performing data transformations. AWS Glue ETL is a serverless ETL service that allows customers to write Spark code and execute it without provisioning resources. Many organizations use Apache Spark to perform their transformations and AWS Glue can be more cost-effective than managing Spark yourself. The transformed data after using AWS Glue can then be read and analyzed using Amazon Athena. AWS Glue ETL charges based on the resources that you use. ETL jobs can scale as Glue ETL provides different hardware types and instances in the Spark cluster to run on.

In this section, we will provide a quick summary of what Glue ETL can do. We will look at two ways to author jobs; that is, using Glue ETL and Glue Studio. Let's look at each one.

Glue ETL

AWS Glue ETL uses Apache Spark with Scala and PySpark, a Python-only runtime for lightweight jobs, and Apache Spark Streaming for steaming jobs. To execute a job, a user would create a script, store it in S3, and register it with an ETL job within Glue. Scripts can be executed with a wide variety of properties to give users flexibility and control. The following screenshot shows a sample script editing screen for a PySpark job within the Glue ETL console:

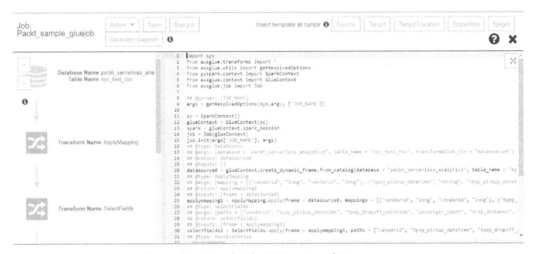

Figure 6.1 – Sample Glue ETL script editing screen

Once the script is ready to be run, the job can be executed using the run job dialog screen, as shown in the following screenshot:

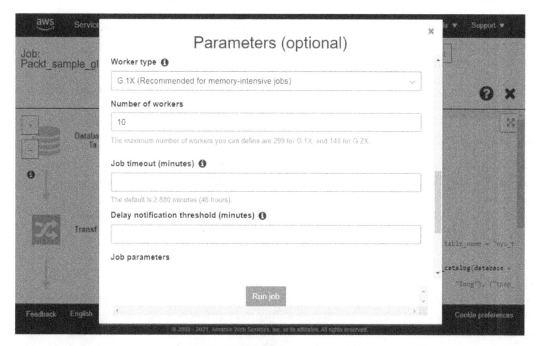

Figure 6.2 – Sample Glue ETL run screen

The Glue console can be used to look at the history of invocations of the job and provide job run information, logs, and other relevant information, as shown in the following screenshot:

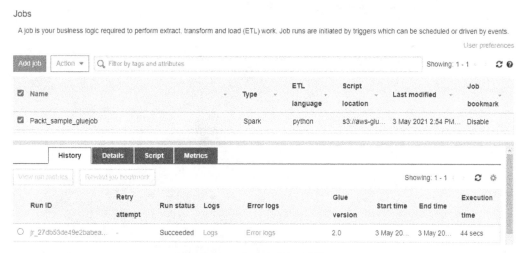

Figure 6.3 – Sample Glue ETL run

Many features make Glue ETL powerful, including Job Bookmarks, which only processes new files when reading a data source, Spark UI to monitor and debug Spark jobs, and publishing job metrics, to name a few.

Now, let's take a look at AWS Glue Studio, which helps with authoring and monitoring Glue ETL jobs.

AWS Glue Studio

AWS Glue Studio is a visual UI that simplifies the process of creating and monitoring Glue ETL jobs. Glue Studio provides enhanced visual editing for Glue jobs and dashboards, which provides job metrics such as running, completed, and failed jobs. It is an ideal tool for non-programmers who are not comfortable with writing code or those that want to do simple transformations.

The visual editor allows users to create complex jobs using mouse clicks instead of writing Spark code. You can piece together three building blocks: sources of data such as S3, RDS databases, Redshift, Kinesis, and Kafka streams; transformations on the data such as joining datasets, renaming, dropping, or filling in empty values in columns; and specifying one or more targets to store the results in various formats. The following screenshot of Glue Studio shows a sample job. Here, we have taken the NYC Taxi dataset and joined it to a location dataset to enrich it by translating the location IDs. We then output the resulting dataset to S3 using the Parquet format:

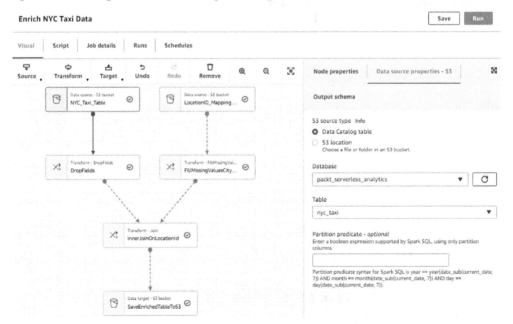

Figure 6.4 – Glue Studio visual editor screen to enrich the NYC Taxi dataset

Once the job has been authored within the visual editor, the source code that implements the execution graph will be auto-generated. The code can then be executed as a regular Glue ETL job on a scheduled basis or automatically triggered by an external event. You are then taken to a dashboard where you can monitor Glue ETL job executions. When you have multiple Glue ETL jobs that run regularly, monitoring and debugging jobs become essential to ensure data is getting generated successfully and on time. Glue Studio has a **Monitoring** tab that shows Glue ETL jobs that have run, their run state, their overall DPU usage to track costs, and other metrics. The following screenshot shows an example dashboard and the available metrics:

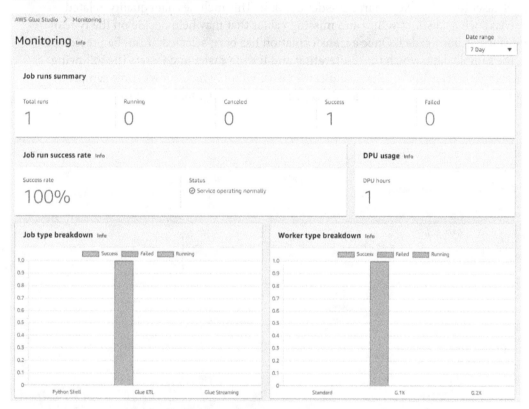

Figure 6.5 – Sample of the Glue Studio Monitoring screen

While AWS Glue Studio builds on top of AWS Glue ETL, Glue also has a separate product that makes it even easier to transform, cleanse, and explore datasets to get them ready for applications and machine learning. Let's take a quick look.

AWS Glue DataBrew

AWS Glue DataBrew is a data preparation and exploration tool that is entirely visual and doesn't require any coding. Unlike Glue Studio, where the visual job editor generates Glue ETL code, Glue DataBrew generates something else call recipes. Recipes are a collection of operations or transformations that are applied to a dataset that can be saved and applied to other datasets.

Glue DataBrew's visual editor provides rich functionality designed to make data preparation simple for all users. When a dataset is loaded into the editor, it will sample the dataset to surface key characteristics about it. This includes data quality-related metrics such as distinct values and missing values that may help decide on the type of transformations needed. Once a transformation has been selected, it can be previewed on the sample data, which makes iterating and testing easier and faster. The following is a sample screenshot of Glue DataBrew's visual editor, which shows some sample transformations on a column called object_name on the sample dataset:

Figure 6.6 – Glue DataBrew sample screen

Glue DataBrew, at the time of writing this book, has over 250 operations and transformations that can be applied to datasets. The transformations range from simple column transformations such as changing data types or renaming columns, data cleaning functions such as changing case on strings, data quality operations such as filling in missing or empty values, and change column structures such as splitting a single column into multiple columns or merging them, to name a few.

Glue DataBrew's pricing is different than Glue ETL's. Glues DataBrew's visual editor charges by the session hour billed per half-hour, which is currently $2.00/hr. When DataBrew executes a recipe, it will use DataBrew execution nodes. Each node has 4 vCPUs and 16 GB of memory and is charged per hour and billed per minute.

Now that we've learned how individual datasets can be transformed and cleansed, let's look at AWS Glues Workflows and how it can piece together multiple transformations that generate data pipelines.

Using AWS Glue Workflows

Glue has many building blocks that can be used together to create what is known as data pipelines. Data pipelines consist of multiple extract, transform, and load jobs that take a complex operation and break them down into manageable parts. Some parts can be reused, run on different execution engines, and executed at other times. The goal is to make pipelines easier to optimize, make them easier to debug and monitor, and then check data quality in different stages to help identify issues earlier.

For example, suppose we are a seller on Amazon.com, and we get raw sales data put into an S3 bucket. We want to transform the data to feed it into a reporting system to generate reports. Before we can generate the reports, we need to cleanse the data, join the data to a product table that translates Amazon product IDs, called ASINs, to product names, join to an inventory table to show how many items we have in stock, and then group all the results by report periods. All these steps can be done within a single job, but our job may run for a long time, and diagnosing data quality issues may be complex. We may also want to save the output of enriched data before grouping the data to generate other reports or share it with another team. It would make sense to break the single job into multiple steps to reuse the job's output.

To manage the order of the job executions and dependencies, we would need an orchestrator to run these jobs and monitor them. This is where Glue Workflows can help. Glue Workflows allows Glue ETL jobs, Glue Crawlers, and **Glue Triggers** to execute in a particular order or workflow. The following screenshot shows a Glue Workflow that can be created to manage the report generation flow we discussed previously. Here, a workflow has been defined using Glue Workflows for the process of report generation:

Figure 6.7 – Glue Workflow of Glue ETL jobs, Glue Triggers, and Glue Crawlers to make a data pipeline

With a workflow defined, you can execute it based on a Glue Trigger. Glue Triggers kick off an action based on job flow dependencies that need to be met to execute the next action in the flow. Glue Triggers can be triggered on a fixed schedule, on-demand, or wait for other tasks to finish, such as Crawlers or Glue ETL jobs. In the preceding example, the workflow triggers are based on a schedule that kicks off the workflow at midnight every night. Once the workflow begins to execute, you can monitor each component, as shown in the following screenshot:

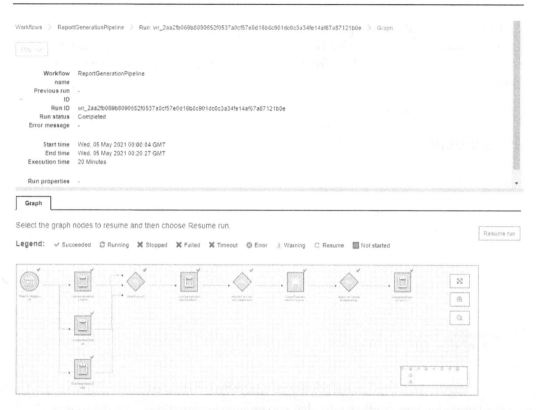

Figure 6.8 – Glue Workflow execution monitoring

In this way, we can monitor the workflow and look over its details as well.

Using AWS Lake Formation blueprints

A standard process that users perform within their data lakes is ingesting data. With a few clicks within the Lake Formation console, you can ingest data from databases, AWS CloudTrail, and load balancer logs. Lake Formation provides **blueprints**, a set of predefined code templates orchestrated with a Glue Workflow, to ingest from these data sources.

Lake Formation provides two types of database blueprints that can extract snapshots of data, or pull data incrementally; that is, data that has been inserted over a certain time. To create a blueprint, select the type from the Lake Formation console, as shown in the following screenshot:

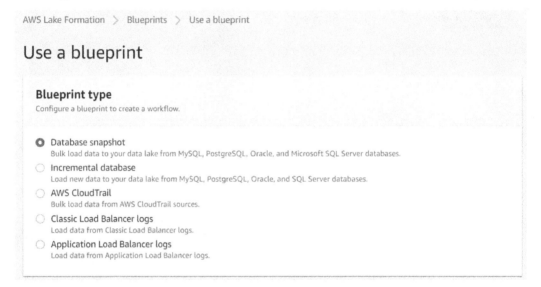

Figure 6.9 – Available blueprint types in Lake Formation

All Lake Formation blueprints require that you provide information about the source, the destination path in S3, and the frequency to pull the data.

Now, let's take a quick look at Glue Crawlers.

Using AWS Glue Crawlers

Glue Crawlers are processes that scan S3 for datasets and register the datasets into Glue Data Catalog. The Crawler reads a sample of the data in the dataset to retrieve or infer the dataset's schema, making registering datasets much easier and less error-prone. We touched on Glue Crawlers in previous chapters, and we take an in-depth look at them in *Chapter 4, Metastores, Data Sources, and Data Lakes*.

Now, let's look at how Lake Formation can help with securing your data lake.

Securing your data lake with Lake Formation

As we mentioned previously, Lake Formation leverages AWS Glue features, including Glue Data Catalog, to simplify creating, accessing, and securing data lakes. Athena uses Glue Data Catalog as its default Metastore and interacts with the service to retrieve metadata to execute queries against tables stored in Glue Data Catalog. Lake Formation adds a security layer on top of Glue tables by eliminating the need to secure individual tables using IAM. When Athena and other AWS analytics services need to access a table, they request permission from Lake Formation, which will authorize based on the calling principal's access policy. The following diagram illustrates this at a high level:

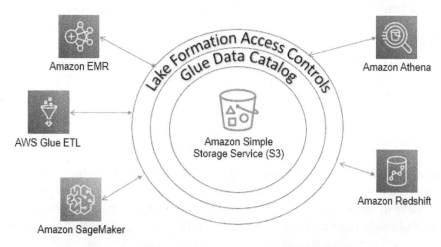

Figure 6.10 – How services interact with Lake Formation

In the following section, we will discuss the benefits of using Lake Formation for authorization and then look at some limitations to consider.

Benefits of using Lake Formation for authorization

Let's look at some of the benefits of using Lake Formation for authorization with Athena and AWS in general.

Finer grained data access controls

Lake Formation authorization occurs at a finer level than what can be achieved using IAM permissions alone. IAM policies can only provide permissions to objects stored in S3 and cannot control what the user can access within the files. This is what we refer to as coarse-grained access control. Lake Formation provides finer-grained access control by allowing us to define policies for subsets of data within an S3 object, namely column- and row-level control. This can be useful for various scenarios. If a dataset contains columns that contain sensitive data, instead of transforming the data to remove these columns, you can leave them in and restrict users to only see those columns containing non-sensitive information. Many times, this is required to meet compliance regulations.

Applying policies at the database, table, and column level

Access policies in Lake Formation are applied to databases, tables, and columns but not S3 paths. This has some benefits in that an administrator does not need to know about the underlying data in S3 when granting and revoking permissions. Tables can be used as an abstraction to the underlying data.

Scalability

Lake Formation permissions do not have a set size limit compared to what is allowed by IAM policies alone. There are limits to the number of inline and managed policies attached to an IAM role or IAM user with IAM policies. Large organizations could reach these limits and would need to develop custom code to generate credentials on the fly or split users into different AWS accounts.

Separating permissions with credentials

One of the leading security benefits of using Lake Formation is that the user running Athena queries does not need to configure access to the underlying data. Instead, when Athena needs access to the data, it sends a request to Lake Formation on behalf of the user to authorize them. If the request is authorized, temporary AWS credentials are provided to access the data. This separates the IAM permissions from the Lake Formation permissions. These temporary credentials are provided to the calling service – in this case, Athena – and not to the user to ensure they can only access data from a trusted service and not directly. When a Glue table is registered with Lake Formation, IAM permissions to S3 and the Glue table can be safely removed. All requests are logged, which can be audited. This flow can be challenging to follow, which is why we have provided the following diagram to help illustrate it:

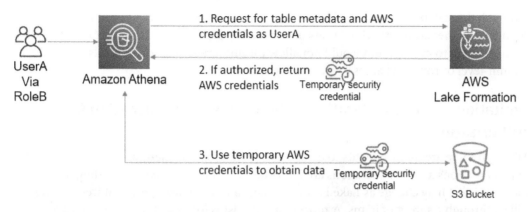

Figure 6.11 – How Athena interacts with Lake Formation to provide access control

For advanced users who use an identity provider to federate access from a directory service such as Microsoft Active Directory, Lake Formation can authorize the federated user and the directory groups they belong to. For example, if **UserA** federates and assumes the role of **RoleB**, then authorization can be done using **UserA** rather than **RoleB**. This is very useful when multiple users are assuming the same role to gain access to AWS services.

Security policies applied consistently across AWS services

Lake Formation provides a central administrative tool to control access to your data from AWS Glue, Amazon Athena, Amazon EMR, Amazon Redshift using Spectrum, and Amazon SageMaker. Access policies that are set in Lake Formation are applied to all Amazon Athena queries, Amazon Redshift queries on tables in S3 using Spectrum, AWS Glue and Amazon EMR Spark jobs, and Amazon SageMaker machine learning exploration using notebooks, pre-processing, and training.

Consistent security across AWS accounts

Many customers have adopted splitting their lines of business or groups using multiple AWS accounts. This allows for use cases where there is a central AWS account that contains the data lake, and different AWS accounts are the producers and consumers of data. Lake Formation allows you to share datasets with other AWS accounts by enforcing permissions on the metadata and data from a central place. Once data has been shared with consumer accounts, users can run queries in Athena against these tables.

Although this can be done using IAM policies, direct data access would need to be provided to other accounts. This results in a more complex set of policies that could be challenging to manage and would not allow for fine-grained access controls such as column-level or row-level access.

Limitations and considerations when using Lake Formation for authorization

Although there are many benefits to using Lake Formation for fine-grained access control, some limitations and considerations are important to understand when deciding to adopt. This list may change as Lake Formation continues to release new features, and we will go through a subset of items. A more complete list is located at https://amzn. to/3nwAvGN. If you have any questions, please contact your AWS representative or AWS support.

Athena query results cannot be managed by Lake Formation yet

When Athena runs a query, the query's results are stored in S3 in the customer's account. When results are requested through Athena APIs, they are read from S3 by Athena and returned to the caller. This ensures that customers have complete control over the resulting data. However, Lake Formation does not currently manage access permissions on S3 paths, but rather only catalog resources such as databases, tables, and columns. For this reason, it is recommended to use another mechanism to limit access to the query results. One solution is to use Athena workgroups to force the query result's location to a particular S3 location, and then employ IAM policies so that the results cannot be read by anyone outside the workgroup.

Athena does not query tables managed by Lake Formation that are encrypted using CSE-KMS encryption

S3 locations that are registered with Lake Formation cannot use **CSE-KMS encryption** with Athena yet. We do not recommend using CSE-KMS if possible, as discussed in *Chapter 5, Securing Your Data*. If this is not possible, then it is not recommended to use Lake Formation for data access controls and to rely on IAM policies instead.

Table partitions data must be located inside the tables directory

In the majority of cases, partition data is stored in a subdirectory inside the table's location. For example, if a table's location is s3://my_bucket/my_table/, then the partitions would be located at s3://my_bucket/my_table/my_partition=val1/ and s3://my_bucket/my_table/my_partition=val2/. If you have a partitioned table where the partition's location is not under the table's location, then Athena with Lake Formation authorization will not work.

Now that we have gone through a subset of limitations, let's look at enabling Lake Formation for data access control with Athena.

Walkthrough to enable Lake Formation for access control

To learn the process of enabling Lake Formation for access control, it is best to go through a walkthrough. This section will go through a sample setup for a new database that will have its access controlled using Lake Formation. We will test the access controls using Athena. If you wish to follow along, you will need to create two IAM users and an S3 bucket that will contain sample datasets. The first user will be given administrative access to Glue and Lake Formation to grant and revoke access to our data lake. A sample IAM policy for this user is available at `https://bit.ly/3er86iv`. The second user will be our Athena user, who will be able to run queries. A sample IAM policy for this user is available at `https://bit.ly/2R87t4B`.

The process that we will be going through will contain four steps as follows:

1. First, we will create and register a data lake administrator.
2. Then, we will register our S3 location with Lake Formation for management.
3. After that, we will grant permissions to our database and tables.
4. Finally, we will test the permissions that we have granted with Athena.

> **Upgrading Production Accounts to use Lake Formation Access Controls.**
>
> If you are looking to upgrade existing AWS accounts and databases, it's strongly suggested that you test the process in a non-production account first and document the steps taken. The upgrade process may look slightly different depending on factors such as data being encrypted, the IAM users/roles, existing policies, and more. The process of upgrading existing databases to use Lake Formation can be a little complicated. However, going through this process should give you a solid understanding of the pieces of Lake Formation that will make the upgrade process easier to navigate.

Creating and registering a data lake administrator

The first step is to register data lake administrators. For this walkthrough, I have created an IAM user named `athena-lakeformation-admin` that will act as our admin. We must select the administrative roles and tasks within the Lake Formation console and then click on the **Choose administrators** button to add our administrator user. Once we've done that, our console should look like this:

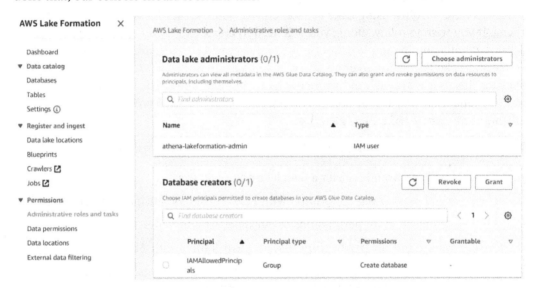

Figure 6.12 – Data lake administrator's screen

Once `athena-lakeformation-admin` has been added, we can switch to that user for the remainder of the interactions with the Lake Formation console. The next step is to register an S3 location with Lake Formation.

Registering an S3 location

The next step is to register our S3 location so that it can be accessed by Lake Formation. This process grants permissions to the Lake Formation service to assume an IAM role so that the service can interact with the data within the S3 location. When an authorization request is made to Lake Formation by an AWS service on behalf of an end user for a dataset, Lake Formation will assume this role and create temporary credentials. For this walkthrough, we will allow Lake Formation to assume a **Service Linked Role**, a type of IAM role that can only be used by AWS services. The AWS service will grant the role with the least amount of privilege to perform actions on your behalf. Once the Service Linked Role has been created, you can view the role in your IAM console and review the permissions that were granted to it. The only scenario when you would not want to use a Service Linked Role is when you want to manually manage permissions or use EMR with Lake Formation.

The following screenshot shows the **Register location** screen, which is where you can register a bucket named `packt-serverless-analytics-888889908458-lakeformation` using the Service Linked Role:

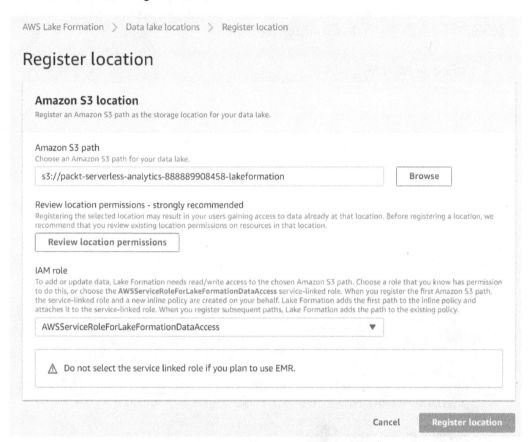

Figure 6.13 – Registering an S3 location with Lake Formation

> **Registering an S3 Location with Encryption Enabled**
>
> If you're registering an S3 location that has encryption enabled, some additional steps must be followed. See `https://amzn.to/3hf39uW` for more information on how to enable encrypted S3 paths.

Before registering a new S3 location, it is good to review the permissions that have already been granted to the S3 location to ensure that the registration process doesn't give permissions to unintended principals.

Now that we have registered an S3 location with Lake Formation, let's grant permissions for our admins to manage datasets in the storage location.

Granting permissions to an S3 location

The next step is to grant permissions to S3 locations for users that we wish to create databases and tables for. There is no other reason to grant users permissions to specific S3 locations. For this walkthrough, we will grant our `athena-lakeformation-admin` user permissions, as shown in the following screenshot:

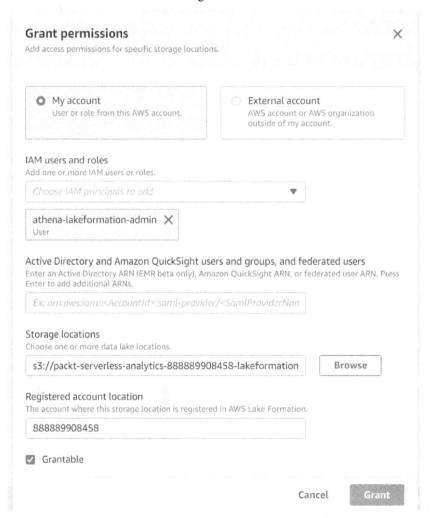

Figure 6.14 – Granting permissions to an S3 location for Lake Formation

The **Grantable** checkbox allows this user to grant other users permissions to this location as well. For example, if we wanted `athena-lakeformation-UserA` to grant permissions to `athena-lakeformation-UserB`, we would set **Grantable** for `athena-lake-formation-UserA`.

With permissions granted to `athena-lakeformation-admin`, let's create our database.

Creating and configuring a database

This step will create a new database called `packt_serverless_analytics_lakeformation` so that we can register tables within it. This database will be configured so that Lake Formation only manages its permissions. The following screenshot shows how to create the database within the Lake Formation console:

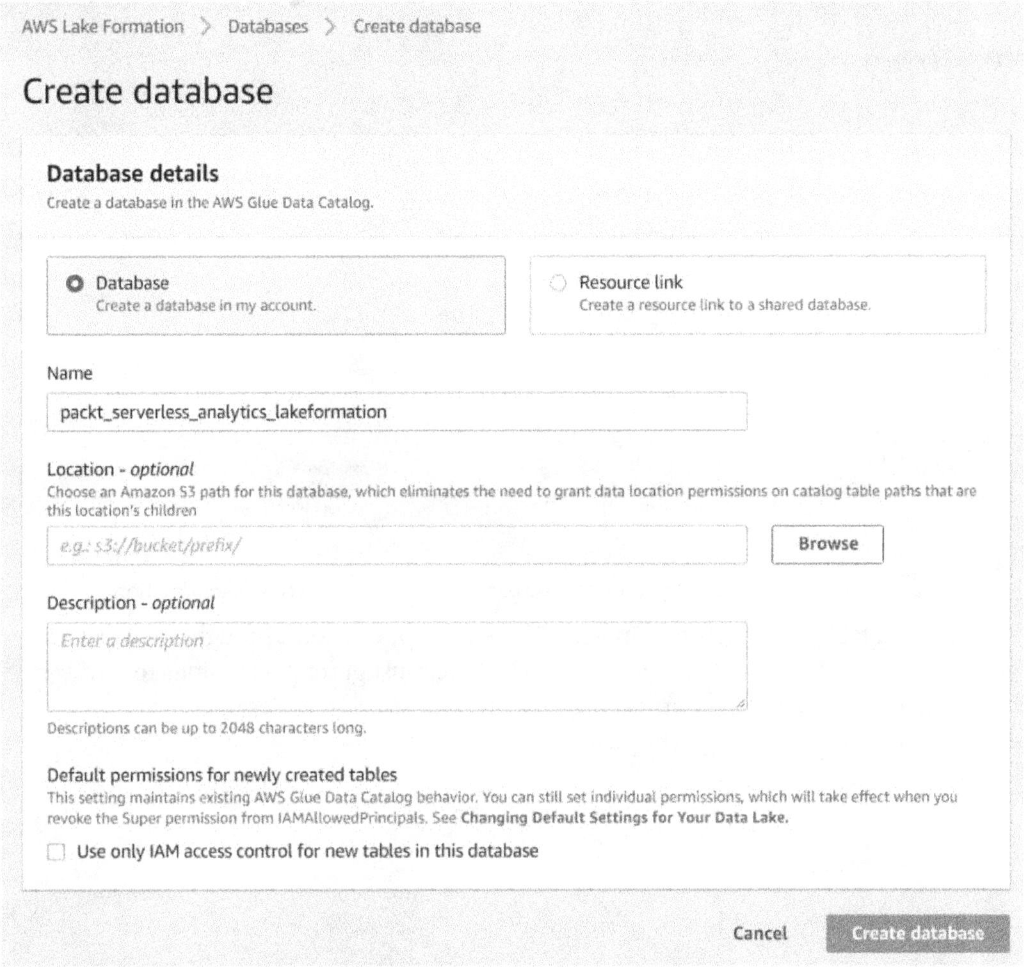

AWS Lake Formation > Databases > Create database

Create database

Database details
Create a database in the AWS Glue Data Catalog.

○ **Database**
Create a database in my account.

○ Resource link
Create a resource link to a shared database.

Name

packt_serverless_analytics_lakeformation

Location - *optional*
Choose an Amazon S3 path for this database, which eliminates the need to grant data location permissions on catalog table paths that are this location's children

e.g.: s3://bucket/prefix/   Browse

Description - *optional*

Enter a description

Descriptions can be up to 2048 characters long.

Default permissions for newly created tables
This setting maintains existing AWS Glue Data Catalog behavior. You can still set individual permissions, which will take effect when you revoke the Super permission from IAMAllowedPrincipals. See **Changing Default Settings for Your Data Lake.**

☐ Use only IAM access control for new tables in this database

Cancel   Create database

Figure 6.15 – Creating a database in Glue Data Catalog for Lake Formation permissions

We want to make sure that we uncheck **Use only IAM access control for new tables in this database**. We want Lake Formation to manage all permissions to our tables within the `packt_serverless_analytics_lakeformation` database.

If you look at the data permissions screen for the database and/or tables within the database and see that the `IAMAllowedPrincipals` principal has permissions, revoke its access. The `IAMAllowedPrincipals` group is a special group within Lake Formation that grants permissions to any IAM principal to interact with this location. Removing it will make Lake Formation the only source for permissions. The following screenshot illustrates this:

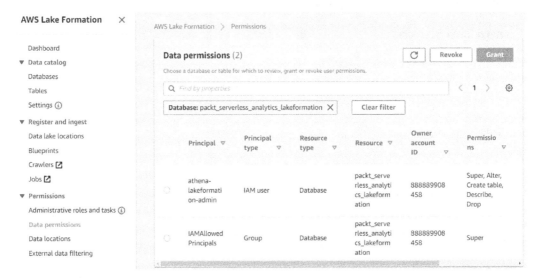

Figure 6.16 – Data permissions for database packt_serverless_analytics_lakeformation

Now, register a table that exists within your S3 bucket. For our walkthrough, we will register our NYC Taxi dataset as `nyc_taxi`. Now, we must grant permissions to our `athena-lakeformation-UserA` to access the database and tables of `packt_serverless_analytics_lakeformation`.

Granting permissions to a user

If we log in as `athena-lakeformation-UserA` and we go to Athena, we will see that the `packt_serverless_analytics_lakeformation` database is not visible, as shown in the following screenshot:

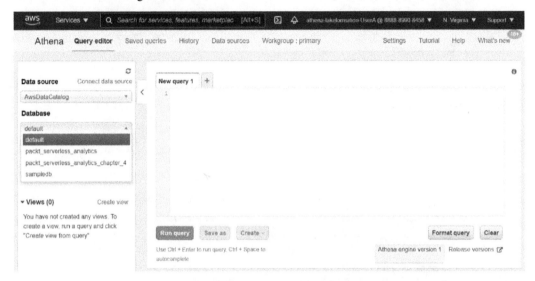

Figure 6.17 – The athena-lakeformation-UserA user's list of databases in
Athena before being granted permission

Let's add permissions for `athena-lakeformation-UserA` by permitting them to describe the database, as shown in the following screenshot:

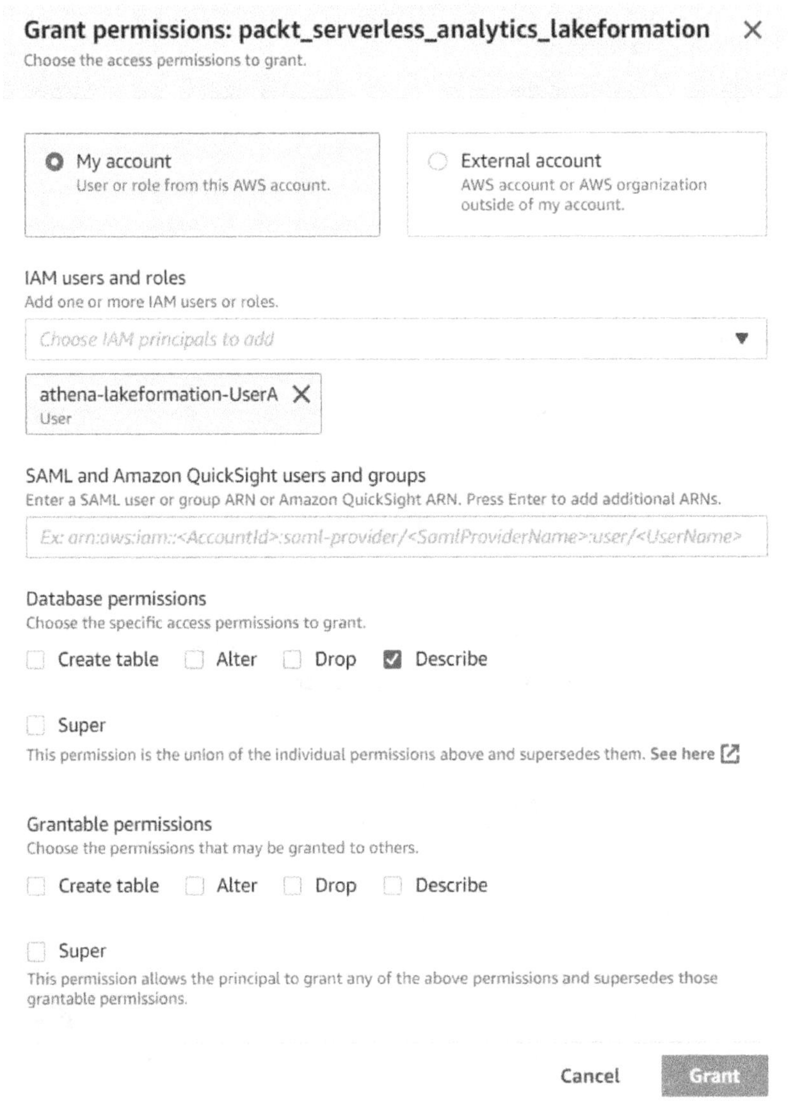

Grant permissions: packt_serverless_analytics_lakeformation   ✕

Choose the access permissions to grant.

○ **My account**
User or role from this AWS account.

○ **External account**
AWS account or AWS organization outside of my account.

IAM users and roles
Add one or more IAM users or roles.

> Choose IAM principals to add ▼

athena-lakeformation-UserA ✕
User

SAML and Amazon QuickSight users and groups
Enter a SAML user or group ARN or Amazon QuickSight ARN. Press Enter to add additional ARNs.

> Ex: arn:aws:iam::<AccountId>:saml-provider/<SamlProviderName>:user/<UserName>

Database permissions
Choose the specific access permissions to grant.

☐ Create table   ☐ Alter   ☐ Drop   ☑ Describe

☐ Super
This permission is the union of the individual permissions above and supersedes them. **See here** ⧉

Grantable permissions
Choose the permissions that may be granted to others.

☐ Create table   ☐ Alter   ☐ Drop   ☐ Describe

☐ Super
This permission allows the principal to grant any of the above permissions and supersedes those grantable permissions.

Cancel   **Grant**

Figure 6.18 – Granting athena-lakeformation-UserA permission to the
packt_serverless_analytics_lakeformation database

This has granted the user to see the database within Athena, as shown in the following screenshot:

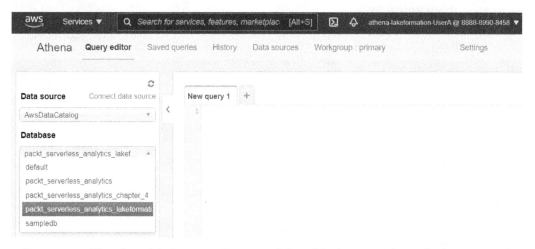

Figure 6.19 – The athena-lakeformation-UserA user's list of databases in Athena after being granted permission

Then, grant the user access to query the `nyc_taxi` table and exclude the `tip_amount` column as it may be sensitive data for the user to query. The following screenshot shows how to grant this permission:

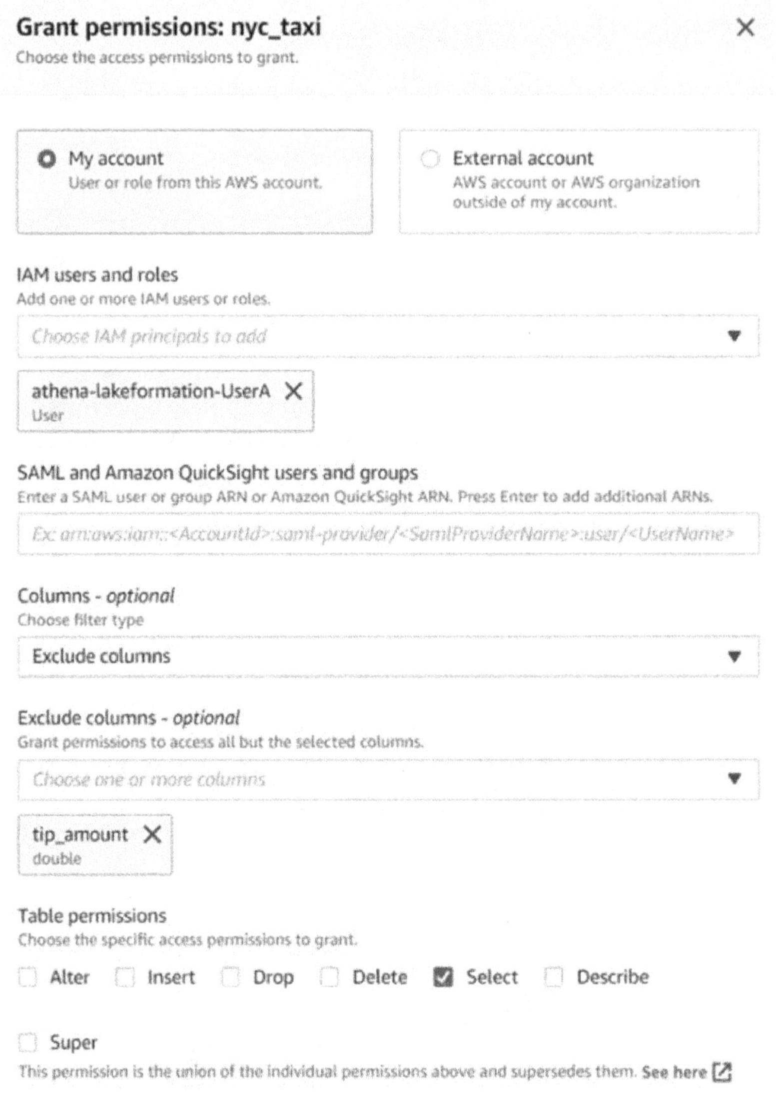

Figure 6.20 – Granting permission to nyc_taxi table to the athena-lakeformation-UserA user, excluding the tip_amount column

> **Note**
>
> If you have a filter on included or excluded columns, you should not select
> **Describe** permissions as you may receive an error message.

After granting these permissions, the user can query the table within Athena but will not
get the `tip_amount` column:

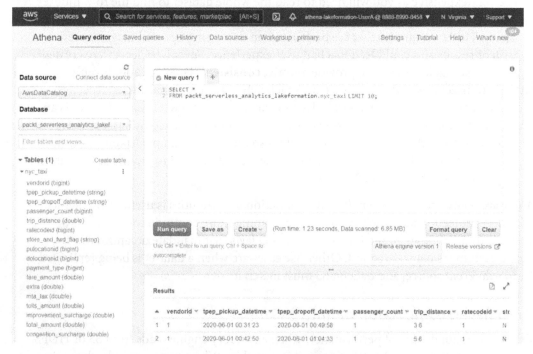

Figure 6.21 – Athena console querying the nyc_taxi dataset with column tip_amount not available

Now that we have enabled Lake Formation for Athena, let's look at governed tables and
how they differ from regular tables.

What AWS Lake Formation governed tables can do for you

Lake Formation introduced a new table format called governed tables. Governed tables
provide many features that aim to solve many of the pain points users have when storing
data on S3. We have an entire chapter, *Chapter 14, Lake Formation – Advanced Topics*,
dedicated to it, but we will summarize many of the benefits here. Let's take a look.

Transactions on tables stored in S3

Distributed filesystems such as Hadoop's Distributed File System and Amazon S3 are excellent choices for storing vast amounts of data and querying them. They also excel at overwriting files and deleting them. However, they were not designed to update and delete data within files. To support this functionality, tools have to download the file, find and update the rows, and then replace the entire file with the new one. This process can be very expensive as you will need to read the whole dataset to find the row. Indexes on primary columns can be added for some file formats to help find which files need replacing. However, queries that are performed on data being updated may not see consistent results and could lead to a bad user experience. Several projects were created to solve these challenges and to provide **atomic, consistent, isolated, and durable (ACID)** transactions to datasets, such as Apache Hive Transactional Tables, Apache Hudi, Apache Iceberg, and Databricks Delta Lake, to name a few. Governed tables is an AWS Lake Formation implementation of similar capabilities backed by a fully managed service. It provides ACID transactions to tables so that users can update and delete files and individual rows through a set of APIs. These store data in S3 to retain the benefits of reliability and scalability that S3 provides.

There are many use cases where having transactional capabilities is useful. Compliance with data protection laws such as GPDR is becoming more common today. This is a growing trend as other countries are introducing laws that mandate organizations to delete customer's data on request. Other use cases are when a dataset is being replicated from a different source, and data on S3 must match it.

Automated compaction of data

One of the main drivers of performance, when any query engine reads from S3 or HDFS, is how data is stored and the data format it is stored in. When customers ask why their queries may not be running as quickly as they think they should be running, the first question I ask is, *how big are the file sizes that are being read?* Most of the time, the files that are being read are tiny, from 10 KB to 10 MB. Having small files can be detrimental to query performance because of the number of round trips an engine must make to read each file. When a file is being read by a query engine, the engine must perform an open file operation to open a stream to the file. Then, the engine performs `GetData` operations to read the stream and closes the stream once it finishes. If the file is tiny, the open file operation can take up to 80% of the time it takes to read data. Having file sizes of a recommended length of between 128 MB and 1 GB dramatically reduces the performance impact of S3 List and Get operations. We'll go through some examples of this in *Chapter 11, Operational Excellence – Maintenance, Optimization, and Troubleshooting.*

AWS Lake Formation governed tables eliminate the issue of small files by automating data compaction by merging small files into larger ones in the background to ensure that data is stored optimally.

Time-traveling queries

Time-traveling queries allow users to execute queries as if those queries were executed at a different time and see what a dataset looked like at that time. This can have multiple applications and use cases. One application is to debug updates to a dataset to see when and how data changed. If an update was done incorrectly, then the transaction that caused the data quality issue can be rolled back. For example, if you have inventory data that gets updated regularly, and a user or customer suggests that the data is incorrect, using time-traveling queries can pinpoint the time when the inaccurate data was updated and the transaction that caused the data to be incorrect.

Row-level filtering

Row-level filtering is a data access feature that allows administrators to grant permissions at the row level for a dataset. There are many applications where this capability is useful. This is best illustrated with an example. Suppose there is a compliance rule in which a user can only access rows of data that match the geographical region from where they are accessing the data. An administrator may set a policy that allows company users residing in Germany to only access data that maps to records for German customers. Users from Germany can perform queries and only get data from their own country. Another example would be with lines of business. For example, a clothing company can allow salespeople from the footwear line of business to only access data for the brands they manage and not see data related to swimwear.

Some customers have implemented this type of behavior by taking a table and breaking it up into different tables representing a particular slice of the data they wish to manage access. However, this is not a scalable solution. If a user has access to multiple data dimensions, they will need to join the tables to get a complete picture.

Now, let's summarize what we went through in this chapter.

Summary

In this chapter, you learned what AWS Glue and AWS Lake Formation provide when building and maintaining data lakes on AWS. We then focused on Lake Formation's ability to provide fine-grained access controls and the benefits and limitations of this. We also went through a sample process of enabling Lake Formation access controls for a new database and how it works within Athena. Lastly, we touched on Lake Formation governed tables, what they are, and how they can solve many issues with storing datasets on a distributed filesystem. There are more advanced features of Lake Formation, and we will dive deeper into governed tables in *Chapter 14, Lake Formation – Advanced Topics*.

In the next part of this book, we will get our hands dirty by using Amazon Athena in various settings ranging from ad hoc data analysis, using Athena to build ETL pipelines, and building applications that use Athena. We'll also take some time to cover how you can troubleshoot and tune common Athena issues in the pursuit of operational excellence.

Further reading

To learn more about the topics that were covered in this chapter, take a look at the following resources:

- AWS Lake Formation resources, including blog posts and demo videos: `http://amzn.to/394z9x7`

- Registering an encrypted Amazon S3 location – AWS Lake Formation: `https://amzn.to/3hf39uW`

- Registering an Amazon S3 location in another AWS account – AWS Lake Formation: `https://amzn.to/3baTVfI`

- Limitations of using Lake Formation security with Athena: `https://amzn.to/3nwAvGN`

Section 3: Using Amazon Athena

This section is all about getting our hands dirty using Amazon Athena in various settings, ranging from ad hoc data analysis, ETL pipelines, and embedded in your own applications. We'll also take some time to cover how you can troubleshoot and tune common Athena issues in your pursuit of operational excellence.

This section consists of the following chapters:

- *Chapter 7, Ad Hoc Analytics*
- *Chapter 8, Querying Unstructured and Semi-Structured Data*
- *Chapter 9, Serverless ETL Pipelines*
- *Chapter 10, Building Applications with Amazon Athena*
- *Chapter 11, Operational Excellence – Maintenance, Optimization, and Troubleshooting*

7
Ad Hoc Analytics

Welcome to *Part 3* of *Serverless Analytics with Amazon Athena*! In the preceding chapters, you learned how to run basic Athena queries and established an understanding of key Athena concepts. You then connected to a data lake that you built and secured. Along the way, you've been learning how to organize and model your data for use by Athena. Now that you have much of the prerequisite knowledge for using Athena, we once again shift our focus. The next few chapters will revisit many of the concepts you've already learned as you work through four of the most common use cases that lead customers to choose Athena for their business.

We begin right here, in this chapter, by unraveling both what it means to run ad hoc analytics queries as well as why the industry seems to have an insatiable appetite for running such queries. We'll also go through building a template for how you can adopt Athena and its related tooling within your organization as part of a complete ad hoc analytics strategy.

In the subsequent sections of this chapter, we will cover the following topics:

- Understanding the ad hoc analytics hype
- Building an ad hoc analytics strategy
- Using QuickSight with Athena
- Using Jupyter Notebooks with Athena

Technical requirements

Wherever possible, we will provide samples or instructions to guide you through the setup. However, to complete the activities in this chapter, you will need to ensure you have the following prerequisites available. Our command-line examples will be executed using **Ubuntu**, but most types of Linux should work without modification, including Ubuntu on Windows Subsystem for Linux.

You will need an internet connection to access GitHub, S3, and the AWS console.

You will also require a computer with the following:

- A Chrome, Safari, or Microsoft Edge browser installed
- The AWS CLI installed

This chapter also requires you to have an **AWS account** and an accompanying IAM user (or role) with sufficient privileges to complete this chapter's activities. Throughout this book, we will provide detailed IAM policies that attempt to honor the age-old best practice of "least privilege." For simplicity, you can always run through these exercises with a user that has full access. Still, we recommend using scoped-down IAM policies to avoid making costly mistakes and learning more about using IAM to secure your applications and data. You can find the suggested IAM policy for this chapter in the book's accompanying GitHub repository, listed as `chapter_7/iam_policy_chapter_7.json`, here: `https://bit.ly/2R5GztW`. The primary changes from the IAM policy recommended for *Chapter 1, Your First Query,* include the following:

- The addition of QuickSight permissions. Keep in mind that an administrator will be required to create your QuickSight account and also enable QuickSight to access Athena and S3. These permissions were too broad for us to feel comfortable adding them to the chapter's IAM policy.
- SageMaker notebook permissions.
- IAM role manipulation permissions used to create a SageMaker role for your notebook.

Understanding the ad hoc analytics hype

If you are lucky, you may not be aware of the buzzword levels of hype surrounding ad hoc analytics. Fortunately, there are strong fundamentals behind the increasing level of interest and importance placed on having good tooling for ad hoc analytics. In a moment, we'll attempt to form a proper definition of ad hoc analytics, but not before we run a time travel query of our own to set the stage for what we now know as ad hoc analytics.

As a society, we've been collecting data since the advent of commerce. In the era before modern big data technologies, the business intelligence landscape was a very different place. Most data capture and entry was a manual affair, frequently driven by government accounting and auditing requirements. Particularly savvy companies were tracking their own, non-accounting-related **Key Performance Indicators (KPIs)**, but these exercises were often short-lived and targeted at achieving specific outcomes. It is essential to understand the difference between the past data landscape, where information was scarce, and today, where data availability is not the most common limiting factor.

While preparing to write this chapter, I looked for examples of companies doing the modern-day equivalent of ad hoc analytics before the advent of big data. How did organizations do this before IoT and cloud computing upended the economics of data capture and retention? In the process, I solved a mystery behind a 25-gallon container of pencils that had been in my parents' garage for nearly 30 years. While helping my father clean out his garage, he asked me how this book was coming along. I told him I was stuck looking for an example of how companies answered questions about their day-to-day operations. Questions such as which products get returned most often or how much productivity is lost in maintenance of old machinery. That's when I finally got the entire backstory to the seemingly endless supply of pencils my father kept behind his ear as a contractor. My grandfather had worked at a large pencil manufacturer back in the 1990s. The story begins with quality control issues that caused poor writing quality and led to entire production batches needing to be scrapped. Folks like my grandfather were working overtime to make up for the production shortfall. Oddly, the more they produced, the lower their yields became.

My grandfather was one of the folks pulled off the production line to aid in quality control. They were already doing periodic quality control. That's how they noticed the issue in the first place. It wasn't enough. They started sampling random pencils from the production line every 5 minutes and tagging them with the date, time, ambient temperature, and ambient humidity. Then they'd sharpen the pencil and write a few words with it to gauge its relative quality before recording the results in a notebook. Each day, the numbers from the various production lines were collated and submitted by USPS to the head office. Eventually, after months of these manual activities and lost production, someone noticed a pattern. When humidity rose above a certain threshold, the quality started to falter, but only toward the end of the week. It turned out that when their production exceeded the on-hand supply of raw materials, they'd get fresh batches of glue delivered. The fresh glue was more sensitive to high humidity. Unfortunately, the manufacturing line's humidity tended to peak at the end of the week, as they were due to receive a new batch of raw materials. This entire investigation was a crude form of ad hoc analysis, and that barrel in the garage was full of the pencils my grandfather had tested but didn't want to throw away.

Potentially charming anecdotes aside, this is a classic example of a long OODA loop. The **OODA loop** shown in the following *Figure 7.1* represents the four stages of sound decision making. You start by **observing** in order to **orient** yourself to the problem at hand before **deciding** on what to do and finally **acting** on that decision. The hallmark of many successful businesses is a short OODA loop because they can react to changing information quickly. The longer it takes you to detect and understand why something bad, or good, is happening, the less likely you can navigate the situation successfully. This can result in missed business opportunities, lost customer confidence, or regulatory impact. The need to shorten the OODA loop has driven the world to capture and retain as much potentially relevant data as possible, fueled in part by improvements in embedded systems that have fueled the IoT boom. Physical businesses, such as the pencil manufacturer we just discussed, can now record hundreds of KPIs in real time for a fraction of what it cost them to measure three variables every 5 minutes a few decades ago. The rapid fall in data acquisition costs has led to compound annual data growth rates above 50% and shifted the OODA loop problem to the right.

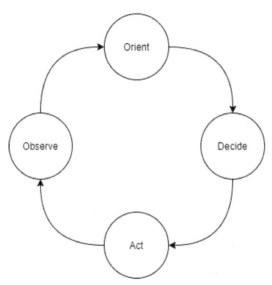

Figure 7.1 – The almightly OODA loop

Fast forward to today, and most organizations can capture more data than they know what to do with. As a result, essential business insights are buried among mountains of uninteresting information that may become useful in the future. Typically, an organization will periodically review KPIs using scheduled reports. These reviews often raise new questions. Why is this off trend? How long has this been happening? At what point will we need to account for that? Identifying unexplained trends is only the first step to generating actionable insights. Once you've observed something interesting, the second step in the OODA loop is to orient yourself to the context that is causing it. To do that, we need to ask follow-up questions of our data. These questions are ad hoc because they are situational and depend on information from previous observations. As a result, only the analytical tasks that go into the observe portion of the OODA loop can be standardized into scheduled reports. The exploratory and root cause research-related queries that often follow are too varied and numerous to be known in advance. There you have the creation of the ad hoc analytics craze.

Organizations that are early on the maturity curve will often establish centralized reporting teams that field requests for both scheduled and ad hoc reports. Reporting teams attempt to bridge a skills gap that has historically existed between the subject-matter experts and query experts. For example, a fashion-savvy merchandiser running an apparel business may not know how to write a `MapReduce` job to identify the emerging trend in unmatched socks. This leads them to miss out on an opportunity to be fully stocked before prices rise as this new style takes off. Or maybe my five-year-old is the only one driving such buying patterns. Organizations often try and bridge this skills gap by creating entire teams dedicated to fielding reporting requests. This model can work for small organizations, but quickly becomes a bottleneck at scale. The unyielding inflow of requests, each of which may spawn more follow-up requests, contributes to high turnover in such teams.

Aside from the scaling challenges, centralized reporting teams add non-obvious friction to the analytics process. Individuals writing the reports may have enough understanding of the data's relationships to properly offer or use techniques such as the approximate query functions we covered in *Chapter 3, Key Features, Query Types, and Functions*. Since they are kept at a distance from the data and tools to query that data, customers become implicitly biased or trained by past reporting experiences. This limits their future asks, creating a cycle that approaches zero utility.

Organizations are looking to tools such as Amazon Athena combined with easy-to-use tools such as QuickSight to democratize access to data. In the next section, you'll explore a possible ad hoc analytics strategy that combines Athena, QuickSight, and Jupyter Notebooks to provide flexible options for a broad spectrum of ad hoc analytics use cases.

Building an ad hoc analytics strategy

As we've seen in our examples, by putting the information in the hands of subject-matter experts, you can make better, faster decisions. Thus, it should be a focal point of any ad hoc analytics strategy to improve the accessibility of data, putting it in the hands of the individuals best suited to interpret the insights it contains. Our first step in forming such a strategy is to remember that while this book will present solutions based on the Athena ecosystem, it is rarely a good idea to lock yourself into any single product or analytics engine. The underlying technologies, pricing models, and supporting tooling will make trade-offs that necessarily favor one use case over others. If something sounds too good to be true, such as a product claiming to be the only analytics system you need, it's probably mediocre at a wide range of things and unlikely to be the best in class for anything. This is part of the philosophy behind AWS's fit-for-purpose database strategy and is equally applicable to analytics. The important things to consider include the following:

- Choosing your storage

- Sharing data

- Selecting query engines

- Deploying to customers

We'll check these out in more detail in the following sections.

Choosing your storage

Let's start our hypothetical ad hoc analytics strategy with storage. Where will you house the data? Will each team store their own data? Suppose for a minute that we avoided being prescriptive about this. After all, we painted the notion of a centralized reporting team as less than ideal. Maybe the same is true for standardizing storage. Different teams may even have different storage needs. Be careful about falling into this trap. The storage system you choose may limit your options for discovering and sharing data across your organization. This can mean the difference between having a ubiquitous data lake and many siloed data ponds. Nearly every **Online Analytics Processing** (**OLAP**) use case can be made better by separating storage and compute with an S3-like object store. Some esoteric use cases may indeed have specialized performance or auditability requirements that make S3 a less-than-ideal choice. You should avoid the temptation to shape your strategy based on outliers that you may never actually encounter. Instead, leave room in your strategy for how you will evaluate, approve, and integrate these exceptions.

If you're following the best practices in earlier chapters, you're likely to arrive at a strategy that treats S3 and your data lake. All data providers will be expected to master their data in S3 using Parquet for structured data and text for unstructured data. The accompanying metadata for these datasets will be housed in AWS Glue Data Catalog. AWS Data Catalog aids in discoverability and sharing since most metadata can be inferred from the S3 data itself. For teams, or systems, that have their own storage, they will be required to maintain an authoritative copy in the data lake on a cadence. This can be done as periodic, incremental additions to S3 or full snapshots that supersede previous versions. In *Chapter 6*, *AWS Glue and AWS Lake Formation*, and *Chapter 14*, *Lake Formation – Advanced Topics,* you learn how Lake Formation helps make it easier to integrate with an S3-backed data lake.

Sharing data

Most of the really interesting use cases for ad hoc analytics will require data from multiple sources. For small organizations, you may be able to get by with handling access requests using IAM policies and organizational processes. However, once you get past a handful of datasets and a couple of consumers, you'll want tooling to support your processes. This is especially true if you deal with sensitive **Personally Identifiable Information** (**PII**) or are subject to GDPR regulations. S3 permissions are limited to enforcing object or prefix (directory) level access. S3 is oblivious to the contents of your objects and their semantic meaning. This means S3 permissions alone cannot restrict sensitive columns or apply row-level filters to prevent someone from reading budget records from a yet-to-be-announced project. If you're taking security seriously, you'll want to make Lake Formation a core part of your analytics strategy. AWS Lake Formation abstracts the details of crafting IAM policies and offers an interface where you can permission your data lake customers on the dimensions they are most familiar with. You simply manage table-, column-, and even row-level access controls from an interface that is designed for analytics use cases. Since Lake Formation is integrated with Amazon Athena, Amazon EMR, and Amazon Redshift, you can avoid the classic problem of having authorization information spread across multiple systems. Lake Formation can even facilitate cross-account data sharing, facilitating the post-acquisition mergers of technology and reducing the strain on your AWS account design.

Selecting query engines

Luckily, our decision to use Amazon S3 to house our data lake means we are not locked into a particular query engine. In fact, we can support multiple query engines with relative ease. You shouldn't take that to mean that it's a good idea to have a dozen different query technologies, only that our choices thus far have derisked the importance of any one product. To begin, you'll want to offer a serverless query engine with a SQL interface, such as Amazon Athena. SQL is a broadly taught and widely understood language. Many of your employees may already be familiar with SQL, and if they aren't, it's easy to get them started. Electing a serverless option always makes it easier to keep costs under control since ad hoc workloads tend to make for challenging capacity planning exercises. This is even more important when your end customers may not be well versed in starting or stopping servers for their queries. Beyond SQL, Athena also offers support for custom data connectors and **User-Defined Functions** (**UDFs**). This can help provide a single query interface even if some of the data may not be in our S3 data lake. While it is beyond the scope of this book, you can complement Athena's SQL interface with Glue ETL or Amazon EMR to add Apache Spark-based query capabilities that can support more sophisticated forms of customization. With Apache Spark, you can introduce your own business logic at every stage of the query. This can become a deeply technical exercise, but our strategy's goal is to lay out a plan. If we encounter these situations, we want a general idea for delivering the required capability.

Deploying to customers

Some customers of your ad hoc analytics offering will be comfortable writing and running SQL directly in the Athena console, but we can do better than Athena's console experience. Your organization will likely want to create repeatable reports and dashboards and even share their ad hoc analysis. They may even want to post-process results using standard statistical libraries. Luckily, Amazon Athena supports JDBC and ODBC connectivity so that you can use a wide range of client applications, such as Microsoft Excel, or BI tools, such as Tableau. For our hypothetical company, we'll support two different ad hoc analytics experiences. The first will be a more traditional experience built on Amazon QuickSight connecting to Athena. This option should be suitable for those experienced with SQL or other BI tooling. It will allow our customers to visualize trends and dig into patterns while minimizing the need for specialized technical skills. We'll also support a more advanced experience in the form of Jupyter Notebooks. Jupyter Notebooks allows authors to mix traditional SQL, statistical analysis tools, visualizations, text documentation, and custom business logic in the form of actual code.

This may seem daunting at first and perhaps only appropriate for developers, but that isn't the case. The collaborative features of notebooks allow you to share and customize analysis in a way that enables you to introduce more powerful tooling with a commensurately steep learning curve.

Now that we've established both the definition and importance of ad hoc analytics, let's see whether we can make our hypothetical strategy a bit more tangible. In the remainder of this chapter, we will walk through implementing this strategy to run ad hoc queries over the data lake we built in previous chapters using Athena, Amazon QuickSight, and SageMaker Jupyter Notebooks.

Using QuickSight with Athena

AWS QuickSight is a data analysis and visualization tool that offers out-of-the-box integrations with popular AWS analytics tools and databases such as Athena, Redshift, MySQL, and others. QuickSight has its own analytics engine called **Spice**. Spice is capable of low-latency aggregations, searches, and other common analytics operations. When combined with a large-scale analytics engine such as Athena, QuickSight can be used for a combination of data exploration, reporting, and dashboarding tasks. This section will briefly introduce you to QuickSight and use it to visualize both our earthquake and Yellow Taxi ride datasets. Since QuickSight itself is a **WYSIWYG (What Ya See Is What Ya Get)** authoring experience with lots of built-in guidance, we won't spend much time walking you through each step in this section. Instead, we will focus on the broad strokes and let you explore QuickSight yourself. Regardless of this simplification, our QuickSight exercise will have multiple steps and take you 15 to 20 minutes to complete. In that process, you'll be tackling the following objectives:

1. Sign up for QuickSight.

2. Add datasets to QuickSight.

3. Create a new analysis.

4. Visualize a geospatial dataset.

5. Visualize a numeric dataset.

6. Explore anomalies in a numeric dataset

Let's dive right in!

Getting sample data

By this point in the book, you've gathered sample data, imported it to S3, and prepared tables for use in Athena half a dozen times. We'll be reusing many of those datasets and tables to save time and focus on the new topics presented in this chapter. In case you skipped previous chapters or just prefer to start with a clean slate, you can download and run our chapter 7 data preparation script using the following commands from AWS Cloud Shell or your preferred terminal environment. The script will download several years of the NYC Yellow Taxicab dataset into an S3 landing zone before reorganizing that data into an optimized table of partitioned Parquet files. It will also download a small dataset containing geospatial data about recent earthquake activity in the US state of California. This script is likely to take 20 minutes to run from AWS Cloud Shell as it encompasses much of the data lake work from the first three chapters. Once the script completes, it may take a few more minutes for the final Athena query it launches to complete and the resulting table to become usable. You can reuse the S3 bucket and Athena workgroups you made in earlier chapters:

```
wget -O build_my_data_lake.sh https://bit.ly/3suTuU8
chmod +x build_my_data_lake.sh
./build_my_data_lake.sh <S3_BUCKET> <ATHENA_WORKGROUP_NAME>
```

Setting up QuickSight

If you aren't already a QuickSight customer, we recommend signing up for the Standard package when following these exercises. Unlike the other services we've used in this book, QuickSight's pricing model is more akin to traditional software licenses, with the Standard plan costing you between $13 and $50 a month per named user. Luckily, if you've never used QuickSight before, you may be able to complete this exercise within the free trial window. Your first time visiting QuickSight, you'll be prompted to sign up. Signing up for and configuring QuickSight requires IAM permissions that are broader than what we typically include in our chapter policies. As such, we recommend using a separate IAM user with administrative access to set up QuickSight. In *Figure 7.2*, we show the key properties you'll need to set when signing up. The most notable is allowing QuickSight to have access to Amazon Athena and Amazon S3. If you attempt to do this without privileged access, you'll encounter issues when you attempt to access Athena in later steps.

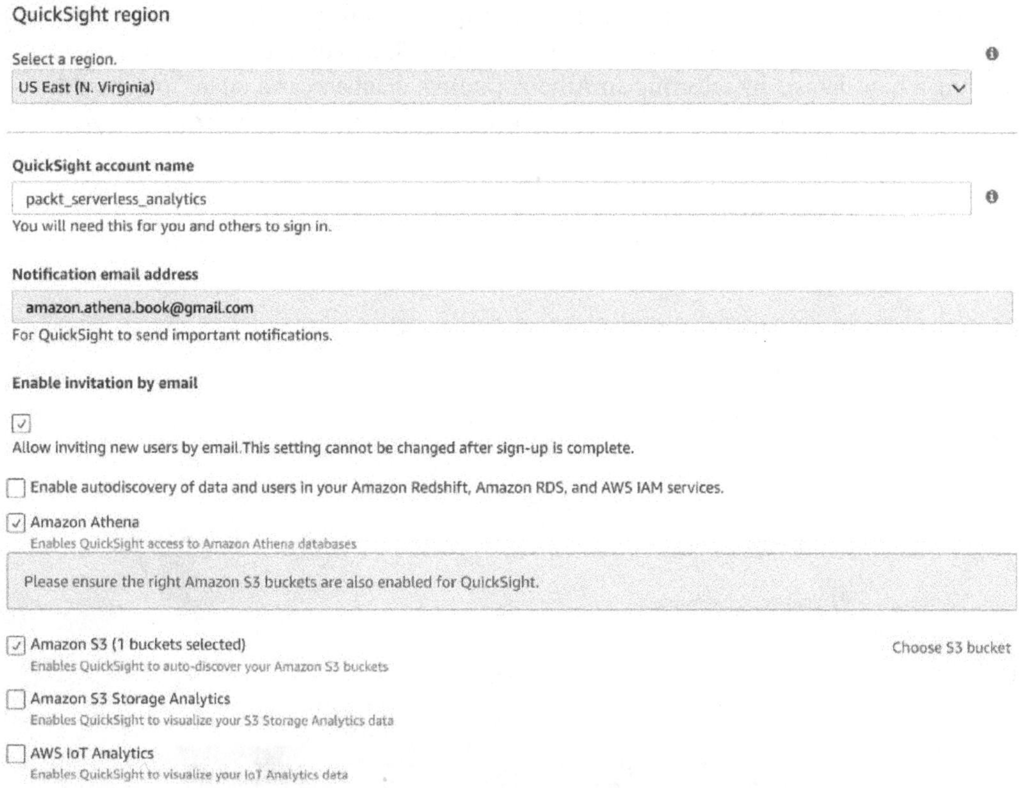

QuickSight region

Select a region.

US East (N. Virginia)

QuickSight account name

packt_serverless_analytics

You will need this for you and others to sign in.

Notification email address

amazon.athena.book@gmail.com

For QuickSight to send important notifications.

Enable invitation by email

☑

Allow inviting new users by email.This setting cannot be changed after sign-up is complete.

☐ Enable autodiscovery of data and users in your Amazon Redshift, Amazon RDS, and AWS IAM services.

☑ Amazon Athena
   Enables QuickSight access to Amazon Athena databases

   Please ensure the right Amazon S3 buckets are also enabled for QuickSight.

☑ Amazon S3 (1 buckets selected)   Choose S3 bucket
   Enables QuickSight to auto-discover your Amazon S3 buckets

☐ Amazon S3 Storage Analytics
   Enables QuickSight to visualize your S3 Storage Analytics data

☐ AWS IoT Analytics
   Enables QuickSight to visualize your IoT Analytics data

Figure 7.2 – Signing up for QuickSight

Once you or an administrator has completed the sign-up process, you'll be able to start analyzing data. Before we create our first dashboard, we'll need to define one or more **datasets** that can be used in our analysis. A QuickSight dataset can be thought of as a table and its associated source connectivity information. For example, we'll be using two datasets from Athena in our exercise. The first will be the `chapter_7_earthquakes` table from the `packt_serverless_analytics` database, with the `chapter_7_nyc_taxi_parquet` table being our second dataset. From the main QuickSight page, we can select datasets to view existing or create new datasets. Even if you've never used QuickSight before, you will have several sample datasets listed as options. When you click **New Datasets**, you'll get to choose from various sources, including Athena. After selecting Athena, a popup will appear asking you to select a workgroup.

If you are following along with the configurations in this book, you should choose the `packt-athena-analytics` workgroup. In *Figure 7.3*, we complete the final steps in adding a new dataset by selecting an Athena catalog, database, and table. You should repeat this process for both the `chapter_7_earthquakes` and `chapter_7_nyc_taxi_parquet` tables.

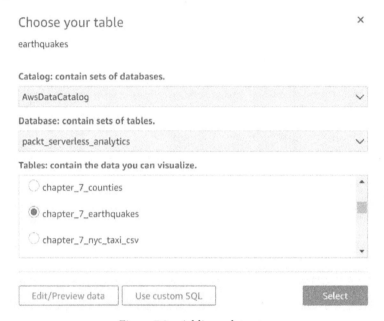

Figure 7.3 – Adding a dataset

Unable to create dataset errors

If you didn't have sufficient administrative rights when you signed up for QuickSight, you might encounter issues adding new datasets. If you see errors related to listing Athena workgroups or accessing the results location, you'll need an administrator to go into the QuickSight settings and re-enable Athena in the **Security and permissions** section.

Once you've created the datasets, you are ready to start analyzing your data. For now, ensure that `chapter_7_earthquakes` is selected. Then you can click on the **New analysis** button on QuickSight's main screen, as shown in *Figure 7.4*. In QuickSight nomenclature, an analysis is a multi-tab workspace with different visualizations and calculated fields. Once you are happy with an analysis, you can publish it as a read-only **dashboard** or share it with other QuickSight users in your organization. Since QuickSight is continuously saving your analysis, you can quickly backtrack from a failed exploration by undoing hundreds of recent edits.

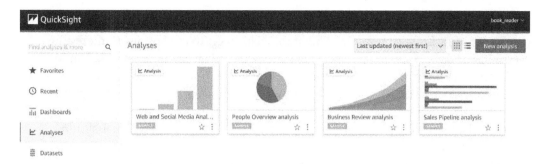

Figure 7.4 – Creating a new analysis

Your new analysis will start with a single blank tab. Let's add a new geospatial visualization to that tab by clicking **Points Map** from the visualization type pallet on the left navigation. Next, we'll select the longitude and latitude fields of the chapter_7_earthquakes dataset we added earlier as our geospatial fields. Since we want to understand the relationship between location, magnitude, and depth, we can use the magnitude and depth columns as our size and color fields, respectively. The ease of use and rich visualizations are where QuickSight shines. *Figure 7.5* shows how, in just a few clicks, we've created our first analysis of earthquakes around the world:

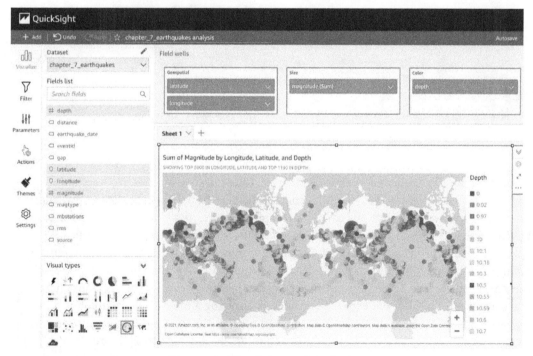

Figure 7.5 – Visualizing earthquake data

Visualizations can help better understand relationships in our data, but ad hoc analysis is all about iterating. Let's see how QuickSight can help us get to the bottom of interesting patterns in our data. If you plan to keep your earthquake data example, go ahead and create a new tab in our analysis for a deep dive into the NYC Yellow Taxi dataset we added earlier. Our first visualization on this new tab will be a combination bar and line graph where we will graph the average `tip_amount` field by year as bars on the chart. For the lines, we will add the count of rides by using the `total_amount` field. The first thing you'll notice is that we seem to have erroneous or incomplete data in our table for many years in the future, and even some in the past are contributing strange data to our graph. Luckily, QuickSight offers a handy tool for filtering data that goes into a visualization. Click on the chart and then on the `year` field's settings in the left navigation pane. From there, you can select **add filter for this field** and use the **Filters** dialog to include only data from 2017, 2018, 2019, and 2020. Once that's done, the graph should automatically refresh and resemble *Figure 7.6*:

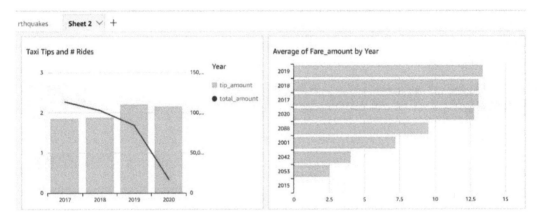

Figure 7.6 – Visualizing yellow taxi ride data

With the noisy data removed, we can see that the total number of rides, represented by the `total_amount` series in our legend, has been trading down for the last 3 years. We mostly ignore 2020 since our data is incomplete. Interestingly, the average tip amount has increased. This would suggest that customer satisfaction is rising. So why could ridership be down? Let's confirm that tips aren't simply growing due to increasing fares by adding a second graph to this tab. In *Figure 7.5*, we added a new bar chart with `year` as the *y* axis and average `fare_amount` as the value. Just as we did for the previous chart, you'll again want to filter out erroneous values by applying a filter to the **year** column in this graph, too. Once the graph renders, you can see that the increase in tips is not tied to a commensurate rise in fares. Customers are consciously choosing to tip more.

QuickSight customization

In *Figure 7.6*, we could not rename the `total_amount` series in the **Taxi Tips and # Rides** chart's legend to something more indicative of the actual value, such as "Number of Rides." While this is a pedantic example of the drawbacks of using WYSIWYG editors, it is indicative of the control you give up when using a tool such as QuickSight. There is an inherent conflict between the myriad of parameters in fully customizable systems and ease-of-use tools such as QuickSight. Please don't take the limited legend customization we've called out here as a reason not to use QuickSight. It's merely an easy-to-convey example of why you're unlikely to find a single tool to satisfy all your customers.

We haven't yet learned why ridership is down. If we really worked for the Taxi and Limousine Commission, we might want to dive deeper into the data and possibly run some A/B testing. Running additional queries along these dimensions could help us understand whether price or other supply and demand factors are playing a role in the decline. This might help confirm the impact of things such as ride-sharing services. For now, we'll put aside our QuickSight analysis and switch to Jupyter Notebooks for the next leg in the ad hoc analysis of our NYC Yellow Taxi dataset.

Using Jupyter Notebooks with Athena

Depending on the proficiency level in querying data, some individuals may consider QuickSight to be more of a **dashboarding** tool that populates results based on pre-set parameters. Individuals looking for a more fluid and interactive experience may feel their needs are better satisfied by a tool designed for authoring and sharing investigations. You're already familiar with the Athena console's basic ability to write queries and display tabular results. Jupyter Notebooks is a powerful companion to analytics engines such as Athena.

In this section, we'll walk through setting up a Jupyter notebook, connecting it to Amazon Athena, and running advanced ad hoc analytics over the NYC Yellow Taxi ride dataset. If you are unfamiliar with SageMaker or Jupyter Notebooks, don't worry. We will walk you through every step of the process so you can add this new tool to your shelf. For the uninitiated, AWS describes SageMaker as the most comprehensive machine learning service around. SageMaker is best thought of as a suite of services that accelerate your ability to adopt and deploy machine learning in any and every situation where it might be useful. That means SageMaker has dedicated tooling for data preparation tasks such as labeling and feature engineering work that may require nuanced techniques for statistical bias detection. You may be wondering what that has to do with Athena and ad hoc analytics. Well, training machine learning models requires good input data.

In many cases, your models are only as good as the inputs to your training. As such, Jupyter Notebooks provides an excellent interface and workflow for exploring data and capturing findings.

To begin, we'll need to create an IAM role that our SageMaker Jupyter notebook will use when interacting with other AWS services such as Athena. You can do this by navigating to the IAM console, selecting the **Roles** section, and clicking the **Create role** button. Once you do that, you'll be presented with the dialog in *Figure 7.7*. Be sure to select **AWS service** as the type of trusted entity and **SageMaker** as the entity, just as we have in *Figure 7.7*:

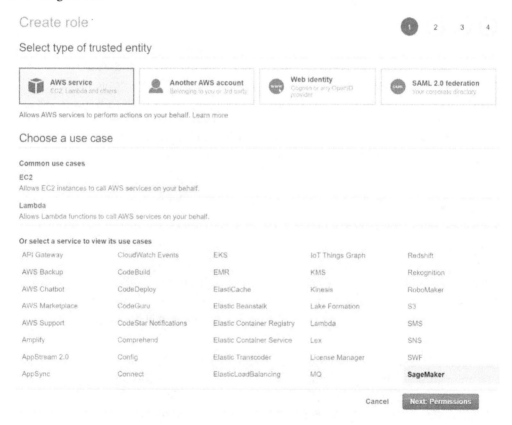

Figure 7.7 – Creating an IAM role dialog

These settings tell IAM that we want to create a role that is explicitly for use with SageMaker. This helps scope down both the types of activities the IAM role can perform and the contexts from which it can be assumed. In the next step, you'll have the opportunity to add the specific policies for the activities we plan to perform using this IAM role.

We recommend adding the `packt_serverless_analytics` policy that we have been enhancing throughout this book and used earlier in this chapter. As a reminder, you can find the suggested IAM policy in the book's accompanying GitHub repository listed as `chapter_7/iam_policy_chapter_7.json` here: `https://bit.ly/2R5GztW`.

Once you've added the policy, you can move on to the **Add tags** step. Adding tags is optional, so you can skip that for now and go to the final step of giving your new IAM role a name. We've recommended naming your new IAM role `packt-serverless-analytics-sagemaker` since this chapter's IAM policy already includes permissions that will allow you to create and modify roles matching that name without added access. If everything went as expected, your IAM role summary should match *Figure 7.8*. If you forgot to attach the `packt_serverless_analytics` policy, you can do so now using the **Attach policies** button highlighted here:

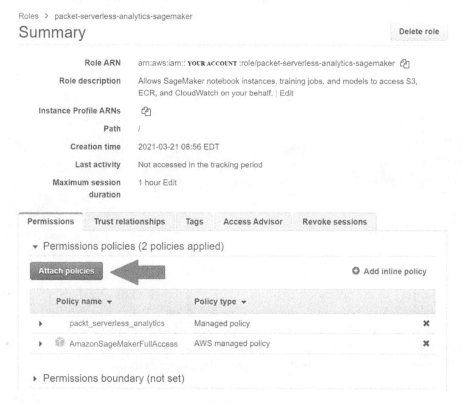

Figure 7.8 – IAM role summary dialog

With our shiny new IAM role in hand, we are ready to start a SageMaker Jupyter notebook and begin exploring the NYC yellow taxi ride dataset with Athena while using handy analysis libraries such as pandas, Matplotlib, and Seaborn. Don't worry if these sound more like species of tropical fish than ad hoc analytics tools. We'll introduce you to these libraries and how they can make your life easier a bit later in this section. On the SageMaker console, you can click on **Notebook** and then the **Notebook instances** section. From there, you can click on **Create notebook instance** to open the dialog in *Figures 7.9* and *7.10*:

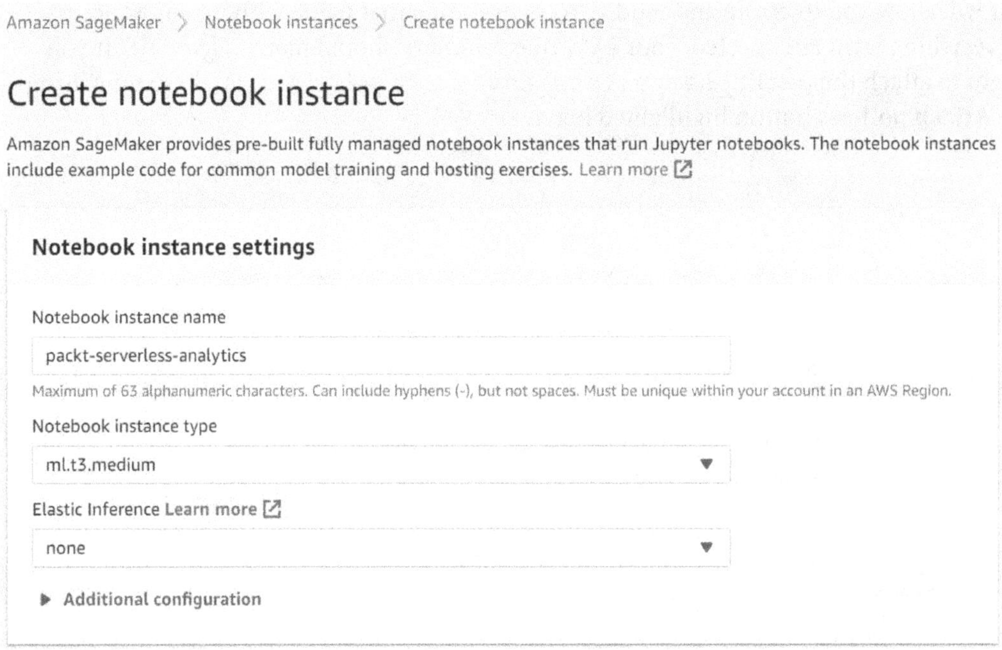

Amazon SageMaker ⟩ Notebook instances ⟩ Create notebook instance

Create notebook instance

Amazon SageMaker provides pre-built fully managed notebook instances that run Jupyter notebooks. The notebook instances include example code for common model training and hosting exercises. Learn more ☑

Notebook instance settings

Notebook instance name

```
packt-serverless-analytics
```

Maximum of 63 alphanumeric characters. Can include hyphens (-), but not spaces. Must be unique within your account in an AWS Region.

Notebook instance type

```
ml.t3.medium                                                              ▼
```

Elastic Inference Learn more ☑

```
none                                                                      ▼
```

▶ Additional configuration

Figure 7.9 – SageMaker Create notebook instance dialog

In the first portion of the notebook creation dialog, you'll pick a name and instance type for your notebook. We recommend naming your notebook `packt-serverless-analytics` since your IAM policy is already configured to grant you the ability to administer notebooks matching that name. Any instance type that is at least as powerful as an ml.t3.medium will be sufficient to complete the exercises in this section.

We recommend the ml.t3.medium because it has a generous *free tier*, which should easily allow you to complete all the exercises at no additional charge. You'll only end up paying for Athena and S3 usage. We won't be doing any heavy machine learning, so you can leave the **Elastic Inference** option on **none**. This option allows you to attach specialized hardware to your notebook that makes the application of machine learning models, also known as inference, significantly faster through the use of AWS's custom inferential chips. Our final step is to set the IAM role that our new notebook instances will use when interacting with other AWS services such as Athena and S3. In *Figure 7.10*, you can see that we used the `packet-serverless-analytics-sagemaker` role we created earlier. Once you've done that, you can leave the remaining options at their default values and create the notebook.

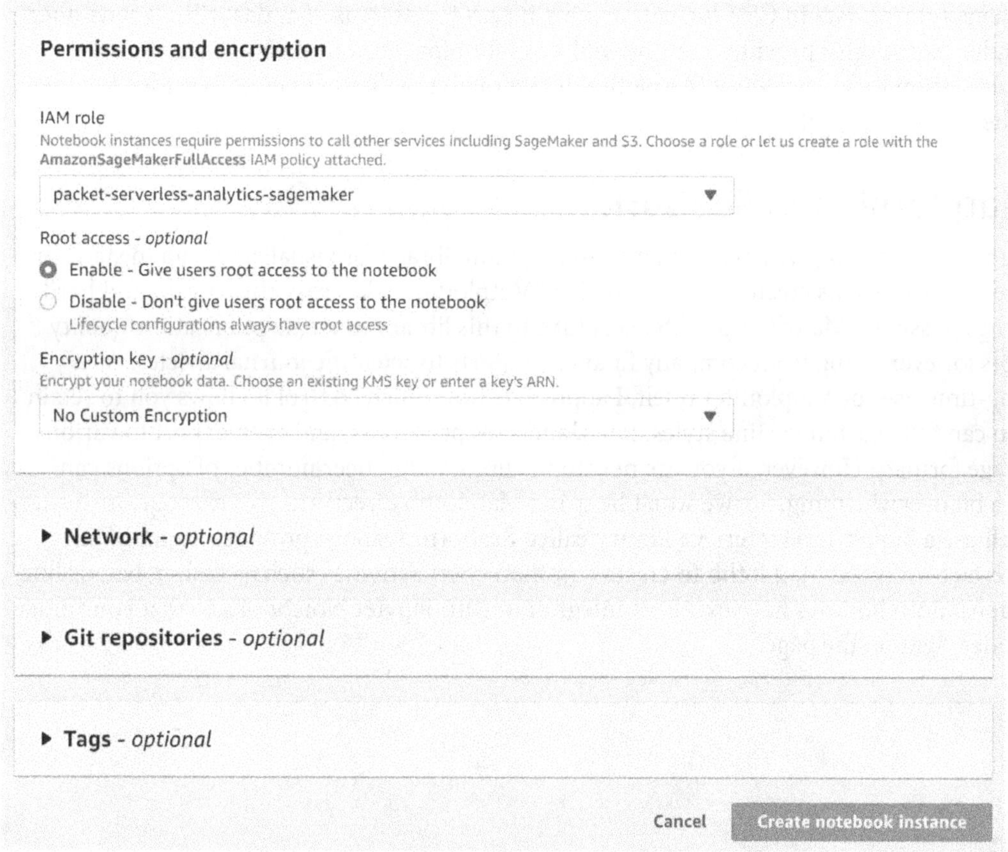

Figure 7.10 – SageMaker Create notebook instance dialog continued

Your new notebook instance will take a few minutes to start. While we wait, let's outline what we're going to do with our notebook and get introduced to the statistical libraries that will help us do even more with Athena.

pandas

pandas is a fast, flexible, and open source data analysis library built on top of the Python programing language. It aims to make working with tabular data such as that stored in spreadsheets or a SQL engine easier. If you're looking for help exploring, cleaning, or processing your data, then pandas is the right tool for you. In pandas, tabular data is stored in a structure known as a *DataFrame*. Out of the box, pandas supports many different file formats and data sources, including CSV, Excel, SQL, JSON, and Parquet. Athena returns data in CSV format, making it easy for us to use pandas with Athena query results. pandas also provides convenient hooks for plotting your data using a variety of visualization tools, including Matplotlib. In a moment, we'll use pandas to bridge between Athena and other data analysis tools.

Matplotlib and Seaborn

Matplotlib is a comprehensive open source Python library for visualizing data in static or interactive plots. Its creators like to say that "Matplotlib makes easy things easy and hard things possible." Many Matplotlib users turn to this library to create publication-quality plots for everything from company financial reports to scientific journal articles. As a long-time user of Matplotlib myself, I appreciate how much control it allows you to retain. You can fully customize line styles, fonts, and axes properties, and even export to various image formats. However, if you are new to the library, the sheer number of options can be a bit overwhelming. So, we won't be using Matplotlib directly in this exercise. Instead, we'll use a higher-level interface library called Seaborn. Seaborn provides a simplified interface for using Matplotlib to create common chart activities such as scatter, bar, or line graphs. Both libraries have excellent integration with Jupyter Notebooks so that your plots render right on the page.

SciPy and NumPy

By now, you can probably guess that both SciPy and NumPy are open source mathematics libraries built in Python. NumPy contains abstractions for multidimensional arrays of data. Such structures can come in handy when applying mathematical operations over an entire column of a table. NumPy also offers highly optimized functions for sorting, selection, applying discrete logic, and a host of statistical operations over these arrays. SciPy builds on the functionality provided by NumPy to create ready-to-use solutions for common scientific and mathematical problems. Later in this section, we will use SciPy's outlier detection algorithm to purge errant data from our Athena results.

Using our notebook to explore

Your notebook instance should just about be ready for use. Let's outline what exploration we'll perform once it's running. The beauty of using Athena from a Jupyter notebook is that you can simply have a conversation with your data and not have to plan it all out in advance. We're itemizing the steps here, so you know what to expect along the way:

1. Connect our notebook instance to Athena.
2. Run a simple Athena query and print the result using pandas.
3. Visualize the result of our simple Athena query using Seaborn.
4. Prune any erroneous data using SciPy for outlier detection.
5. Run a correlation analysis over an aggregate Athena query.

Embedded in these steps is an important cycle. We ask a question by querying Athena. We notice something interesting in the result. We run a follow-up query in Athena. We dissect the result further. This is the ad hoc analytics cycle that differentiates ad hoc analytics from pre-canned reports or dashboards. It has no clear or pre-packaged end. Your next query depends on what you find along the way. This may seem a bit abstract, so we'll make it more concrete by applying this to our NYC yellow taxi dataset.

If you'd like to skip ahead or need added guidance in writing the code snippets we'll be using to run our ad hoc analytics, you can get a prepopulated notebook file from the book's GitHub repository at `chapter_7/packt_serverless_analytics_chatper_7.ipynb` here: `https://bit.ly/3rQKGGI`. GitHub nicely renders the notebook file so that you can see it right from the link. Unfortunately, that makes downloading it for later upload to your SageMaker notebook instance a bit tricky. To get around that, click on the **Raw** view, and then you can perform a **Save as** operation from your browser.

Step 1 – connecting our notebook instance to Athena

From the SageMaker console, go ahead and click the **Open Jupyter** link as shown in *Figure 7.11*. This will open a new browser tab or window connected to your Jupyter notebook instances. Behind the scenes, SageMaker is handling all the connectivity between your browser and what is your own personal Jupyter notebook server.

Figure 7.11 – Opening a Jupyter notebook

Just as we've done in *Figure 7.12*, you'll want to click on **New** and select **conda_python3** for the notebook type. The value may appear at a different position in the dropdown than it does in *Figure 7.12*, so don't be afraid to scroll to find it. This setting determines how our notebook will run the data explorations tasks we are about to write. By selecting **conda_python3**, we are telling Jupyter that it can run our code snippets using Python. Sparkmagic is another common choice if you want to use Apache Spark as your computing platform. For now, we'll stick with Python, but the flexibility of Jupyter Notebooks makes it an excellent choice for any ad hoc analytics strategy. Once you pick the notebook type, yet another browser tab will open with your new notebook. The new notebook file will be named `Untitled.ipynb`, so our first step will be to give it a helpful name by clicking on **File** and then **Rename**.

Figure 7.12 – Creating a new notebook file

Now that you have your notebook ready to use, we'll connect it to Amazon Athena by installing the Athena Python driver. To do this, we'll write the following code snippet in the first **cell** of the notebook. Cells are represented as a free-form textbox and can be executed independently, with subsequent cells having access to variables, data, and other states produced by earlier cells. After executing a cell, its output is shown immediately below it. You can edit, run, edit, and re-run a cell as often as you'd like. You can also add new cells at any time. The entire experience is very fluid, making it perfect for an imperfect exercise such as ad hoc data analysis. Let's put this into practice by running our first cell. Once you've typed the code into the cell, you can either click **Run** or press *Shft + Enter* to run the cell and add a new cell directly below it:

```
import sys
!{sys.executable} -m pip install PyAthena
```

This particular cell will take a couple of minutes to execute, with the result containing a few dozen log lines detailing which software packages and dependencies were installed. You are now ready to query Athena from your notebook.

Step 2 – running a simple Athena query and printing the result using pandas

Go ahead and add a cell to your notebook. This cell will be used to import our newly installed Athena Python driver and the pre-installed pandas library. This is done by typing the first two import statements from the following code snippet. In both cases, we are aliasing our imports to something more convenient. Then we use the connect() function that we imported from pyathena to connect to our Athena workgroup and database using the work_group and schema_name arguments, respectively. You'll also notice that we set the region_name argument to match the AWS Region we've been using for all our exercises:

```
from pyathena import connect
import pandas as pd
conn = connect(work_group='packt-athena-analytics',
          region_name='us-east-1',
          schema_name='packt_serverless_analytics')
```

Still working in the same cell, we can now run our Athena query by using pandas' read_sql() function to read the result of our query into a **DataFrame** as shown in the following code snippet. In this example, we are running a query to get the count of yellow taxi rides by year. On the final line of the cell, we print the first three values from the result. Go ahead and run this cell:

```
athena_results = pd.read_sql("""SELECT year, COUNT(*) as num_
rides
                                FROM chapter_7_nyc_taxi_parquet
                                GROUP BY year
                                ORDER BY num_rides DESC""",
conn)
athena_results.head(3)
```

Viewing the first few rows of the result is great, but we could have done that in the Athena console. We opted for a notebook experience for the ecosystem that included data visualization. That's where Seaborn comes into the picture.

Step 3 – visualizing results using Seaborn

If you didn't already add another cell, go ahead and do that now. In this next cell, we will use Seaborn to graph the number of yellow taxi rides each year as a bar graph. Since this is the first cell that requires Matplotlib and Seaborn, we begin by importing and aliasing these tools. We then conclude this cell by calling Seaborn's barplot function to graph the year and num_rides columns of our DataFrame:

```
from matplotlib import pyplot as plt
import seaborn as sns
seaborn.barplot(x="year", y="num_rides", data=athena_results)
```

But the resulting graph shown in *Figure 7.13* seems a bit odd. There are so many years that we can't even read the *y* axis.

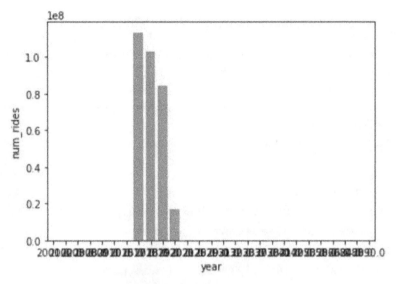

Figure 7.13 – Visualizing data using Seaborn

It seems we have a data quality issue with some rides having erroneous years. In the next cell, we'll use **SciPy** to detect and filter out those outliers.

Step 4 – pruning any erroneous data using SciPy

Our visualization in *step 3* has shown that we have some rides with erroneous start or end values. In our case, our sample dataset only has yellow taxi ride data from 2017, 2018, 2019, and 2020 so any other values must be bad data. In practice, identifying bad data won't always be that easy. It would be useful to have a mechanism for detecting outliers that doesn't require foreknowledge of the dataset. Luckily, SciPy has a set of functions that can help. In our next cell, we'll use SciPy's **stats** module to compute the **zscore** of the num_rides column for each row. A zscore, also known as a standard score, measures how many standard deviations above or below the population mean a value is.

Using the following code snippet as a guide, we start by importing the `stats` module from SciPy. Depending on your version of pandas, you'll want to suppress `chained_assignment` warnings, as we have done. Then we use the `zscore` function from the `stats` module to calculate the zscore for the `num_rides` column. This function returns a DataFrame with as many rows as the input column. pandas DataFrames make it easy to add a new column to our Athena result and fill it with the calculated values from our new DataFrame. We do that by assigning the result to a new column in our original DataFrame. We conclude this cell by printing the results DataFrame to see the zscores alongside our original values:

```
from scipy import stats
#surpressing warning related to chained assignments
pd.options.mode.chained_assignment = None


zscore = stats.zscore(athena_results['num_rides'])
athena_results['zscore']=zscore
print(athena_results)
```

When you are ready, go ahead and run this cell. *Figure 7.14* shows the first few results from the output. As expected, the bulk of the rides are in the four years we loaded into our data lake, but we've also got data from 2088, 2058, and a few other years that are far in the future. Interestingly, SciPy generated negative zscores for all the rows with erroneous years. This is because the ride counts for those years are so far from the population mean. Let's add another cell and repeat our visualization after filtering by zscore.

year	num_rides	zscore
2017	1.13E+08	3.274275
2018	1.03E+08	2.93152
2019	84397884	2.341594
2020	16847996	0.176508
2038	4	-0.3635
2058	3	-0.3635
2088	2	-0.3635
2042	1	-0.3635

Table 7.1 – zscore values

This cell will be short, thanks to pandas' shorthand for filtering a DataFrame. We select the subset of the `athena_results` DataFrame where the zscore column is greater than zero and assign the result to a new `athena_filtered` DataFrame. We then repeat our earlier plot command to produce a new bar chart:

```
athena_filtered = athena_results [athena_results['zscore'] > 0]
seaborn.barplot(x="year", y="num_rides", data=athena_filtered)
```

After running this cell, we get a much more reasonable chart, like the one in *Figure 7.15*. Even with all the erroneous data points removed, we can still see a clear downward trend in the number of yellow taxi rides beginning in 2018. Some of this may be attributed to the rise of ride-sharing services such as Uber, or there may be other factors at play.

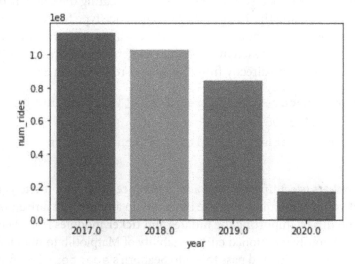

Figure 7.14 – zscore values

Running a correlation analysis

In our final notebook cells, we'll attempt to use the average tip amount as a proxy for customer satisfaction. We'll then check whether using the tip amount is a flawed proxy for customer sentiment by looking at how the tip amount correlates with other metrics such as trip speed and time of day. Add a new cell and run a new Athena query to get the average fare amount, average tip amount, and total rides grouped by day, as we've done in the following code snippet:

```
athena_results_2 = pd.read_sql("""
    SELECT date_trunc('day',
        date_parse(tpep_pickup_datetime,'%Y-%m-%d
%H:%i:%s')) as day,
```

```
    COUNT(*) as ride_count,
    AVG(fare_amount) as avg_fare_amount,
    AVG(tip_amount) as avg_tip_amount
    FROM chapter_7_nyc_taxi_parquet
    GROUP BY date_trunc('day',
        date_parse(tpep_pickup_datetime,'%Y-%m-%d %H:%i:%s'))
    ORDER BY day ASC""", conn)
```

In the same cell, we'll then calculate the zscore of the `ride_count` column so that we can again filter out the outliers. Since this query gathers daily data, we adjust our zscore threshold to `-1` to allow for a broader range of valid values. Once you've included the code from this following snippet, you can run the cell. Executing the cell may take a minute or two if you are using the `ml.t3.medium` instance type for your notebook instance. This is because the notebook needs to retrieve all results from Athena using Athena's results API. As we discussed in an earlier chapter, Athena's results API is not as performant as reading the data directly from the Athena results file in S3:

```
zscore2 = stats.zscore(athena_results_2["ride_count"])
athena_results_2['zscore']=zscore2
athena_filtered_2= athena_results_2[athena_results_2['zscore']
> -1]
```

Once the cell completes executing, you can add another cell that we'll use to generate a scatter plot that varies color and point size based on tip amount and fare amount, respectively. We do this by importing the **mdates** and **ticker** modules from Matplotlib. Then we use the previously mentioned customizability of Matplotlib to manually set a wide aspect ratio for our plot and pass this into Seaborn's `scatterplot` function. You can see the full detail of how to configure the plot in the following code snippet. We conclude the cell by customizing the frequency and format of our graph's *y* axis using the `set_major_locator()` and `set_major_formatter()` functions of our `plot` object:

```
import matplotlib.dates as mdates
import matplotlib.ticker as ticker

fig, ax = pyplot.subplots(figsize= (16.7, 6.27))
plot = seaborn.scatterplot(ax=ax, x="day", y="ride_count",
size="avg_fare_amount", sizes=(1, 150), hue="avg_tip_amount",
data=athena_filtered_2)
```

```
plot.xaxis.set_major_locator(ticker.MultipleLocator(125))
plot.xaxis.set_major_formatter(mdates.DateFormatter('%m/%Y'))
plt.show()
```

When run, the cell produces the chart in *Figure 7.15*. At a glance, we can see that while the daily number of rides is indeed trending down, the average tip amount is actually increasing even though the average cost of a ride is relatively flat. This suggests that customer satisfaction is not a likely reason for the reduction in yellow taxi rides. For completeness, we'll still carry out a correlation analysis of our key metrics to better understand the relationships in our data.

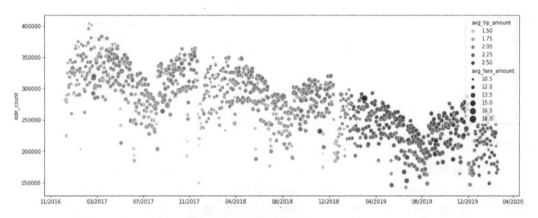

Figure 7.15 – Plotting the ride count versus the average tip amount versus the
average fare amount over time

Let's add one final cell to our notebook. We'll start this cell by running an Athena query to get hourly averages for ride duration, distance, fare, tip, and the number of rides. We conclude the cell by calling the pandas `corr()` function to calculate the correlation between all the columns in our results DataFrame:

```
athena_results_3=pd.read_sql("""SELECT
    max(hour(date_parse(tpep_pickup_datetime,
                    '%Y-%m-%d %H:%i:%s'))) as hour_val,
    avg(date_diff('second',
            date_parse(tpep_pickup_datetime, '%Y-%m-%d
%H:%i:%s'),
            date_parse(tpep_dropoff_datetime, '%Y-%m-%d
%H:%i:%s')))
    as duration,
    avg(trip_distance) as trip_distance,
```

```
        avg(fare_amount) as fare_amount,
        avg(tip_amount) as tip_amount,
        count(*) as cnt
from chapter_7_nyc_taxi_parquet
WHERE year=2018
group by date_trunc('hour', date_parse(tpep_pickup_
datetime,'%Y-%m-%d %H:%i:%s')) """, conn)
```

```
athena_results_3.corr()
```

pandas' `corr()` function implements several techniques for calculating a correlation matrix. By default, it uses the Pearson method to determine the covariance between two variables and then divide that factor by the product of the two variables' standard deviations. Covariance refers to the tendency for two variables to increase or decrease, with the relationship between height and age of students being a simple example of highly correlated variables. The Pearson method can only capture linear relationships between variables and has a range of 1 for highly correlated, 0 for uncorrelated, and -1 for inversely correlated.

Figure 7.16 shows the output of the correlation matrix outputted by our final cell. Interestingly, `tip_amount` is not correlated to duration. This goes against every movie you've seen where someone jumps in a taxi and offers a big tip to run every red light. In fact, `tip_amount` is most correlated with `trip_distance`. The relationship, or lack thereof, between the time of day (`hour_val`) is another surprise. You would think that ride duration would spike during peak commute times, but the lack of correlation between `hour_val` and duration suggests otherwise even though `ride_count` is highly correlated to the time of day. If we were continuing our ad hoc analysis of this dataset, our next step would be to look at how duration manages to be unaffected by ride count, a seemingly obvious traffic volume indicator.

	hour_val	duration	trip_distance	fare_amount	tip_amount	cnt
hour_val	1.000000	0.005754	-0.343937	-0.063703	0.269899	0.691085
duration	0.005754	1.000000	-0.011370	0.012235	0.019973	0.017925
trip_distance	-0.343937	-0.011370	1.000000	0.270629	0.327298	-0.721152
fare_amount	-0.063703	0.012235	0.270629	1.000000	0.210351	-0.150394
tip_amount	0.269899	0.019973	0.327298	0.210351	1.000000	0.190958
cnt	0.691085	0.017925	-0.721152	-0.150394	0.190958	1.000000

Figure 7.16 – DataFrame correlation values

In this section, we managed to run multiple Athena queries, targeting different slices of data, and pivot our ad hoc analysis based on findings along the way. We did all that while staying in one tool, our notebook. The tools we used are capable of much more than the simple explorations we undertook. Hopefully, this exercise has demonstrated why they would be a powerful addition to any ad hoc analytics strategy.

Summary

In this chapter, you got hands-on with the first of Athena's four most common usages – ad hoc analytics. We did this by looking at the history of business intelligence and learning about the OODA loop. Ad hoc analytics shortens the OODA loop by making it easier to use data to observe and orient yourself to the situation. The increased accessibility of data ultimately leads to the heightened situational awareness required for making sound decisions. With clarity of data behind your decisions, your organization will be less likely to waste time before acting on those choices. A short OODA loop also helps you react to poor decisions or calculated risks such as A/B tests.

The OODA loop isn't a new concept, and it's not the catalyst of the rising importance of ad hoc analytics. Instead, the proliferation of data has made it necessary for every decision maker in your organization to have access to critical business metrics at a moment's notice. We saw how some organizations attempt to meet this need through centralized reporting teams that bridge the skills gap between subject-matter experts that understand the semantic meaning of the data and the technical expertise required to access the data itself.

Athena shrinks the skills gap by hiding much of the complexity behind a SQL façade. Basic SQL knowledge is becoming increasingly common even in non-technical roles. Complimentary tools such as QuickSight further democratize access to data by providing a more guided experience. Jupyter Notebooks rounds out the strategy by providing an escape valve for advanced users and data scientists to use popular libraries with their data.

In *Chapter 8, Querying Unstructured and Semi-Structured Data*, you'll learn about another typical Athena use case. Querying loosely structured data is a challenging undertaking and partly the result of traditional SQL tables and schemas being too rigid and ill-equipped to support the pace of software evolution.

8

Querying Unstructured and Semi-Structured Data

Many of the world's most valuable datasets are loosely structured. They come from application logs, which don't conform to any standards. They come from event data generated by a system that users interact with, such as a web server, which stores how users navigate an organization's website. They can also come from an analyst generating spreadsheets on a company's financial performance. This data is usually stored and shared in a semi-structured format to make it easier for others to consume. Some query engines have evolved to fully support this semi-structured data.

When talking about **structured**, **semi-structured**, and **unstructured** data, there are many different definitions out there. For this book, structured data is stored in a specialized data format where the schema and the data it represents are one to one. The data is serialized to optimize how the data is read, written, and analyzed. An example is a relational database. Semi-structured data is when the data format follows a specific format, and a schema can be provided to read that data. For example, the JSON, XML, and CSV file formats have rules on how they are parsed and interpreted. Still, the relationship to a schema or table definition is loose. Unstructured data is data that does not follow a particular data model. Examples of unstructured data include application logs, images, text documents, and more.

In this chapter, we will learn how Amazon Athena combines a traditional query engine and its requirement for an up-front schema with extensions that allow it to handle data that contains varying schemas or no schema at all.

In this chapter, we will cover the following topics:

- Why isn't all data structured to begin with?
- Querying JSON data
- Querying arbitrary log data

Technical requirements

For this chapter, if you wish to follow some of the walk-throughs, you will require the following:

- Internet access to GitHub, S3, and the AWS Console.
- A computer with Chrome, Safari, or Microsoft Edge installed.
- An AWS account and an accompanying IAM user (or role) with sufficient privileges to complete this chapter's activities. For simplicity, you can always run through these exercises with a user who has full access. However, we recommend using scoped-down IAM policies to avoid making costly mistakes and learn how to best use IAM to secure your applications and data. You can find a minimally scoped IAM policy for this chapter in the book's accompanying GitHub repository, listed as `chapter_8/iam_policy_chapter_8.json` (https://bit. ly/3hgOdfG). This policy includes the following:
  - Permissions to create and list IAM roles and policies:
    - You will be creating a service role for an AWS Glue Crawler to assume.

- Permissions to read, list, and write access to an S3 bucket.

- Permissions to read and write access to Glue Data Catalog databases, tables, and partitions:

  - You will be creating databases, tables, and partitions manually and with Glue Crawlers.

- Access to run Athena queries.

Why isn't all data structured to begin with?

Data is generated from everywhere at all times within computer systems. They power our applications and reports and help us make sense of the world and our decisions that impact it. Data that's produced from an application that manages financial portfolios tells us how much risk the instruments in the portfolio are at. Websites can generate click data to tell a story, such as how customer's behavior changes when an update is made to a website. Retail businesses produce sales transactions and marketing data to determine how sales are affected by marketing campaigns. Amazon's user traffic information on individual products can train machine learning models to make recommendations to users who showcase products that they didn't even know they wanted. For this data to be helpful, it must be accessible to data engineers and machine scientists to produce even greater value from them.

However, not all data is created equally. If we take a hypothetical online store that sells everything from A to Z, sales information can be saved in structured data stores such as relational databases. User traffic and click data can be stored as text files in S3 in CSV files. Item description data can be retrieved through a RESTful API and saved as JSON data. The format and structure of the data are usually chosen based on how best to represent the information and how downstream applications consume that information. Usually, this data is not stored in a database system because this tends to be expensive. Hence, they are pruned of older data to keep costs low and performance high. Having data in a semi-structured format makes sharing data very easy. The data usually conforms to open standards, such as CSV, JSON, and XML. It is also estimated that 80-90% of current applications produce non-structured data. So, it makes sense to not change the existing applications but have our query engine read the data directly or ETL the data for Athena to read.

The remainder of this chapter will show you how to query a variety of semi-structured and unstructured data sources using Athena. We will use a fictitious retail business that sells widgets. This retail business wants to perform analytics with data that is produced by various systems. The following table outlines the system and type of data that is generated:

Data Source	Data Format	Description
Customers	JSON	Contains information about customers, such as their names and addresses
Sales	CSV	Contains transactions of customers purchasing widgets
Marketing	TSV	Contains information about a marketing campaign
Inventory	CSV	Contains the number of widgets left in our inventory
Website clicks	Text	Contains information about how customers are using the website

Table 8.1 – Data types and descriptions

The sample data files can be found on GitHub (`https://bit.ly/3wlJSwV`). Let's query these datasets.

Querying JSON data

JSON is a prevalent data format. It can be described as a lightweight version of XML and has many similarities with it. The file format is text-based, contains field names, along with their values, and supports advanced data types such as structures and arrays. A structured data type is a group of columns that are stored and referred to by their structure names and column names. This allows for logically similar columns to be grouped; for example, the structure of a customer's address that contains a street name, street number, city, state, and more. Arrays allow a single row to have a field containing zero or more values that can be referenced by an index number. An example list would be a list of addresses for a customer. JSON supports a mixture of arrays and structures. You can have an array of structures or a structure with an array field within it.

When using Athena, JSON files have to be of a particular format. Athena requires that JSON files must contain a single JSON object on separate lines within a file. If there are multiple objects on the same line, only the first object will be read, and if an object spans multiple lines, it will not throw an error. If the file format does not conform to what is compatible with Athena, then the data will need to be transformed (see *Chapter 9, Serverless ETL Pipelines*).

Now, let's look at some sample queries and read our customer's table.

Reading our customer's dataset

The following is a sample JSON record from our customer's table (formatted to be easier for a human to read but not Athena!):

```
{
  "customer_id": 10,
  "first_name": "Mert",
  "last_name": "Hocanin",
  "email": "mert@somedomain.com",
  "addresses": [
    {
      "address": "63 Fairview Alley",
      "city": "Syracuse",
      "state": "NY",
      "country": "United States"
    }
  ]
}
```

Here, we can see that this record has five fields: customer_id, first_name, last_name, email, and addresses. The addresses field is an array that contains a structure that contains four fields.

To register this table in our catalog, we can run a Glue crawler. But if we want to create this table using a CREATE TABLE statement (available at https://bit. ly/3yna5eV), it would look like this:

```
CREATE EXTERNAL TABLE customers (
  customer_id INT,
  first_name STRING,
  last_name STRING,
  email STRING,
  addresses ARRAY<STRUCT<address:STRING,city:STRING,
state:STRING,country:STRING>>,
  extrainfo STRING
)
ROW FORMAT SERDE
  'org.openx.data.jsonserde.JsonSerDe'
STORED AS INPUTFORMAT
```

```
   'org.apache.hadoop.mapred.TextInputFormat'
OUTPUTFORMAT
   'org.apache.hadoop.hive.ql.io.HiveIgnoreKeyTextOutputFormat'
LOCATION
   's3://<S3_BUCKET>/chapter_8/customers/';
```

Let's view some sample data from this table by running `SELECT * FROM customers LIMIT 10`:

▲	customer_id ▾	first_name ▾	last_name ▾	email ▾	addresses ▾
1	1	Rori	Struss	rstruss0@somedomain.com	[{address=20532 Debra Place, city=Boston, state=MA,
2	2	Maribeth	Myers	mmyers1@somedomain.com	[{address=4800 Montana Terrace, city=Anniston, state=
3	3	Pearle	Merrell	pmerrell2@somedomain.com	[{address=855 Upham Junction, city=Minneapolis, state=

Figure 8.1 – Sample data from the customers table

The results look as expected. Now, let's query the table and see how many customers have primary addresses in each state. We will assume that the first address in the address list is their primary address, so we can run the following query:

```
SELECT
    addresses[1].state AS State,
    count(*) AS Count
FROM customers
WHERE cardinality(addresses) > 0
GROUP BY 1 ORDER BY 2 DESC LIMIT 5
```

The results will look as follows:

▲	State ▾	Count ▾
1	TX	11
2	CA	10
3	FL	7
4	NY	7
5	VA	6

Figure 8.2 – Query results to show the top five numbers of customers by US state

With arrays, we can reference the element by using square brackets (*addresses[1]*). Since this returns a structure, we can reference the field by its name (*.state*). So, putting this together, we can specify the first address's state by writing `addresses[1].state`. Now, let's look at how we can parse fields that contain JSON data.

Parsing JSON fields

There are cases where some fields contain a string that contains JSON as a payload. This is sometimes done to make the payload completely abstract. Only the readers of the payload would understand the data in it. Our customer's table has a field called *extrainfo* containing JSON. In this section, we will describe an unlimited shipping program called the Pinnacle program. When we run `SELECT customer_id, extrainfo FROM customers WHERE extrainfo is not null LIMIT 5`, we get the following results:

	customer_id ▾	extrainfo ▾
1	1	{"is_pinnacle_customer":"true","pinnacle_id":"12423"}
2	5	{"is_pinnacle_customer":"true","pinnacle_id":"543433"}
3	11	{"is_pinnacle_customer":"true","pinnacle_id":"544333"}
4	15	{"is_pinnacle_customer":"true","pinnacle_id":"667645"}
5	23	{"is_pinnacle_customer":"true","pinnacle_id":"2342322"}

Figure 8.3 – The extrainfo field within the customers table

So, what can we do with this JSON data in the field? Athena (and PrestoDB/TrinoSQL) supports a JSON data type and a variety of built-in functions that allow us to interact with the JSON data easily without parsing or transforming the data. There are two JSON functions that are really useful: `json_extract` and `json_extract_scalar`. These functions take a string and a JSON path and return the JSON data type or a string. These functions extract any field within the JSON object, regardless of how nested the data may be. For example, if we run `SELECT json_extract_scalar(extrainfo, '$.is_pinnacle_customer')` `FROM customers where extrainfo IS NOT NULL`, we would get the following result:

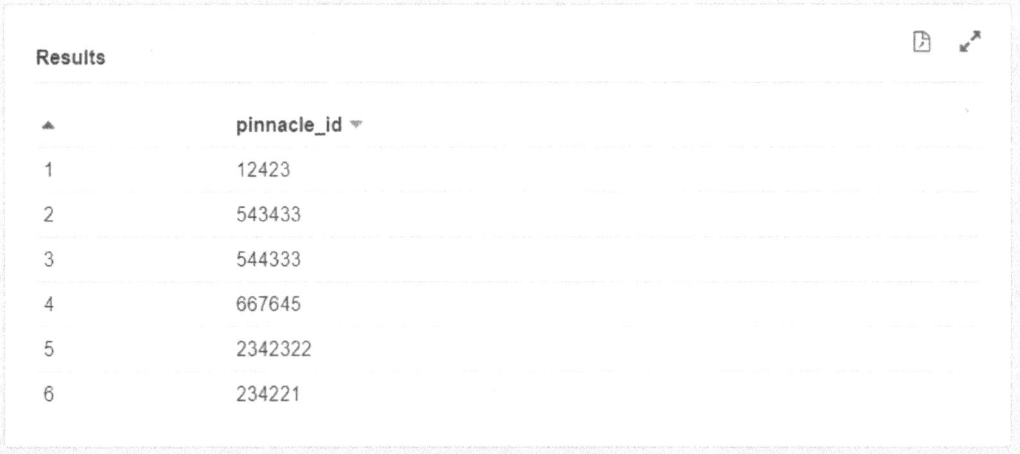

	pinnacle_id ▼
1	12423
2	543433
3	544333
4	667645
5	2342322
6	234221

Results

Figure 8.4 – json_extract_scalar function example

Let's look at some things we should consider when reading JSON data.

Other considerations when reading JSON

Let's take a look at some other things you should consider when reading JSON.

Schema updates with JSON

One of the benefits of using JSON is that fields can be added and removed from the records without it impacting Athena's ability to read the table. Since the files contain field names and their values, any fields that are not present in the files are ignored. If there is a row that doesn't include the field, then a null is returned. This is useful as data evolves and new fields are added and removed. Additionally, the ordering of fields does not impact the ability to read the data.

JSON SerDe comparison

Athena provides two different SerDes to be able to read JSON data. Each SerDe has slightly different functionality, so it's important to compare the two. In the preceding CREATE TABLE statement, we specified org.openx.data.jsonserde. JsonSerDe. The other SerDe is the Hive JSON SerDe. Our recommendation is to use the OpenX version. It contains some beneficial properties that can help read JSON that the Hive SerDe does not have.

When specifying the following properties, they need to be specified in the SerDe properties of the table, like so:

```
CREATE EXTERNAL TABLE customers (
    ... Table columns
)
ROW FORMAT SERDE
   'org.openx.data.jsonserde.JsonSerDe'
WITH SERDEPROPERTIES (
  "property1" = "value1",
  "property2" = "value2"
)
... Rest of the table attributes like INPUTFORMAT, LOCATION,
etc
```

Let's look at some of the useful OpenX JSON SerDe properties:

- **Mapping**: The OpenX JSON SerDe has a property that allows you to map a field name within the JSON file to a column name within your table definition. This can be useful if a field in your JSON file cannot be used within your table definition, such as if a keyword is used. For example, if you have a *timestamp* field name in your JSON file, you won't create a column called timestamp because it is a reserved keyword. Instead, you can map the *timestamp* field to a column named *ts* by specifying the WITH SERDEPROPERTIES ("mapping.ts" = "timestamp") SerDe property.

- **Case Insensitivity**: By default, the OpenX JSON SerDe will compare field names found in JSON files and column names in your catalog in a case-insensitive way. For most cases, this behavior is ideal as it will reduce the likeliness of errors being caused because of the case. However, in some rare cases, this may not be wanted as two field names may conflict if they only have case differences. For example, if your JSON file contains a field called *time* and *Time*, then it will seem like there is a duplicate field in the file, and it will be rejected as it will be deemed malformed. To get around this, we can leverage the mappings feature and turn off case insensitivity. For the time fields, we can use the `WITH SERDEPROPERTIES ( "mapping.time1" = "time", "mapping.time2" = "Time", "case.insensitive" = "FALSE" )` SerDe property.

- **Periods in Field Names**: If your JSON files contain field names with periods in them, then Athena won't read their data. To get around this, we must set `dots. in.keys` to true. Turning this property on will convert all the periods into underscores. For example, if you had a field in your JSON file named `customers. name`, then SerDe will translate this into `customer_name`.

Now that we have learned how to read JSON, let's look at how we can query CSV and TSV.

Querying comma-separated value and tab-separated value data

The **comma-separated value (CSV)** and **tab-separated value (TSV)** formats are some of the oldest data formats. They have lasted the test of time. They are heavily used today in many legacy systems and even among heavy Microsoft Excel users. Their main advantages versus other formats are their simplicity, their popularity, and that most spreadsheet applications can open them. CSV and TSV data also map very closely to tables within a database, where you have rows and columns of data. CSV and TSV files are text-based. Field values are separated by a delimiter, usually commas or tabs, and rows are separated by newlines. You can find examples of CSV files at `https://bit. ly/2TQY8z5` and `https://bit.ly/3h1G919`. We will use them as example datasets. Let's look at an example.

Reading a typical CSV dataset

Reading CSV and TSV data within Athena is simple, and in most cases, it is very straightforward. For most use cases, we can use the built-in delimited text parser. Let's take a look at the CREATE statement for our sales table (this can be found at https://bit.ly/2TQG73W):

```
CREATE EXTERNAL TABLE sales (
    timestamp STRING,
    item_id STRING,
    customr_id INT,
    price DOUBLE,
    shipping_price DOUBLE,
    discount_code STRING
)
ROW FORMAT DELIMITED
    FIELDS TERMINATED BY ','
    ESCAPED BY '\\'
    LINES TERMINATED BY '\n'
LOCATION 's3:// <S3_BUCKET>/chapter_8/sales/'
TBLPROPERTIES ('serialization.null.format'='',
               'skip.header.line.count'='1')
```

We can set two table properties here: skip.header.line.count and serialization.null.format. The skip.header.line.count property tells the parser to skip the first line in the CSV file as it contains the column names or the header row. The serialization.null.format property tells the parser to treat empty columns as nulls. Now that we have defined our sales data, let's take a look at some sample data, as shown in the following screenshot:

Results

	timestamp ▼	item_id ▼	customer_id ▼	price ▼	shipping_price ▼	discount_code ▼
1	2020-05-24T02:27:34Z	4	50	19.36	1.43	
2	2020-05-24T08:02:05Z	5	43	13.39	1.51	
3	2020-05-24T22:49:15Z	1	72	11.58	2.17	
4	2020-05-25T19:05:42Z	5	4	16.85	1.67	54162-269
5	2020-05-26T10:49:05Z	3	82	16.44	1.53	

Figure 8.5 – Sample data from the sales dataset

If your data contains a string field containing a comma, you can deal with it in two ways. First, you can escape the comma by using the specified character provided by the ESCAPED BY value. The second would be to surround the field with quotes, but you will need to use the OpenCSVSerDe parser for Athena to parse quoted fields correctly. We'll look at OpenCSVSerDe in more detail later. Now, let's learn how to read TSV files.

Reading a typical TSV dataset

TSV files are similar to CSV files, except tabs are used as delimiters between field values rather than commas. Tabs are less likely to be contained within string fields, so they are sometimes convenient to use rather than commas and escape unintentional commas. If you have tried to open a CSV file with Microsoft Excel and the columns do not align with these unexpected commas, you will understand that they can be challenging to fix.

In our example, we have a table representing marketing campaigns that contains a starting timestamp that represents the start of a marketing campaign, an ending timestamp that represents the end of a marketing campaign, and a description of the campaign. Suppose the marketing department provides this data as an export from Microsoft Excel that delimited the fields by tabs. To register the table, the CREATE TABLE statement (available at https://bit.ly/3xZVIwU) will look very similar to the CSV table, as shown in the following statement:

```
CREATE EXTERNAL TABLE marketing (
   start_date STRING,
   end_date STRING,
   marketing_id STRING,
   description STRING
)
ROW FORMAT DELIMITED
      FIELDS TERMINATED BY '\t'
      ESCAPED BY '\\'
      LINES TERMINATED BY '\n'
LOCATION 's3:// <S3_BUCKET>/chapter_8/marketing/';
```

You'll notice that the delimiter is \t, which represents a tab. The sample data is shown in the following screenshot:

Results

	start_date ▾	end_date ▾	marketing_id ▾	description ▾
1	2021-01-30 11:00:00Z	2021-01-30 18:00:00Z	193223	Extra adverstisement on search engine.
2	2021-06-19 00:00:00Z	2021-06-19 23:59:59Z	543222	A test marketing compaign, in order to observe im
3	2020-12-14 00:00:00Z	2020-12-14 23:59:59Z	432543	Pinnacle Day.
4	2020-07-01 00:00:00Z	2020-07-01 23:59:59Z	537734	Advertise on Canada Day.
5	2020-12-25 00:00:00Z	2020-12-25 23:59:59Z	796456	Advertise on Christmas Eve.

Figure 8.6 – Sample data from the marketing dataset

You will notice a comma in the second row that we did not need to escape. Now that we have a dataset for sales, customers, and marketing information, we can do some simple analytics from data that could have come from three different systems or sources. Let's look at a quick example.

Example analytics query

Suppose that we wanted to know how effective our marketing campaigns were by looking at the number of sales on days with campaigns versus days that do not. Additionally, we want to know the states that the sales were coming from. The following is a sample analytics function that can achieve that (this can be found at `https://bit.ly/3gXwf1u`, which also contains a breakdown of the query):

```
SELECT
    date_trunc('day', from_iso8601_timestamp(sales.timestamp))
as sales_date,
    CASE WHEN marketing.marketing_id is not null then TRUE else
FALSE END as has_marketing_campaign,
    SUM(1) as number_of_sales,
    histogram(CASE WHEN cardinality(customers.addresses) > 0
THEN customers.addresses[1].state ELSE NULL END) as states
FROM
    sales
LEFT OUTER JOIN
    marketing
ON
    date_trunc('day', from_iso8601_timestamp(marketing.start_
date))
    = date_trunc('day', from_iso8601_timestamp(sales.
```

```
timestamp))
LEFT OUTER JOIN
    customers
ON
    sales.customer_id = customers.customer_id
GROUP BY 1, 2 ORDER BY 3 DESC
```

This Athena query would produce the following results:

Results

	sales_date ▼	has_marketing_campaign ▼	number_of_sales ▼	states ▼
1	2021-01-30 00:00:00.000 UTC	true	9	{TX=1, AZ=1, GA=1, VA=2, CO=2, CA=2}
2	2020-12-04 00:00:00.000 UTC	true	8	{IL=1, IN=1, OH=1, MI=1, CA=2, DC=1}
3	2020-09-07 00:00:00.000 UTC	true	7	{NY=2, ID=2, CA=2}
4	2020-08-03 00:00:00.000 UTC	false	7	{MA=2, TX=2, FL=1, GA=1}
5	2020-07-08 00:00:00.000 UTC	false	7	{IN=1, NY=3, VA=1, NJ=1}

Figure 8.7 – Results of the analytics query

Here, we can see that for our top three results, days that had marketing campaigns produced the most sales and that most of our sales came from California. This information can help inform future marketing campaigns as well as inventory when marketing campaigns are run. This was just a warmup; we will look at more cases and explain how to do this type of analytics in *Chapter 7*, *Ad Hoc Analytics*. Now, let's learn how to read inventory data.

Reading CSV and TSV using OpenCSVSerDe

So far, we have looked at using the default version of SerDe to parse CSV and TSV files. However, another SerDe that we should look at deals with CSV and TSV files called OpenCSVSerDe. This SerDe compares to the default SerDe in a few crucial ways. First, it supports quoted fields, meaning that values are surrounded by quotes. This is usually done when the fields contain the same delimiter values, which are then ignored until the quote values are found. However, if there are quote values, those need to be escaped as well. The second difference is that all columns are treated as STRING data types, regardless of the table definition, and need to be implicitly or explicitly converted into the actual data type. The following is a sample CSV file from our inventory dataset that illustrates when OpenCSVSerDe should be used:

```
"inventory_id","item_name","available_count"
"1","A simple widget","5"
"2","A more advanced widget","10"
```

```
"3","The most advanced widget","1"
"4","A premium widget","0"
"5","A gold plated widget","9"
```

If we used the default SerDe, the `inventory_id` and `available_count` data would need to be specified as a string, and all field values would be returned with quotes, as shown in the following screenshot:

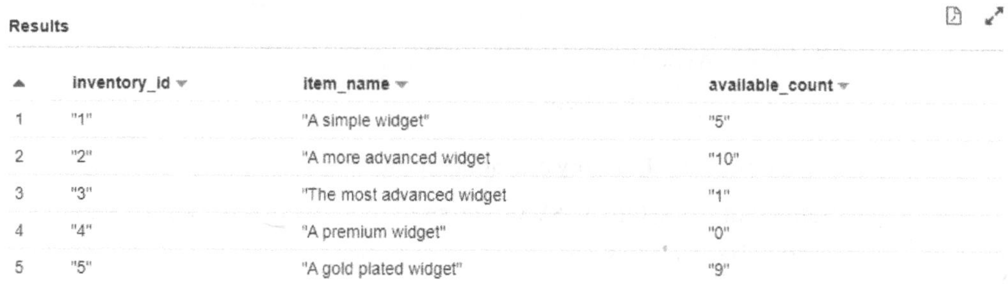

Results

▲	inventory_id ▼	item_name ▼	available_count ▼
1	"1"	"A simple widget"	"5"
2	"2"	"A more advanced widget	"10"
3	"3"	"The most advanced widget	"1"
4	"4"	"A premium widget"	"0"
5	"5"	"A gold plated widget"	"9"

Figure 8.8 – Reading the inventory dataset using the default SerDe

When the data is returned, as shown in the preceding screenshot, it would be tough to use. Using OpenCSVSerDe will solve this issue, as shown in the following CREATE TABLE statement (which is available at `https://bit.ly/35UsP9k`):

```
CREATE EXTERNAL TABLE inventory (
    inventory_id BIGINT,
    item_name STRING,
    available_count BIGINT
)
ROW FORMAT SERDE 'org.apache.hadoop.hive.serde2.OpenCSVSerde'
WITH SERDEPROPERTIES ("separatorChar" = ",", "escapeChar" =
"\\")
LOCATION 's3://<S3_BUCKET>/chapter_8/inventory/'
TBLPROPERTIES ('skip.header.line.count'='1')
```

Using OpenCSVSerDe will give us the following results:

Results

▲	inventory_id ▾	item_name ▾	available_count ▾
1	1	A simple widget	5
2	2	A more advanced widget	10
3	3	The most advanced widget	1
4	4	A premium widget	0
5	5	A gold plated widget	9

Figure 8.9 – Inventory data using OpenCSVSerDe

For more information about using OpenCSVSerDe, see Athena's documentation, which is located at `https://amzn.to/3isnvzr`.

Now that we have learned how to read CSV and TSV data that's been generated from Microsoft Excel or another source, let's dive into reading arbitrary log data.

Querying arbitrary log data

One very common use case for system engineers or software developers is to scan log files to find a particular logline. This may be to find when bugs have occurred, gather metrics about how a specific system performs, how users interact with a system, or to diagnose user or customer issues. There is a vast amount of useful and valuable data that comes out of log data. It's a great idea to store application log data in data to be mined in the future. Many of the AWS services are already pushing log data into S3 or can easily be configured. Athena's documentation provides templates for reading many of these services' log files, which can be found at `https://amzn.to/3dJzt6H`. Let's learn how Athena can be used to quickly and easily scan through logs stored on S3.

Doing full log scans on S3

Many logs are pushed to S3. Reading through those log files can be difficult and/or time-consuming when stored on S3. If those logs need to be read to look for problems, issues, or some kind of event, you may download the files and then run a grep command. This could take a lot of time because the download from S3 is done using a single computer, so it is limited by a single computer's network and CPU resources. You could spin up a Hadoop cluster and attempt to read the logs in parallel, but that requires expertise in using Hadoop and the time it takes to create and configure the cluster.

Athena can scan your log files in a parallel and easy way and return lines in log files that match search criteria. Let's go through an example. Anyone who has used Amazon EMR before knows that the application logs of Apache Hive, Apache Spark, and other applications are pushed to S3. When a particular Spark or Hive job fails, finding the specific log file that caused the failure may be difficult. Using Athena, we can search for the failure and log out the files that contained those failures. To do this scanning, we will rely on the default version of SerDe that Athena provides, which we looked at in the *Querying comma-separated value and tab-separated value data* section. But the trick here is to specify a delimiter that is very unlikely to exist in our log files. Let's look at `CREATE TABLE`:

```
CREATE EXTERNAL TABLE emrlogs (
   log_line string
)
ROW FORMAT DELIMITED
    FIELDS TERMINATED BY '|'
    LINES TERMINATED BY '\n'
LOCATION
   's3://<S3_BUCKET>/elasticmapreduce/j-2ABCDE34F5GH6'
```

Since the pipe character is unlikely to be in EMR logs, the `log_line` field will contain the value of a single logline. Then, we can submit queries while looking for a specific text. For example, we can use `regexp_like` to specify a regex to search for:

```
SELECT log_line FROM emrlogs where regexp_like(log_line,
'ERROR|Exception')
```

This query will print the entire line. Although this can be useful, we can also specify a hidden column that gives us the path of the file that the row was found in:

```
SELECT log_line, "$PATH" FROM emrlogs where regexp_like(log_
line, 'ERROR|Exception')
```

The $PATH field is very useful as it will give us the path that the logline was found in to download the file or files and take a closer look. The $PATH field can also be put in the WHERE clause to search for a particular application, EC2 instance ID, or EMR step ID. The following screenshot shows the example query output from the previous query:

	log_line ▼	$PATH ▼
1	Caused by: ExitCodeException exitCode=143:	s3://aws-logs-XXXXXXXXXX-us-east-1/j-106EEVKIB88X9/i-0ea08c8003c460588/hadoop-yarn/nodemanager-ip-10-0-0-227.log.gz
2	Caused by: ExitCodeException exitCode=143:	s3://aws-logs-XXXXXXXXXX-us-east-1/j-106EEVKIB88X9/i-0ea08c8003c460588/hadoop-yarn/nodemanager-ip-10-0-0-227.log.gz
3	Caused by: ExitCodeException exitCode=143:	s3://aws-logs-XXXXXXXXXX-us-east-1/j-106EEVKIB88X9/i-0ea08c8003c460588/hadoop-yarn/nodemanager-ip-10-0-0-227.log.gz
4	Caused by: ExitCodeException exitCode=143:	s3://aws-logs-XXXXXXXXXX-us-east-1/j-106EEVKIB88X9/i-0ea08c8003c460588/hadoop-yarn/nodemanager-ip-10-0-0-227.log.gz
5	Caused by: java.lang.InterruptedException	s3://aws-logs-XXXXXXXXXX-us-east-1/j-106EEVKIB88X9/i-0ea08c8003c460588/hadoop-yarn/nodemanager-ip-10-0-0-227.log.gz
6	Caused by: java.lang.InterruptedException	s3://aws-logs-XXXXXXXXXX-us-east-1/j-106EEVKIB88X9/i-0ea08c8003c460588/hadoop-yarn/nodemanager-ip-10-0-0-227.log.gz
7	Caused by: java.lang.InterruptedException	s3://aws-logs-XXXXXXXXXX-us-east-1/j-106EEVKIB88X9/i-0ea08c8003c460588/hadoop-yarn/nodemanager-ip-10-0-0-227.log.gz
8	<h2>HTTP ERROR: 404</h2>	s3://aws-logs-XXXXXXXXXX-us-east-1/j-10WFFA6AT6SM8/i-030875b499532e0b4/daemons/instance-state/instance-state.log-202
9	<h2>HTTP ERROR: 404</h2>	s3://aws-logs-XXXXXXXXXX-us-east-1/j-10WFFA6AT6SM8/i-030875b499532e0b4/daemons/instance-state/instance-state.log-202

Figure 8.10 – Running a Grep search on EMR logs using Athena

This way of using Athena can be applied to any text-based log files and can make it quick and easy to scan logs stored on S3. But what if we wanted to scan log files that are a little more structured to filter based on fields? This is where using Regex and Grok SerDes can help.

Reading application log data

Athena has two built-in SerDes that allow you to parse log data that follows a pattern. They then map the pattern to different columns within a table to query many types of log files. These two SerDes are the Regex SerDe and the Grok SerDe. With both of these SerDes, you provide an expression that Athena will use to parse each line of your text file and map the expressions to columns in your table.

Regular expressions, or regexes for short, are commonly used within many programming languages and editors to provide a search expression to find or replace a particular pattern. We won't go into how to write regular expressions, but if you are familiar with how to write regular expressions, then the Regex SerDe can be useful. The good news is that for many types of application logs, Athena's documentation provides the expressions so that they're ready to use to parse some of the most popular log types, such as Apache web server logs (see https://amzn.to/3xqrNhO) and most AWS services (see https://amzn.to/3dJzt6H). If you do want to create regular expressions for other log types, then we recommend using an online regular expression evaluator to test your expressions, which can help.

The Grok SerDe was built based on Logstash's grok filter. This SerDe takes in a Grok expression that is used to parse log lines. Grok expressions can be seen as extensions of regexes as Grok expressions are built using regexes, but regex expressions can be named and referenced. With named expressions, you can put them together to express a full logline in a more human-readable format. An added benefit is that Logstash has many built-in, ready-made expressions that you can use. The list is available at `https://bit.ly/3hEqq8n`. Let's look at an example of how this works.

Let's take our fictional company. They have a web server that outputs when visits occur, which page they visited, and referred them. Some example rows are as follows:

```
1621880197 59.73.211.164 http://www.acmestore.com/ https://www.
yahoo.com
1597343145 50.13.226.237 http://www.acmestore.com/ https://www.
google.com
1617872146 32.2.141.225 http://www.acmestore.com/product/4
https://www.duckduckgo.com
1621960907 65.105.235.14 http://www.acmestore.com/product/1
https://www.google.com
```

We have the time in epoch format, the visitor's IP address, the page that was visited, and the referrer (usually a search engine). Looking at the pre-built grok expressions, the preceding code can be expressed as `"%{NUMBER:time_epoch} %{IP:source_addr} %{URI:page_visited} %{URI:referrer}?"`. Let's create the table and query it using the following `CREATE TABLE` (available at `https://bit.ly/3xjRxMD`):

```
create external table website_clicks (
   time_epoch BIGINT,
   source_addr STRING,
   page_visited STRING,
   referrer STRING
) ROW FORMAT SERDE
   'com.amazonaws.glue.serde.GrokSerDe'
WITH SERDEPROPERTIES (
    'input.format'='%{NUMBER:time_epoch} %{IP:source_addr}
%{URI:page_visited} %{URI:referrer}?'
   )
STORED AS INPUTFORMAT
   'org.apache.hadoop.mapred.TextInputFormat'
```

```
OUTPUTFORMAT
    'org.apache.hadoop.hive.ql.io.HiveIgnoreKeyTextOutputFormat'
LOCATION
    's3:// <S3_BUCKET>/chapter_8/clickstream/';
```

The SerDe we specified is `com.amazonaws.glue.serde.GrokSerDe`, and we put it in the Grok expression via the `input.format` SerDe property. Now, if we query the table, we will get the following results:

Results

	time_epoch ▼	source_addr ▼	page_visited ▼	referrer ▼
1	1619373564	176.53.241.205	http://www.acmestore.com/product/1	https://www.yahoo.com
2	1620345099	52.96.144.79	http://www.acmestore.com/product/3	https://www.acmestore.com/search?s=awesome
3	1609494659	74.208.254.89	http://www.acmestore.com/product/5	https://www.duckduckgo.com
4	1600553532	159.68.111.113	http://www.acmestore.com/product/1	https://www.duckduckgo.com
5	1616796863	87.233.147.184	http://www.acmestore.com/product/1	https://www.acmestore.com/search?s=awesome

Figure 8.11 – Running a query against a table using Grok SerDe

Now that we can parse and query application logs, let's summarize what we have learned so far.

Summary

In this chapter, we explored the different ways in which we can query unstructured and semi-structured data. This data that comes from applications, databases, or even Microsoft Excel can be queried using Athena. We looked at two of the most commonly used file formats used by legacy and source systems, JSON and CSV/TSV, and how to determine which SerDes to use to parse them. We then looked at the Regex and Grok SerDes to help us parse log files that conform to some patterns, such as Log4J logs. Using these SerDes, we can query and derive value.

The next chapter will examine how we can take unstructured and semi-structured data and transform it into a more performant and cost-effective format, such as Apache Parquet or Apache ORC.

Further reading

To learn more about what was covered in this chapter, take a look at the following resources:

- Athena's documentation on the OpenCSVSerDe documentation: `https://docs.aws.amazon.com/athena/latest/ug/csv-serde.html`.

- Athena's documentation on the Grok SerDe: `https://docs.aws.amazon.com/athena/latest/ug/grok-serde.html`.

- Grok: `https://www.elastic.co/guide/en/logstash/7.13/plugins-filters-grok.html`.

- Athena's documentation on the Regex SerDe: `https://docs.aws.amazon.com/athena/latest/ug/regex-serde.html`.

- Athena's templates for consuming AWS Service logs: `https://docs.aws.amazon.com/athena/latest/ug/querying-AWS-service-logs.html`.

- Athena's supported SerDes: `https://docs.aws.amazon.com/athena/latest/ug/supported-serdes.html`.

9

Serverless ETL Pipelines

In the previous chapter, you learned how to tame unstructured or loosely structured data using **Athena** to manipulate logs, **JavaScript Object Notation (JSON)**, and other types of machine-generated data. In this chapter, we'll continue with the theme of controlling chaos by using automation to normalize newly arrived data through a process known as **extract, transform, load (ETL)**. We start with a brief explanation of ETL, and once we've established a basic understanding of ETL processes, we will move on to best practices and common pitfalls of using Athena for ETL.

As with most of the chapters in this book, we'll then get hands-on by designing and implementing a serverless ETL pipeline. More precisely, we'll implement the serverless ETL pipeline discussed in *Chapter 2, Introduction to Amazon Athena*. In that chapter, we described a fictional hedge fund with a propensity for trading widely shorted meme stocks. Their equally fictional yet surprisingly realistic compliance department needed a way to automatically process newly arrived trade reports from across the company and use the data to update the company's risk management system in near real time. By combining Athena with Lambda and using **Simple Storage Service (S3)** event notifications as a trigger, we can build a robust, cost-effective, and completely serverless ETL pipeline.

In the subsequent sections of this chapter, we will learn about the following topics:

- Understanding the uses of ETL

- Deciding whether to ETL or query in place

- Designing ETL queries for Athena

- Using Lambda as an orchestrator

- Triggering ETL queries with S3 notifications

Technical requirements

Wherever possible, we will provide samples or instructions to guide you through the setup. However, to complete the activities in this chapter, you will need to ensure you have the following prerequisites available. Our command-line examples will be executed using **Ubuntu**, but most Linux flavors should work without modification, including Ubuntu on **Windows Subsystem for Linux** (**WSL**).

You will need internet access to GitHub, S3, and the **Amazon Web Services** (**AWS**) console.

You will also require a computer with the following installed:

- Chrome, Safari, or Microsoft Edge browser

- The AWS **Command-Line Interface** (**CLI**) installed

This chapter also requires you to have an **AWS account** and an accompanying **Identity and Access Management** (**IAM**) user (or role) with sufficient privileges to complete this chapter's activities. Throughout this book, we will provide detailed IAM policies that attempt to honor the age-old best practice of "least privilege." For simplicity, you can always run through these exercises with a user that has full access. Still, we recommend using scoped-down IAM policies to avoid making costly mistakes and learning more about using IAM to secure your applications and data. You can find the suggested IAM policy for this chapter in the book's accompanying GitHub repository listed as `chapter_9/iam_policy_chapter_9.json`, here: `https://bit.ly/2RklBaW`. The primary additions from the IAM policy recommended for past chapters include the following:

- Lambda invoke and execution role changes

- S3 event notifications access

- CloudWatch Logs and Metrics access

Understanding the uses of ETL

In the most literal terms, ETL refers to a procedure with three conceptual phases that begin with reading data from a source system and end with a derivative of the original data being stored into a target system. In between these deceptively simple steps sits the most important facet of ETL, the transformation from the source system's semantic and physical schema to the domain model expected by the target system. In this step, we are essentially integrating source and target systems that may represent data differently.

Much of the academic literature on ETL points to the expansion of data warehousing concepts in the 1970s as its origin. It was a time when businesses rapidly adopted databases and found themselves with multiple data repositories, often using incompatible formats. Sounds familiar? Fast forward to today, and not much has changed aside from the date. The ability to integrate data from siloed or incompatible systems continues to be a key enabler for many **business intelligence (BI)** functions.

Traditional data warehouses were born of this era. They frequently served as the data integration point for everything from mainframes to spreadsheets. Data warehouses and ETL often play key roles in mergers and acquisitions, with ETL processes forming a beachhead for the more challenging technology integration effort to follow. Over time, the number of new data formats, databases, and source systems has led to an order-of-magnitude increase in data sprawl. Naturally, ETL is as popular as ever. Even the recent emergence of federated query engines such as Athena and their ability to query data in place hasn't done much to change the popularity of ETL processes. Tools such as Athena have led to an evolution of ETL from a primarily offline operation to a near-real-time integration.

For many years, businesses depended on ETL pipelines as a mechanism to get a consolidated view of critical business data. Today's ETL processes have evolved beyond the original charter that spawned the term and now include reactive pipelines, modularization of complex data processing, and even elements of performance optimization through pre-computed materialized views. Let's take a closer look at common ETL use cases.

ETL for integration

In our first ETL use case, we focus on scenarios where the goal is to enable two or more systems that have no direct mechanism for exchanging data to do so through a translation layer—namely, an ETL pipeline. Customers often run into this situation when they have a mix of systems from different vendors or that were developed in-house for various purposes over several years. Then, suddenly, due to an emerging regulatory requirement, an acquisition, or the pursuit of greater efficiency between two previously independent divisions, you need these applications to operate with an understanding of the whole.

To illustrate our point, we'll again use a fictional company. Suppose we worked for an e-commerce company that had purchased systems for managing its product catalog, pricing, and inventory tracking. Unfortunately, each piece of software came from a different vendor. As our company has grown, we've realized that we can improve the customer experience by ensuring these systems can work together. The following diagram shows how our product catalog system could avoid disappointing customers when searching for out-of-stock items and have extended lead times. Similarly, our pricing system could offer lower prices or promotional prices for items with too much inventory or that have achieved lower marginal cost due to the sheer volume we'll sell:

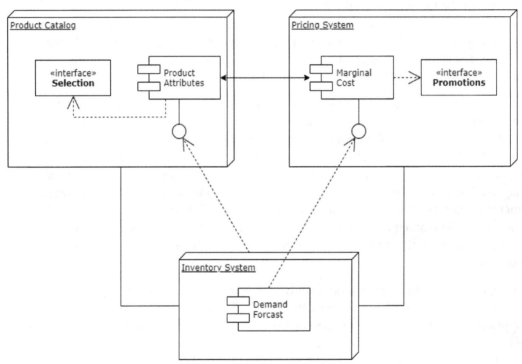

Figure 9.1 – Signing up for QuickSight

We'll need a way to load availability data into the product catalog, which only accepts JSON files. Similarly, we'll need a way to ingest unit costs into our pricing system, which only supports MySQL bulk loads. The inventory system was not built to export this kind of information. Luckily for us, it is built on a Postgres database, and the information we want is available in two easy-to-query tables. With these primitives, it is possible to create an ETL pipeline that integrates these systems as follows. Using Athena's Query Federation capabilities, we run an extract query against the inventory system's Postgres database and insert it into a **comma-separated values (CSV)**-formatted table backed by S3. The second step in our ETL pipeline triggers a bulk load into the pricing system's MySQL database using Aurora MySQL's load from the S3 feature. The next step in our pipeline is to run another Athena query that converts our temporary CSV table to JSON for use by the product catalog system. In our final step, we trigger the product catalog system's load from the S3 feature to pull the inventory availability data into the catalog.

Using ETL for system integrations such as the one we just discussed may not win many *Design of The Year* awards, but it is often a straightforward way to get disparate systems to work together. The main downside to such integrations tends to be latency since bulk load operations may impact the performance of the live system if done too often. You may be able to increase the frequency of the pipeline, and thus the freshness, but extracting and loading only the records that have meaningful change since the last run. This can help the average case but may not be viable if your dataset is so large that even the changed portion is too large to export frequently.

ETL for aggregation

One of the most common uses for the ETL pattern is consolidating information from across organizations into a single location for ease of reporting. As we saw in *Chapter 7, Ad Hoc Analytics*, a prerequisite for shortening the **OODA (observe-orient-decide-act) loop** is the accessibility of information. If your data is scattered across the organization, it can be impossible to answer even basic questions. In the following diagram, we again use our fictional e-commerce company to illustrate the utility of ETL by reviewing. This color-coded **entity-relationship diagram** (**ERD**) conveys where each table is stored with cross-table relationships or foreign key references, depicted by lines. Whenever a connection crosses storage systems, the line is dotted:

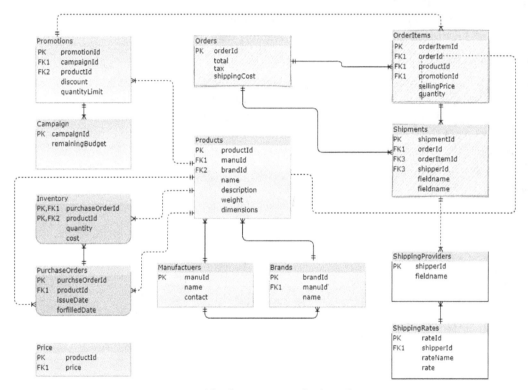

Figure 9.2 – Siloed e-commerce database diagram

> **ERDs**
>
> ERDs illustrate how entities within a particular domain are interconnected. This approach was initially developed in 1976 by Peter Chen to aid database design and development. Since then, their usage has expanded to include any context in which it may be helpful to understand both the universe of entities, their key properties, and how those properties define relationships between the entities. In *Figure 9.2*, we used a very basic ERD to illustrate the entities in our fictional e-commerce company.

As is the case with most businesses and the software systems they run on, our e-commerce example has evolved over time. The team separated concerns into five different microservices with accompanying datastores, including promotions, inventory, product catalog, pricing, order management, and shipping. While this is undoubtedly a reasonable level of decomposition that makes it easier to develop and maintain the system, it complicates even the most basic BI tasks.

For example, a campaign manager would be unable to understand which other items were purchased when a customer buys a promotional item at the sale price. This information is commonly used to determine the *lift*. A lift is a measure of the sales boost given to neighboring, related products when something goes on sale. A deal on cellphones is likely to generate additional purchases or lift for chargers and screen protectors. This information may be the difference between canceling an underperforming promotion or canceling an advertisement that only appears to be underperforming due to a lack of data.

This is where ETL can help. It is common for AWS customers to create ETL jobs for each system they may need to report across. The jobs extract the data from the source system, normalize the data types and semantics of the data, and finally load the data into a data lake. Once the data is aggregated into the data lake, it can be queried with ease from many tools, including Athena.

ETL for modularization

In the previous section, we saw how ETL could aggregate data from a modularized system. Here, we'll see how ETL itself can be a tool for modularization. This most commonly comes into play when you have a complex (multi-step) or long-running offline computation that you'd like to break down into more manageable parts. Allowing the calculation to unfold as bitesize steps can even improve reliability because you can avoid rerunning the entire process if one stage fails. Instead, you simply rerun the failure step and all the yet-to-be-run downstream steps. Let's look at an example.

Figure 9.3 depicts a four-step ETL pipeline that uses eight jobs to generate a seasonal buying plan for our fictional e-commerce company. We run a single job to find all the seasonal items in our product catalog in **Step 1**. **Step 2** has two jobs: one calculates the current inventory **sell-through rate** (**STR**) for the season items, and the other produces key weather-related inputs for the demand forecasting exercise that follows in **Step 3**. **Step 3** is the most complex and has been broken down into four independent jobs before flowing into the final step, which produces a buying plan for the upcoming season.

You can view the diagram here:

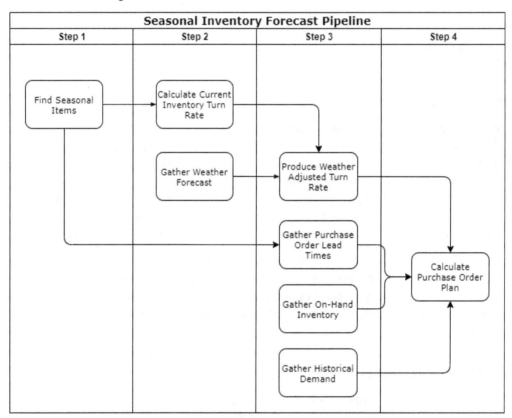

Figure 9.3 – Modular inventory forecast pipeline

Even if we had all this data in a single data store such as an S3-baked data lake, it would still be advantageous to break this process up into smaller units.

ETL for performance

Using ETL to optimize performance, usually of reporting systems, is yet another old pattern emerging with a new twist. In the last decade, we've seen a rapid expansion in the computational capabilities of query engines. Dozens or hundreds of nodes working together can achieve incredible data scan rates against an S3 data lake; for example, a typical Athena query against a well-structured S3 table can easily exceed a 200 **gigabytes per second (Gbps)** data scan rate. Increased query scale reduced the need for pre-computed aggregations or materialized views. This new class of query engines could compensate for increasing data sizes or misaligned data model access patterns with raw horsepower.

Unsurprisingly, the growth in business data has caught up with even the most advanced query engines, and the need to balance cost, latency, and performance has resurfaced, although it is also probably fair to say this balancing act never stopped being relevant. Many customers use ETL jobs to pre-compute common aggregations or generate materialized views that move costly operations such as joins into offline processes. Pre-computing enables query engines such as Athena to return results more quickly, use fewer resources, or incur lower costs per query.

Imagine your data lake has a table with customer orders from the last 10 years. Every line item in an order corresponds to a row in your table. For a successful company, such a table might have millions or billions of rows. If your most common access patterns look at weekly, monthly, or quarterly trends for a product or a category of products, you may benefit from generating aggregate tables. Using ETL jobs to pre-compute aggregate or rollup tables could improve cost and performance by two or three orders of magnitude in this example. The same concept can be applied to other costly operations such as joins. Building on this example, suppose we want to run a sales report by product manufacturer. This requires us to join our sales table with attributes from our product catalog and manufacturer tables. Joining these tables as part of our report can add tens of minutes to the query. If the person crafting the query is a **Structured Query Language (SQL)** novice or is using a BI tool that abstracts the SQL itself, you can easily end up with a long-running query that exhausts the query engine's memory and fails. As an alternative, you can use an ETL job to pre-join frequently used attributes into the sales table as a one-time effort and avoid the expensive join on the more frequent, ad hoc queries. Next, we'll see when it may make sense to avoid ETL by querying the data in place.

Deciding whether to ETL or query in place

The distinction between ETL and querying in place is blurred when using a service such as Athena. In the preceding sections, we reviewed common ETL use cases. In this section, we'll unpack the details that should go into deciding when the downsides of querying in place tilt the scale in favor of ETL. You might be curious why we've deliberately framed the choice as defaulting to querying in place. The reason is simple and comes to us courtesy of John Gail, who in 1975 theorized, "*A complex system that works is invariably found to have evolved from a simple system that worked. A complex system designed from scratch never works and cannot be patched up to make it work. You have to start over with a working simple system.*" In many ways, querying the data in place can be viewed as the most straightforward starting point. Athena's scalability reduces the need to curate your data model to your access patterns highly. In *Chapter 12, Athena Query Federation*, we'll see how federated queries lessen the need to extract data into your data lake at all.

Even though reducing the need to ETL or otherwise prepare your data for querying is a central part of Athena's mission to simplify querying your data, Athena doesn't completely eliminate the need for ETL. All of the preceding use cases for ETL still apply when using Athena, but the point at which they become relevant shifts. For most customers, performance becomes the primary factor, with cost a distant second. The actual threshold will vary based on your use case and latency needs. In the next section, we'll see how the ETL jobs' implementation also changes when using Athena.

Designing ETL queries for Athena

This section highlights workload traits and design considerations that Athena customers sometimes overlook creating ETL pipelines. Many of the items we are about to discuss are not specific to Athena. We'll be sure to note the ones that do stem from idiosyncrasies in the way Athena works. Generally speaking, there are no differences between regular Athena queries and those intended for use in an ETL pipeline. All of the performance suggestions covered in *Chapter 2, Introduction to Amazon Athena,* apply, and all the same Athena features are applicable across ad hoc analytics, ETL, and other use cases.

Don't forget about performance

Since ETL is not expected to be an interactive process, it allows us to run more time-consuming operations than we might otherwise. Just because ETL is typically viewed as an offline or asynchronous process that doesn't have a human sitting at a screen waiting for a response doesn't mean you can ignore performance. A good way to think about ETL performance is that all the same metrics as latency and cost matter but the scale shifts. You might not want to exceed a 30-second response time with an interactive query, but you might target 30 minutes with an ETL query. In the case of Athena, you'll want to pay special attention to the amount of memory your ETL jobs require. Joins, window functions, and high-cardinality distinct operations all have an amplifying effect on your query's peak memory demand. You may recall from previous chapters that Athena's version of Presto has limited but growing support for spilling query memory to disk. This capability reduces but does not eliminate the likelihood that your query will encounter Athena's **Query exhausted resources at this scale factor** error message. As you break down your ETL process into stages, keep the memory requirements top of mind and consider breaking up a complex query into multiple steps, sub-queries, or over independent time periods to reduce peak memory.

End-to-end (E2E) latency, sometimes referred to as data freshness, is the second performance dimension to be aware of. Often, customers will focus on the runtime of individual queries but lose sight of the total time it takes for their ETL pipeline to complete. In *Figure 9.4*, we have a dependency tree for an ETL pipeline from our fictitious e-commerce company. We've highlighted the sales extract job in red because it represents a chokepoint in our pipeline. Individually, each job runs quickly and meets or exceeds our expectations given the nature of the work being done, yet our customers routinely complain that their reports arrive late and can't be used in the weekly sales meeting.

You can view the dependency tree here:

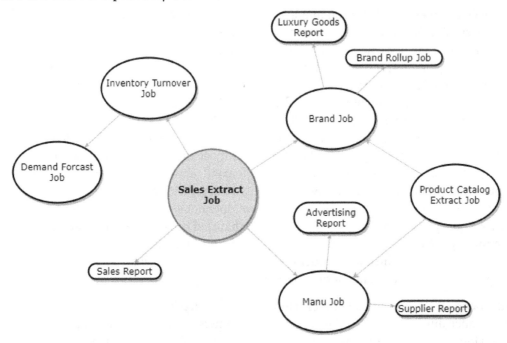

Figure 9.4 – ETL pipeline with a chokepoint

When designing this pipeline, we didn't factor in the time of day when the required inputs become available. In particular, the sales extract job cannot start running until a critical input from our credit card processing vendor arrives. This data tends to arrive on time on only 50% of days. Unfortunately, the sales extract job is an input to every other part of our pipeline. A common strategy in these cases is to break up the pipeline to separate the late-arriving data and the downstream items that actually depend on it. This may seem obvious, but it's common to combine simple operations to reduce the total number of jobs you must manage. Another option is to find an alternate source or use estimated values on occasions when the critical dependency arrives later than expected. This decouples you from the late source at the expense of accuracy. Depending on the nature of your data, this strategy may not be feasible. However, if the late-arriving dataset was updated product color information that rarely changes, it may be preferable to depend on a previous day's data.

Begin with integration points

Identifying how you will extract data from your sources and then load the result into your target systems may not seem like the most straightforward part of your ETL design, but you should start there anyway. Conceptually, the extract and load usually require little design, but it's not because they are trivial parts of ETL. These operations are often the most constrained and have the fewest options for you to choose from when designing your ETL process. This is precisely why we emphasize that you start by understanding your options for extracting and loading data.

Do your sources support incremental exports or only bulk snapshots? If your source supports incremental exports, you may be able to speed up your ETL jobs while also reducing costs. You'll also likely be taking on some extra complexity to handle cases where you need to backfill missing or incorrect data. Conversely, suppose your sources can only provide bulk snapshots that amount to a full export of the dataset. In that case, you can build a simpler pipeline with less error handling and reconciliation work. The downside is that you need to transfer more data than in the incremental model. This can increase costs as well as E2E runtime. You might even find yourself adding a stage in the pipeline to produce your own incremental feed of the source to reduce storage and compute costs for downstream queries and systems. There is no magic formula to decide when to use incremental or snapshot-style extract operations; the performance and feature set of the source system will likely dictate your options. Knowing this upfront will save you from rewriting or restructuring the transform phase of your ETL process.

An identical but reversed set of constraints applies to the load phase of your ETL pipeline. Does the target system support bulk loads? What happens to the performance of the system while you are loading new data? You may have limited or no control over the behavior of the target system that the results of your ETL pipeline will flow into. For example, if the end of your ETL pipeline is a table in an S3 data lake, then you've got a pretty broad set of options. The most challenging thing you'll likely need to handle is recovering from a partially failed job or having to restate erroneous results. On the other hand, if your target system is a MySQL instance (or any **relational database management system (RDBMS)**, to be honest), you'll want to think carefully about what happens to other queries when you begin to bulk load new data. If performance begins to degrade, you might even need to restructure your ETL process to produce smaller results. This can have implications for the original business purpose of the jobs, in addition to the underlying technologies.

It's always a good idea to disambiguate the things you don't control. Speaking of control, next, we'll look at what is managing or orchestrating each step in your ETL process.

Use an orchestrator

Athena is an excellent choice for many ETL queries, but until the service adds support for running queries on a schedule or in response to an event, you'll need to pair Athena with an orchestrator. You may be familiar with the concept of orchestration if you've worked with large, multi-step ETL pipelines in the past. For the lucky individuals who haven't had to organize 1,000 ETL jobs with tangled dependencies and conflicting column definitions, we'll take a moment to better explain what an orchestrator does.

Suppose you have an ETL query that you'd like to run after the last shipping truck leaves your warehouse for the day. You want this query to run after the shipping system has exported the day's shipping summary. How do you do that? Well, a naïve approach would be to schedule your query to run after midnight. After all, no shipments can go out for that day if the day is over. Unfortunately, the shipping system runs periodic maintenance, which can delay the availability of the summary data that our query depends on. It would be easier if our query could be aware of the completion of its dependency and trigger as soon as the data was available. That is where an orchestrator comes in. An orchestration system such as AWS Data Pipeline Amazon Managed Workflows for Apache Airflow can watch for a condition to be satisfied and trigger an action such as an Athena query. That condition can be the arrival of a file in S3, the completion of a previous query that subsequent queries depend on, or simple time-based schedules.

The complex, multi-query scenario we just laid out seems like a reasonable candidate for a dedicated orchestration tool. What may be less obvious is that even a single query ETL pipeline needs an orchestrator when using Athena. Until Athena adds a mechanism to schedule queries or reacts to events, we'll need to have something kick off our ETL queries. Later in this chapter, we'll use a simple Lambda function to orchestrate a simple serverless ETL pipeline.

Using Lambda as an orchestrator

An AWS Lambda function is an ideal orchestrator for simple ETL processes that run for 15 minutes or less and can be triggered by an event stream. If the number of steps, dependencies, or runtime grows, you'll want to consider using a more fully-featured orchestrator, such as **AWS Managed WorkFlows** for **Apache Airflow**. Putting that aside, building your own, simpler, serverless ETL pipeline with Lambda as an orchestrator is a great way to learn what to look for in a good orchestrator.

In this section, we'll precisely do that. Imagine we work for a fictitious hedge fund that is reeling from the great meme stock uprising of early 2021. Due to recent market volatility, the firm's risk management department is requiring trading desks across the company to report their recent trades on an hourly basis. Unfortunately, each trading desk uses different specialized trading software with no common interface for data extraction. Luckily, the trading desks can produce a CSV file with their trades and push the file to AWS S3 every hour. Our ETL process will use these files as input and feed them into the risk management processes at the end. The information in these files is time-sensitive, but the different trading systems will require varying amounts of time to generate and publish the hourly file. For that reason, we'll aspire to make this ETL process event-driven instead of working on a naïve hourly schedule. Let's look at the steps we'll need to complete as part of this ETL process, as follows:

1. Trigger an event when new trade files arrive in our S3 bucket.

2. Import the new trade data into the risk management data lake's trades table.

3. Publish an updated trade summary for each stock symbol with a nonzero net position or number of shares owned to our risk management monitoring process.

Depending on the data sizes involved, this is a fairly short ETL process. For simplicity, we'll assume that the E2E process takes well below the 15-minute runtime limit of AWS Lambda. The remainder of this chapter focuses on building a working version of this ETL process using S3 event notifications to trigger an AWS Lambda function that acts as an orchestrator.

Creating an ETL function

In order for our Lambda function to interact with Athena, S3, and the other services that our serverless ETL process needs, we'll have to first create a Lambda execution role in IAM. Since the creation of such a role requires broad IAM privileges, we omitted it from the chapter's IAM policy. You should use a privileged account or have your account administrator create the role for you. The following screenshot shows an example of how to configure the execution role for our Lambda function.

We recommend providing the role with this chapter's IAM policy and using `packt-serverless-analytics-lambda` as the name of this new role since the chapter's IAM policy already grants you the `iam:PassRole` permission on that name. If you choose a different name for your Lambda execution role, you may be unable to assign that role to the Lambda function we make in the next step unless you update the chapter's IAM policy:

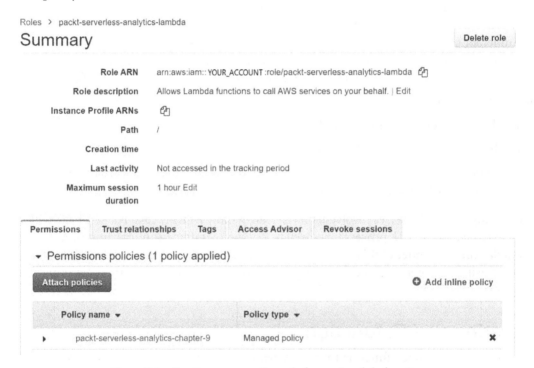

Figure 9.5 – Creating an execution role for our Lambda function

Now that we have our Lambda execution role set up, we can use the Lambda console to create a new Lambda function. If you are unfamiliar with AWS Lambda, it is arguably the genesis of the serverless movement. With AWS Lambda, you define functions, literal fragments of code, which can be invoked from various contexts, including S3 events. In the past, you'd have had to write an entire web service or **Remote Procedure Call** (**RPC**) service to do this, but AWS Lambda removes the need to manage any infrastructure or write any boilerplate service code. In the following screenshot, you can see just how easy it is to create a function in AWS Lambda. You simply provide a name for your function, pick a runtime, and select an existing role that will be used to provide the function with access to other AWS resources. The IAM policy for this chapter is set up to use `packt-serverless-analytics-etl` as the name of your function. Our ETL code will be written in Python, so you should select **Python 3.8** or later as the runtime for your function. AWS Lambda also supports Java, Node.js, and other runtimes, which you can try later:

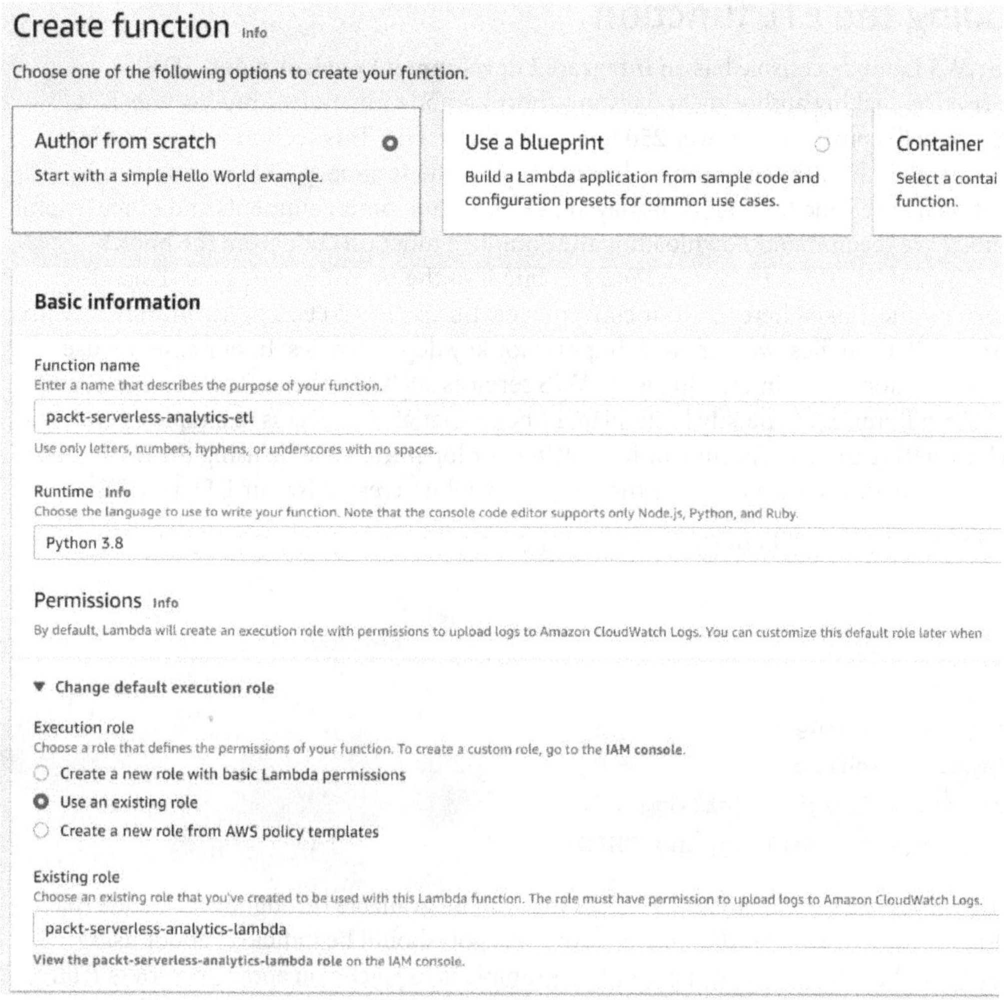

Create function Info

Choose one of the following options to create your function.

Author from scratch ⦿	Use a blueprint ○	Container
Start with a simple Hello World example.	Build a Lambda application from sample code and configuration presets for common use cases.	Select a contai function.

Basic information

Function name
Enter a name that describes the purpose of your function.

```
packt-serverless-analytics-etl
```

Use only letters, numbers, hyphens, or underscores with no spaces.

Runtime Info
Choose the language to use to write your function. Note that the console code editor supports only Node.js, Python, and Ruby.

```
Python 3.8
```

Permissions Info

By default, Lambda will create an execution role with permissions to upload logs to Amazon CloudWatch Logs. You can customize this default role later when

▼ **Change default execution role**

Execution role
Choose a role that defines the permissions of your function. To create a custom role, go to the **IAM console**.

○ Create a new role with basic Lambda permissions
⦿ Use an existing role
○ Create a new role from AWS policy templates

Existing role
Choose an existing role that you've created to be used with this Lambda function. The role must have permission to upload logs to Amazon CloudWatch Logs.

```
packt-serverless-analytics-lambda
```

View the packt-serverless-analytics-lambda role on the IAM console.

Figure 9.6 – Creating an AWS Lambda function

Lambda function timeout

By default, AWS Lambda functions use a 3-second timeout. This means that after 3 seconds, AWS Lambda will terminate calls to your function, even if the code is still running. Our serverless ETL example will typically complete in 30 seconds or less. To avoid unnecessary errors and troubleshooting, be sure to increase the timeout of your newly created Lambda function. We recommend using a maximum of 15 minutes for this exercise as you are unlikely to exceed the AWS Lambda free tier in this chapter. You can update this setting from the **Configuration** tab of your function in the Lambda console.

Coding the ETL function

The AWS Lambda console has an **integrated development environment (IDE)** experience, making authoring and testing short Lambda functions a breeze. Our ETL function will consist of just over 250 lines of Python code. This section will go line by line to explain how the function orchestrates our trade summary ETL process. While functional, the code fragments displayed in this section omit comments and other helpful context. We recommend downloading the complete function code from the book's accompanying GitHub repository. You can find it in the `chapter_9/etl_lambda_func.py` file, linked here for your convenience: `https://bit.ly/3wbAZp4`. As with all sound Python files, we start with imports for key dependencies. In our case, we use the `boto3` library for interacting with AWS services such as Athena. The `time`, `os`, and `logging` libraries are mostly boilerplate imports that give us access to simple operations such as getting the current time or formatting our log lines. We'll be using the `hashlib` library to create unique names for the temporary tables created by our ETL process.

The code is illustrated in the following snippet:

```
import time
import boto3
import os
import logging
import hashlib
logger = logging.getLogger()
logger.setLevel(logging.INFO)
```

Next, we declare several global resources that will be available throughout the code that follows. When writing production-quality code, you should be judicious about using global variables. Since the purpose of this example is to teach you about serverless ETL and not ideal Pythonic design, we're using global variables to keep things simple. Our first two global variables, ATHENA and CLOUDWATCH, are `boto3` clients for the respective services. You'll notice that when creating these clients, we didn't provide any credentials or region information. When invoking our function code, AWS Lambda injects credentials and region details into environment variables that `boto3` understands. This magic makes it very easy to get started with Lambda but can confuse folks when they run this code where these environment variables aren't automatically provided. The remaining variables act as configuration for our ETL process, conveying which Athena workgroup to run the queries in, which database and table names to import data to, and lastly, where our ETL data should be stored. Be sure to update these settings to match your environment if you haven't been following the suggested names in this and previous chapters.

The code is illustrated in the following snippet:

```
ATHENA = boto3.client('athena')
CLOUDWATCH = boto3.client('cloudwatch')
WORKGROUP = "packt-athena-analytics"
DATABASE = 'packt_serverless_analytics'
BASE_TABLE = 'chapter_9_trades'
ETL_LOCATION = 's3://<YOUR_S3_BUCKET>/chapter_9/'
```

With the dependencies and configuration out of the way, we can get to the body of the ETL process. The `lambda_handler` Python function is the main entry point that AWS Lambda calls when it wants to invoke our code. AWS Lambda sets the event and context parameters for each call. In our case, the `event` will contain details of the S3 object that was uploaded into our ETL `import` folder and acts as the trigger for our ETL process. We'll see how to set up an S3 event stream in a later section. Each line in the body of the `lambda_handler` Python function represents a step in our ETL process. We've designed the function this way both to improve maintainability and make it easier for you to follow along. These steps are modeled as helper functions that appear later in the code. We'll go over the steps briefly before looking at the code for each step in more detail.

First, we extract the **Uniform Resource Identifier (URI)** of the S3 object that was uploaded to our ETL `import` folder using our `event_to_s3_uri` helper. Before calling the `make_temp_import_table` helper function, we use the `ensure_trade_table_exists` helper function to set up our ETL tables. This only needs to be run one time, but as you'll see later, we used a `CREATE IF NOT EXISTS` query to cut down on the number of steps to get your serverless ETL process up and running. Once we've verified that our ETL tables are ready to go, we use the `make_temp_import_table` helper function to create a temporary table pointed at the newly arrived `s3_object` element. Athena doesn't have the concept of a temporary table, so you'll notice that we later use a `drop_table` helper function to delete the temporary table we created. But before doing that, we call the `import_data` helper function to transform the data in the newly arrived S3 object to a form that can be stored in our data lake tables. The details of the transformation are hidden in the `import_data` helper function that we'll look at shortly.

Lastly, we use the `update_trade_summary` and `publish_trade_summary` helpers to refresh our system's view of the world. By recalculating trade summary data by stock symbol and then publishing the summary values to CloudWatch Metrics, our hedge fund's risk management team can author alerts on these values. Those alerts can notify them of violations or trigger additional Lambda functions to take automated action.

The code is illustrated in the following snippet:

```
def lambda_handler(event, context):
    s3_object = event_to_s3_uri(event)
    ensure_trade_table_exists(DATABASE, BASE_TABLE, ETL_
LOCATION)
    import_table = make_temp_import_table(DATABASE, s3_object)
    import_data(DATABASE, BASE_TABLE, import_table)
    trade_summaries = update_trade_summary(DATABASE, BASE_
TABLE)
    publish_trade_summary(trade_summaries)
    drop_table(DATABASE, import_table)
    return {}
```

Ignoring the highly reusable helper functions and import statements, our entire ETL process is about 20 lines of Python code. It's difficult to get more straightforward than that. We can now dig into the helper Python functions we used to abstract reusable bits of the ETL process. Starting with event_to_s3_uri, you can see that this helper function extracts the S3 bucket and object key from the event that triggered our Lambda function. The function also trims off the actual object name from the key. We'll see why this is required in a later step. The schema of the event object is rather complex, but we'll show you an easy way to test your Lambda function later in this chapter.

The code is illustrated in the following snippet:

```
def event_to_s3_uri(event):
    record = event['Records'][0]
    s3_bucket = record['s3']['bucket']['name']
    s3_key = record['s3']['object']['key'].rsplit('/', 1)[0]
    return "s3://" + s3_bucket + "/" + s3_key
```

After the `event_to_s3_uri` helper function extracted the S3 bucket and object key that triggered the event, we used the `ensure_trade_table_exists` helper function shown in the next code snippet to check, and if need be, create a table our ETL process will ultimately load into. We used a few anti-patterns here in the interest of time. Firstly, we hardcoded the schema of our table in the function. It would be better if we abstract this away from the code and provide it as a separate configuration file or, better yet, use a CloudFormation template to handle setting up our data lake so that our ETL function doesn't need to perform this check. The other important thing to note here is the schema of our table. It has six fields, including the stock symbol, the date of the trade, the price, and the number of shares traded. The `year` and `month` fields are used as partition dimensions. We use the `replace` feature of Python strings to substitute our database and table name into the query before using the `run_query` helper function to execute the query in Athena. We'll look at that Python helper function next.

The code is illustrated in the following snippet:

```
def ensure_trade_table_exists(database, table_name, location):
    base_table_query = """CREATE EXTERNAL TABLE IF NOT EXISTS
        'DATABASE'.'TABLE_NAME'(
        'symbol' string,
          'trade_date' string,
          'price' double,
          'num_shares' bigint)
        PARTITIONED BY ('year' bigint, 'month' bigint)
        STORED AS PARQUET
        LOCATION 'S3_LOCATION'
        tblproperties ("parquet.compression"="SNAPPY");
    """.replace("TABLE_NAME", table_name)\
    .replace('DATABASE', database)\
    .replace('S3_LOCATION', location + table_name)

    run_query(base_table_query, 120)
```

The `run_query` helper function provides a convenient wrapper over the `boto3` Athena client and simply sets up the `request` object using some of the global variables we defined at the start of our Python code. It also adds helpful logging to make troubleshooting issues easier when they inevitably arise. You'll also notice that the helper makes use of yet another helper function called `wait_for_query`. The combination of these two helpers simplifies how our ETL process interacts with Athena's asynchronous query execution model. Usually, you'd want to avoid synchronously waiting for your Athena ETL queries to finish. Listening for the CloudWatch event that Athena generates when your query transitions from running to complete is a far more scalable approach. Since this may be your first time designing a serverless ETL process with AWS Lambda, we've opted to limit the event-driven flow to S3 event notifications and rely on a synchronous model for interacting with Athena.

The code is illustrated in the following snippet:

```
def run_query(query, wait_seconds = 0):
    logger.info('run_query: Preparing to run query %s', query)
    query_id = ATHENA.start_query_execution(
        QueryString=query,
        QueryExecutionContext={'Database': DATABASE},
        WorkGroup=WORKGROUP
    )['QueryExecutionId']
    logger.info('run_query: Started query with id: %s', query_
id)
    query_result = wait_for_query(query_id, wait_seconds)
    logger.info('run_query: Query result: %s', query_result)
    return [query_id, query_result]
```

As we saw in the `run_query` helper, the `wait_for_query` Python helper function is used as an adapter from the asynchronous programming model provided by the `boto3` Athena client and our desire for a simpler, albeit less scalable, synchronous model. In the synchronous model, our code runs an Athena query and then waits for it to complete instead of exiting and using a query completion event to wake our code up. The helper function accomplishes this by calling Athena, retrieving the status of our query in a loop, and sleeping between each check. Once the query moves to a terminal state, either succeeded or failed, the loop condition is met, and `wait_for_query` returns control to its caller. The function also takes an optional timeout that represents the maximum number of seconds it will wait for the Athena query to reach a terminal state.

The code is illustrated in the following snippet:

```
def wait_for_query(query_id, max_wait_seconds = 5):
    state = 'RUNNING'
    while (state in ['RUNNING', 'QUEUED'] and max_wait_seconds
> 0):
        query_execution = ATHENA.get_query_execution(
                            QueryExecutionId = query_id)
        try:
            qexec = query_execution['QueryExecution']
            exec_status = qexec['Status']
            state = exec_status['State']
            if state == 'FAILED':
                reason = exec_status['StateChangeReason']
                raise RuntimeError(query_id, reason)
            elif state == 'SUCCEEDED':
                return qexec['ResultConfiguration']
['OutputLocation']
        except KeyError:
            pass
        time.sleep(1)
        max_wait_seconds = max_wait_seconds - 1
    return False
```

The next step in our ETL process is the call from `lambda_handler` to the `make_temp_import_table` helper. This function's purpose is to create a temporary table pointing to the newly arrived S3 object that triggered our Lambda function via S3 event notifications. You may recall that when we looked at the `event_to_s3_uri` helper, we noted that it trims off the actual object name such that `import_location` is actually the folder containing the new object. This was done because an Athena table cannot point directly to an object. Instead, it must point to a prefix or folder. You'll need to keep this in mind when you upload test data by ensuring you upload it to a subfolder in our `import` directory.

The helper itself uses a hash of `import_location` to create a unique name for our temporary table and then runs an Athena query to create an AWS Glue Data Catalog table. We again took a shortcut by hardcoding the table definition in our Lambda code. In practice, you'll want to codify this as a configuration file or possibly a CloudFormation template. The essential pieces to remember are that the ETL process expects the `import` files to be CSV files with a header and a four-column schema consisting of the symbol, trade date, price, and number of shares.

The code is illustrated in the following snippet:

```
def make_temp_import_table(database, import_location):
    hash_import_location = hashlib.md5(import_location.
encode())
    table = 'chapter_9_import_' + hash_import_location.
hexdigest()
    temp_table_query = """
        CREATE EXTERNAL TABLE IF NOT EXISTS 'DATABASE'.'TABLE_
NAME' (
            'symbol' string,
            'trade_date' string,
            'price' double,
            'num_shares' bigint)
        ROW FORMAT DELIMITED
            FIELDS TERMINATED BY ','
        STORED AS INPUTFORMAT
            'org.apache.hadoop.mapred.TextInputFormat'
        OUTPUTFORMAT
            'org.apache.hadoop.hive.ql.io.
HiveIgnoreKeyTextOutputFormat'
        LOCATION
            'S3_LOCATION'
        TBLPROPERTIES (
            'areColumnsQuoted'='false',
            'columnsOrdered'='true',
            'delimiter'=',',
            'skip.header.line.count'='1',
            'typeOfData'='file')
    """.replace("TABLE_NAME", table)\
    .replace('DATABASE', database)\
```

```
    .replace('S3_LOCATION', import_location)

    run_query(temp_table_query, 120)
    return table
```

Once all the preparatory work to validate the extract portion of the ETL process has been completed, our Lambda function is ready to transform and load the new data. It does this by calling the `import_data` helper. As the name suggests, this Python helper function runs an Athena query to read from the temporary `import` table and transform the data into a format that can be stored in our data lake for use by downstream processes. This ETL process may look familiar since we used a similar procedure in *Chapter 3, Key Features, Query Types, and Functions*, to manually transform and load incremental data for our NYC Yellow taxi cab rides. Again, we use a hardcoded query, this time of the `INSERT INTO` variety, to transform and load the newly arrived data into our data lake's `trades` table.

The code is illustrated in the following snippet:

```
def import_data(database, target_table, source_table):
    import_data_query = """
        INSERT INTO DATABASE.TARGET_TABLE_NAME
        SELECT
            symbol,
            trade_date,
            price,
            num_shares,
            year(date_parse(trade_date, '%Y-%m-%d %H:%i:%s')) as
year,
            month(date_parse(trade_date, '%Y-%m-%d %H:%i:%s')) as
month
        FROM
            DATABASE.IMPORT_TABLE_NAME
    """.replace("TARGET_TABLE_NAME", target_table)\
    .replace('DATABASE', database)\
    .replace('IMPORT_TABLE_NAME', source_table)

    run_query(import_data_query, 120)
```

With the new trade data added to our data lake, we're nearly at the end of our serverless ETL orchestration code. Aside from cleaning up our temporary table and resources, we have one final load operation to perform. Using the update_trade_summary helper function, we run an Athena query to calculate the net holdings for each symbol our traders have bought and sold. We also use a HAVING clause to ensure we only return symbols that have a nonzero position. Put another way, the query will only return stock symbols where we are holding a positive (long) or negative (short) position. The update_trade_summary function concludes by returning an iterator over the results of the Athena query. This begins to get at the roots of why our faux hedge fund's risk management team asked us to create this serverless ETL process in the first place. Next, we'll see how our ETL process helps automate the handling of risky positions by downstream systems.

The code is illustrated in the following snippet:

```
def update_trade_summary(database, table_name):
    summary_query = """SELECT symbol, sum(num_shares)
                        FROM DATABASE.TABLE_NAME
                        GROUP BY symbol HAVING sum(num_shares)
!= 0
    """.replace("TABLE_NAME", table_name)\
    .replace('DATABASE', database)
    query_id = run_query(summary_query, 120)[0]
    paginator = ATHENA.get_paginator('get_query_results')
    return paginator.paginate(QueryExecutionId=query_id,
                        PaginationConfig={'PageSize':
1000}
    )
```

To demonstrate that our ETL process can integrate with live systems via **application programming interface** (**API**) calls, not just data lake queries, our Lambda function uses the publish_trade_summary helper function to publish trade summaries to Cloudwatch Metrics. It does this by walking the provided results iterator and using a boto3 client for Cloudwatch Metrics to publish the number of outstanding shares by a symbol. Our fictional risk management team can then author alarms with custom thresholds to alert them when a risk policy is violated. We could easily call a risk management API instead of CloudWatch Metrics.

The code is illustrated in the following snippet:

```
def publish_trade_summary(trade_summaries):
    for trade_summary in trade_summaries:
        for row in trade_summary['ResultSet']['Rows'][1:]:
            symbol = row['Data'][0]['VarCharValue']
            num_shares = float(row['Data'][1]['VarCharValue'])
            CLOUDWATCH.put_metric_data(
                MetricData=[{ 'MetricName': 'POSITION',
                    'Dimensions': [ {'Name': 'SYMBOL','Value':
symbol},],
                    'Unit': 'None',
                    'Value': num_shares
                },],
                Namespace='RISK/SUMMARY'
            )
    return num_summaries
```

The final step in our Lambda function is to clean up our temporary resources. In this case, we drop our temporary import table, as illustrated in the following code snippet:

```
def drop_table(database, table):
    drop_table_query ="""
        DROP TABLE DATABASE.'TABLE_NAME';
    """.replace("TABLE_NAME", table).replace('DATABASE',
database)

    run_query(drop_table_query, 120)
```

Now that we've completed coding our ETL function, we are ready to test it and configure S3 to trigger our Lambda function whenever a new trade report CSV file is uploaded to our import directory.

Testing your ETL function

At the top of the AWS Lambda function development screen, you'll see buttons for deploying and testing your function. The **Test** button has a dropdown that allows you to define one or more test events that Lambda will generate to trigger and test your function. The following screenshot shows the test event configuration screen that will guide you through defining a test event. Since we'll trigger our ETL process from an S3 event, you should pick the s3-put template from the provided examples. Simply change the S3 bucket and object key to match an actual test CSV file we'll upload to S3 next:

Figure 9.7 – Creating a test event

You can use the following sample trade data to create a test trade report in S3 corresponding to the test event you just configured. Be sure to put this test object in a subdirectory of the imported prefix you plan to use. For example, `S3://<BUCKET_NAME>/chapter_9/import/trade_desk_1_8am/trades.csv` would be a good upload location for trade desk 1's 8 a.m. trade file. In the next section, we'll configure S3 to send event notifications to our Lambda function any time an object is added to the import director:

```
symbol,trade_date,price,num_shares
GME,2021-01-01 00:41:22,240.00,1000
GME,2021-01-01 01:41:22,260.00,200
GME,2021-01-01 02:41:22,460.00,-200
GME,2021-01-01 03:41:22,560.00,-800
```

Next, we'll set up an automatic trigger for our ETL queries.

Triggering ETL queries with S3 notifications

Due to its low cost, high reliability, and seemingly infinite scalability, Amazon S3 is often at the center of many cloud architectures. In 2014, this led the S3 team to add the ability to trigger events for operations on your objects. These events can be filtered by bucket, prefix, and operation type with possible destinations, including **Simple Queue Service (SQS)**, **Simple Notification Service (SNS)**, and Lambda. You may also be interested to know that S3 does not charge for this feature. You'll only pay for the associated SQS, SNS, or Lambda usage for processing the events.

As we said earlier, we want our ETL process to react to the arrival of new data without the need to wait or poll. This reduces latency and increases data freshness for time-sensitive workloads such as our trade summary reports. The integration between S3 events and AWS Lambda also automatically handles re-driving failed events, simplifying our error handling. To begin, navigate your browser to the S3 console and select the bucket you'll be using for this exercise. In the **Properties** tab of your bucket, you'll find an **Event notifications** section. Clicking on **Create event notification** will pull up the dialog shown in the following screenshot:

Create event notification

The notification configuration identifies the events you want Amazon S3 to publish and the destinations where you want Amazon S3 to send the notifications. Learn more ⤢

General configuration

Event name

```
packet-serverless-analytics-chapter-9-new-data
```

Event name can contain up to 255 characters.

Prefix - *optional*
Limit the notifications to objects with key starting with specified characters.

```
chapter_9/import/
```

Suffix - *optional*
Limit the notifications to objects with key ending with specified characters.

```
.csv
```

Figure 9.8 – S3 event notification: General configuration

You can pick any name for the event configuration, as it's mostly used for documentation. The **Prefix** and **Suffix** fields should match the location you plan to use for this exercise. We recommend using the location provided in *Figure 9.8*. Next, we'll specify which S3 actions should generate an event. As shown in the following screenshot, we only need **Put** events:

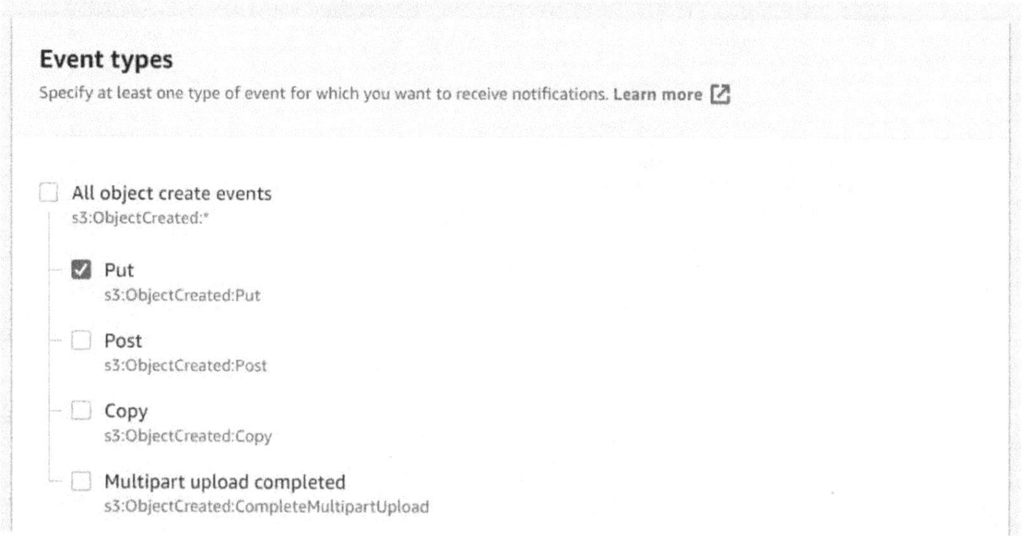

Figure 9.9 – S3 event notification: Event types

Lastly, you'll need to configure the destination that S3 should use for matching events. After selecting the Lambda function destination type, as shown in the following screenshot, a dropdown with available Lambda functions will appear. Find the ETL Lambda function you created earlier and save your changes:

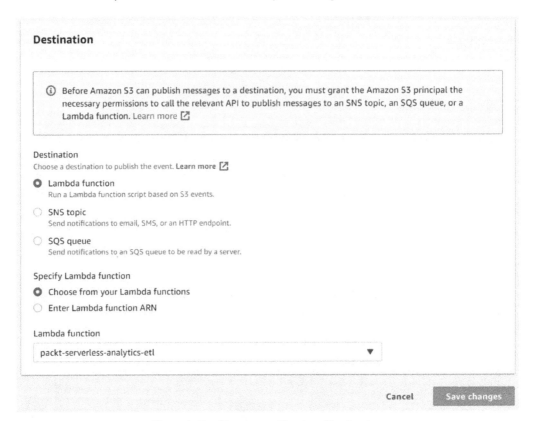

Figure 9.10 – S3 event notification: Destination

Now, you can upload a trade report CSV file to trigger our completed ETL process! You can use the same sample trade data from our S3 test event in the Lambda console, or you can use the provided trade file in the book's accompanying GitHub repository, found here: https://bit.ly/3f5DALJ. Unlike our earlier tests, which used the AWS Lambda development console, this E2E test will write its logs to Cloudwatch Logs, and the trade summaries will be published to Cloudwatch Metrics. You can navigate your browser to the Cloudwatch Logs console to find both. You can also run an Athena query to see if the new trades from the report file you uploaded were added to the chapter_9_trades table.

Summary

In this chapter, you learned about common usages of the ETL pattern, including integration, aggregation, modularization, and performance. The integration patterns offer a lowest-common-denominator approach to connecting disparate systems, even if they have no native support for integrating with each other. ETL for aggregations helps produce a **single source of truth** (**SSOT**) for getting a view of data across your estate. This is a common pattern for creating data lakes that work with services such as Athena. Modularization is an approach for using ETL to break up monolithic processes that are difficult to maintain or operationally prone to failure. Lastly, ETL for performance is a technique that moves expensive or time-consuming processing out of the live query path by either creating materialized views or running other pre-computations of anticipated workloads.

Armed with this knowledge of ETL design patterns, you reviewed key criteria for designing ETL queries for use with Athena. Deciding if you can skip the *extract* phase and use Athena Query Federation to query the data in place can help avoid unnecessary ETLs. When federation isn't a viable option, you saw that scale, integration points, and orchestration need to be factored into your ETL design.

The chapter concluded by putting what we learned into practice to build a serverless ETL pipeline with AWS Lambda and S3 event notifications. In *Chapter 10, Building Applications with Amazon Athena*, you'll continue putting what you've learned into practice by seeing how to build other types of applications with Athena.

10
Building Applications with Amazon Athena

Up to this point in the book, we've primarily been focusing on getting a feel for Athena as a product and what you can do with it. In this chapter, we're going to look at it from a slightly different angle and see how we can build our own product that leverages Athena. There are many things to consider when doing this, with the simplest being, how are we even going to call Athena? Previously, we've either used the AWS Console, the AWS CLI, and also occasionally the Athena Python SDK, but what other options are there? In terms of connecting to Athena, what should we consider? What security features are there for connecting? And finally, how do we make sure we continue to leverage Athena in the most performant and cost-effective way? These are all questions that we are going to try to answer throughout this chapter. This chapter will also be a nice reminder of how subjective writing software really is. A lot of decisions come down to personal preference, so I will do my best to present the facts and it will be up to you to decide which ones you care about the most.

In the forthcoming sections, we will cover the following topics:

- Connecting to Athena
- Best practices for connecting to Athena
- Securing your application
- Optimizing for performance and cost

Technical requirements

Wherever possible, we will provide samples or instructions to guide you through the setup. However, to complete the activities in this chapter, you will need to ensure you have the following prerequisites available. Our command-line examples will be executed using **Ubuntu**, but most Linux flavors should work without modification, including Ubuntu on Windows Subsystem for Linux.

You will need internet access to GitHub, S3, and the AWS Console.

You will also require a computer with the following:

- A Chrome, Safari, or Microsoft Edge browser installed
- The AWS CLI installed

This chapter also requires you to have an **AWS account** and an accompanying IAM user (or role) with sufficient privileges to complete this chapter's activities. Throughout this book, we will provide detailed IAM policies that attempt to honor the age-old best practice of "least privilege." For simplicity, you can always run through these exercises with a user who has full access. Still, we recommend using scoped-down IAM policies to avoid making costly mistakes and learning more about using IAM to secure your applications and data. You can find the suggested IAM policy for Chapter 10 in the book's accompanying GitHub repository listed as `chapter_10/iam_policy_chapter_10.json` here: `https://bit.ly/3zM54wG`. The primary additions from the IAM policy recommended for past chapters include the following:

- Adding SNS topic permissions for topics beginning with `packt-*`
- CloudTrail permissions for trails beginning with `packt-*`
- EventBridge permissions for managing rules

Connecting to Athena

So, you're ready to get started on your application built on top of Athena. You've got some initial data models prepared and registered within Athena and you want to start querying the data. Now how do you do that? If you've been following along with all of the exercises in this book, we've primarily interacted with Athena either directly through the AWS Console or the AWS CLI. If you have read *Chapter 7, Ad Hoc Analytics*, then you did get a small preview of the Athena Python SDK. So, your other options include using a **JDBC Driver**, an **ODBC Driver**, or, more generally, the AWS SDK, which is available in many languages (for a full list, see `https://amzn.to/3BgXrQc`).

So, before we figure out which one is right for you, let's go over what some of these options are. The SDK should be pretty straightforward; it's a language-native implementation for interacting with AWS's many APIs. But what about JDBC and ODBC; what are those?

JDBC and ODBC

JDBC, or **Java Database Connectivity**, is a Java database abstraction API. It is oriented primarily around interacting with relational, SQL-based databases (though there are some JDBC drivers out there for NoSQL databases as well). Essentially, it provides a standard mechanism for Java developers to connect to different database technologies by using the exact same (or very similar) code.

ODBC, or **Open Database Connectivity**, provides the same functionality as JDBC but is written in C, and so is intended for use in C, C++, C#, and so on.

The way both of these technologies work is that there is the common API, which is what the developers will be using directly in their code, and then there are drivers, which are the actual underlying implementation of the API. Both technologies allow for the dynamic loading of drivers, so as long as the driver is available to the running process, they can be used together. Let's take a look at a couple of examples of what using each one would look like. In both examples, we are going to connect to a MySQL database containing a table named `awesome_packt_table` with the data below and run a simple query against them.

Title	Publisher	Year of Publication
Serverless Analytics with Amazon Athena	Packt	2021

Table 10.1 – awesome_packt_table data

We will run the following query:

```java
import java.sql.Connection;
import java.sql.DriverManager;
import java.sql.SQLException;

public static void main(String args[]) {
  // Notice we are using DriverManager, DriverManager is able
to
  // determine that we want to use the mysql driver by way of
the
  // "jdbc:mysql" in the url

  // Create the connection in a try-with-resources to auto
close
  // when we are done
  try (conn =
      DriverManager.getConnection("jdbc:mysql://localhost/
test_db?" +

"user=packt&password=supersecure")) {
    // Statement is the object that will accept our query
    stmt = conn.createStatement();

    // And here we execute! Again putting results in the try
with
    // resources so it closes when we're done
    try (results = stmt.executeQuery("SELECT title, publisher,
publish_year FROM awesome_packt_table")) {
      while (results.next()) {
        // Returns "Serverless Analytics with Amazon Athena"
        result.getString("title");
        // Returns "Packt"
        result.getString("publisher");
        // Returns 2021
        result.getInt("publish_year");
```

```
      }
    }
  }
}
```

Code 10.1 – Sample JDBC code

And now let's take a look at what this would look like for ODBC:

```
using Microsoft.Data.Odbc;


static void Main(string[] args) {
  // Unlike Java, there is no DriverManager, simply instantiate
  // a new connection and indicate the driver in that
connection.
  string MyConString = "DRIVER={MySQL ODBC 3.51 Driver};" +
          "SERVER=localhost;" +
          "DATABASE=test_db;" +
          "UID=packt;" +
          "PASSWORD=supersecure;" +
          "OPTION=3";
  // Same as in Java, auto close when we are done
  using (OdbcConnection connection =
                      new OdbcConnection(connectionString)) {
    OdbcCommand MyCommand =
            new OdbcCommand("SELECT title, publisher, publish_
year FROM awesome_packt_table", connection);
    using (OdbcDataReader Reader = command.ExecuteReader()) {
      while (Reader.Read()) {
        // Returns "Serverless Analytics with Amazon Athena"
        Reader.GetString(0);
        // Returns "Packt"
        Reader.GetString("publisher");
        // Returns 2021
        Reader.GetInt32("publish_year");
      }
    }
```

```
    }
}
```

Code 10.2 – Sample ODBC code

Now let's say we've moved our table out of MySQL and loaded it into Athena. Let's say everything else remains the same – the table name (assuming a default catalog) and column names are all identical. All we would have to do is change the following:

```
try (conn =
    DriverManager.getConnection("jdbc:awsathena://" +
                        "AwsRegion=[AWS_REGION];" +
                        "User=[AWS_ACCESS_KEY];" +
                        "Password=[AWS_SECRET_KEY];" +
                        "S3OutputLocation=[OUTPUT]") {
```

Code 10.3 – Migrating to Athena JDBC

We can do the same for ODBC as follows:

```
string MyConString = "DRIVER={Simba Athena ODBC Driver};" +
            "AwsRegion=[AWS_REGION];" +
            "AuthenticationType=IAM Credentials;" +
            "UID=[AWS_ACCESS_KEY];" +
            "PWD=[AWS_SECRET_KEY];" +
            "S3OutputLocation=[OUTPUT]";
```

Code 10.4 – Migrating to Athena ODBC

And that's it! All we did was change the connection strings for both drivers to match that of Athena and the drivers themselves have done the heavy lifting of understanding how to interact with MySQL versus Athena.

For the sake of completion, let's quickly discuss what you would have to do with the Athena SDK to accomplish the same query. You would start by instantiating your Athena client. Depending on where this is running (for example, on AWS provided compute such as EC2 or Lambda), you'd either use the default credentials provider, or you'd supply the credentials as in the two preceding figures. Then you would call `StartQueryExecution` with the query string and also provide a result location, which would be the same as `Output` above. Next, you would call `GetQueryExecution` repeatedly in a loop until the query completes, and finally, when it's done, you would call `GetQueryResults`.

Which one should I use?

The reality is that there is no perfect answer to this question; it kind of comes down to preference. Obviously, some decisions will be made for you depending on your tech stack; for instance, you probably won't use the JDBC driver, which is Java-specific, if you are writing your application in Python. You'd just go ahead and use the Python SDK. But let's say you've chosen Java as your application language, what now? Well, this is really where it gets a bit more subjective. There are pros and cons to both, so it's really up to you which ones matter most. First, let's get one thing out of the way; the implementation of the JDBC driver (and ODBC driver) utilizes the respective SDK implementations, so there's no difference in performance.

In general, the decision between the API abstraction options and the SDK centers around convenience versus flexibility. The convenience of JDBC/ODBC comes in a few different forms. Firstly, if your organization is one of many that already heavily uses those abstractions, then this would certainly fit in nicely with your stack. Also, if you think there's a chance that you might be switching data storage options, then this makes that painless, as we showed above (or perhaps you are switching to Athena, as we did above). And finally, the code can, in some cases, be more succinct when using JDBC/ODBC. The general call flow that we described previously ends up being around 100 lines of code, versus the 25 or so that we wrote for JDBC/ODBC. The abstractions provide easy mechanisms for getting the correct data type that you need for a value (refer to the preceding examples where we have `getString` and `getInt`), whereas with the Athena SDK, everything is returned as a string and it's your responsibility to convert it into whatever underlying type it is.

So then, why bother with the SDK? Well, if you have very long-running queries, you may not want to be constantly occupying a thread while waiting for the query to complete. Some queries could run for hours and that's a pretty significant waste of resources. That's not an option with JDBC/ODBC. There are some libraries that make them operate in an "async-like" fashion, but underneath there is always a thread that is fully taken while it waits for the query to complete. Below, we're also going to talk about how instead of polling for query execution status, we can actually integrate with AWS EventBridge to get push notifications for when a query execution status changes. Again, that is not something you can accomplish with abstractions. There is also always the possibility, since the JDBC and ODBC drivers depend on the SDK, that they may not immediately get any new features, or at least not as quickly as the SDK itself will. So, these sorts of things are where the SDK really shines in its ability to allow you to interact with Athena exactly how you want to.

With this information that we just covered, you now have the means to decide which option for interacting with Athena is right for you. For the remainder of this chapter, we're going to focus on making sure you are getting the most out of your usage of Athena. The metrics by which you track whether you are being as optimal as possible are going to depend on your circumstances – whether your goal is to have the lowest possible AWS bill, or whether it's to have a blazing fast application, that's up to you. My goal is that you leave this chapter with the necessary tools in your toolbox to accomplish your goals.

Best practices for connecting to Athena

In this section, we're going to go over some things to consider when connecting to and calling Athena, including idempotency tokens and query tracking.

Idempotency tokens

I know this statement may come as a huge surprise to you, but perfect software does not exist. It's going to fail. There's a reason why there are so many different options out there for monitoring the operational status of an application. And among the infinitesimal category of possible failure scenarios, they can be narrowed down to two large categories – safe to retry and not safe to retry. It's that second category we will be focusing on in this section. More specifically there is a subcategory of *not safe to retry* that can quickly be summarized as ¯_(ツ)_/¯ – you have no clue whether it is safe to retry; you know something happened, but exactly what happened is a complete mystery.

Thankfully Athena (and many other services) has a nice mechanism for handling these very scenarios. They are called **idempotency tokens**. To be idempotent, an operation has to be able to guarantee that if repeatedly given an identical request, the operation will return an identical response. Surprisingly, there is a decent amount to unpack from such a simple statement. What defines an identical request? What defines an identical response? Those can be sort of subjective things. For example, an absolute value is an idempotent operation. It is always true that $|x| == |x|$. So, the request in that case is "x," and the response is always, well, the absolute value of x.

Now let's take a real-world example. Say you are going to buy coffee and you pay with your credit card; based on the definition of idempotency, for that single transaction, you could be charged the exact same amount twice, and get the exact same behavior twice (having a charge on your credit card), and that would qualify as "idempotent," if the request is simply defined as the "amount to charge" and the response is "charge successful." But that would not make you very happy, would it? That was rhetorical; of course it wouldn't! So instead, the request is defined as the combination of "amount to charge" AND a unique identifier for that transaction and the response is that that transaction successfully charged that amount exactly once. Now, if the coffee shop tries to send two identical requests containing that amount and the unique identifier, your credit card company will know not to take the second one as it was probably sent in error. Et voilà, we've arrived at an idempotency token! That transaction ID, in this case, is acting as the idempotency token; it is saying that if you see this ID twice and you've already successfully processed it, disregard any further attempts to process it. And that's exactly how they work in Athena.

In Athena, they are called `ClientRequestTokens` and they are only supported by some APIs (essentially any in which it could be undesirable to retry an identical request). `StartQueryExecution` is the one we are going to focus on, but another that is supported is `CreateNamedQuery`, because named queries are uniquely identified by an "ID," but that ID is not supplied at creation; it is generated as part of the creation process, so a retry without an idempotency token would result in two identical named queries being created with different IDs.

To better understand why we care about `ClientRequestTokens` in the context of the `StartQueryExecution` API, let's look at a couple of sample call flows.

In the first sequence below, *Figure 10.6*, you can see that no ClientRequestToken was provided. Athena successfully begins the execution of the query on a cluster but fails to return the response to the customer. The customer assumes it failed and reruns the query. Because there is no ClientRequestToken, Athena assumes it's a new query and runs it again. Now the customer has incurred double charges, which, much like the coffee scenario, is not desirable!

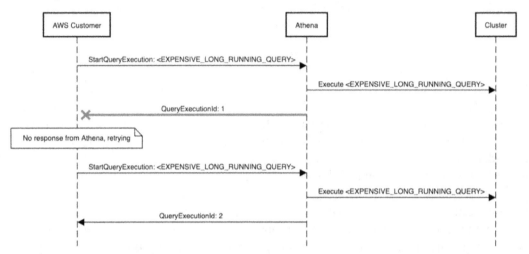

Figure 10.1 – Retrying a query against Athena without ClientRequestToken

But now, in the next sequence, *Figure 10.7*, you see that we ran the same query, experienced the same failure with Athena, but this time the customer supplied a ClientRequestToken. So, when the customer goes to retry, Athena is able to determine that it actually did successfully execute the previous request, and simply returns to the customer the identical response that it attempted to return in the previous call. Yay, we only paid once for our one cup of coffee!

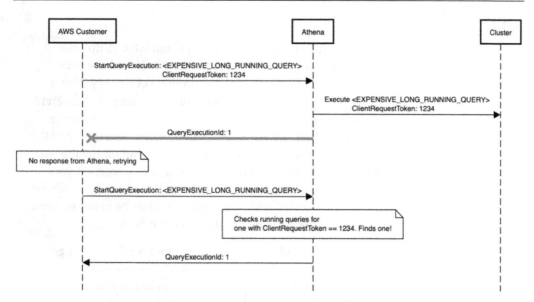

Figure 10.2 – Retrying a query against Athena with ClientRequestToken

Hopefully, now you have an idea of the importance of idempotency and
`ClientRequestTokens`. And now for the best news of all! If you are using the AWS
SDK (or JDBC/ODBC driver, since, as we discussed previously, those rely on the AWS
SDK) or the AWS CLI, then you actually don't have to do anything to leverage this feature!
The AWS SDK/CLI will automatically populate the `ClientRequestToken` in your
requests, which means that if that request gets retried, for whatever reason, it will be
idempotent!

Query tracking

Pivoting away from failure handling and retries, let's talk a bit about what to do once you've successfully started execution of a query. Throughout this book, we have been leveraging the GetQueryExecution API to monitor the progress of a query. This is fine, but as your usage of Athena scales up, you are going to run into a couple of different issues. Firstly, tuning your application to poll at the right frequency can be a challenge. You don't want to poll so infrequently that you are adding unnecessary time on top of the query execution, particularly if you have queries that execute quickly, but on the flip side, you don't want to poll so frequently that you are consuming a ton of resources on your end (threads, I/O sockets, and so on) and also Athena API limits. Limits can generally be increased, but of course, there's a limit to that limit, and wouldn't it just be better to avoid having to deal with that? Well, some more good news! There is a way to do that!

Athena publishes any changes in the status of a query execution to **AWS EventBridge**. AWS EventBridge is a managed event bus service that allows AWS customers to process events produced by other systems (either AWS services or anyone else) in real time utilizing a push model. The way it works is that you configure a **rule** that tells EventBridge, for a given scenario and event, to forward that event to a **target**. There is also a second type of rule, which is a scheduled rule, so rather than reacting to an event, it triggers a target on some sort of schedule, either a cron job or an arbitrary time rate (for example, once an hour). For our purposes, we are going to focus on the first type of rule, which is the event-based rule.

So, let's run through a quick example of how to get set up with an EventBridge rule for Athena queries. To keep things simple, we're going to set it up so that we send an email based on Athena status changes. In this case, our target will be an SNS topic, which will, in turn, forward any messages it receives to an email we configure.

> **Note**
>
> Everything we are doing should fall well within the Free Tier limits, so assuming your account still qualifies, this next section should not cost you anything!

Step 1 – Setting up the SNS topic

Navigate to the SNS console, find topics, and then select **Create topic**. On the **Create topic** page, select the **Standard** type of topic, and then give it a descriptive name. Then you can leave the rest blank, go ahead, and scroll and click on the **Create topic** button.

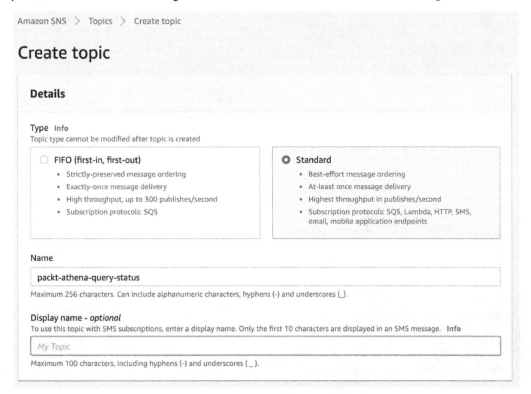

Figure 10.3 – Creating an SNS topic

You should be taken to the newly created topic's **Details** page. Look for a button now that says **Create subscription**. On that page, find the **Protocol** dropdown, and select **Email**. Enter the email you wish to receive the notifications where it says **Endpoint**. Then again, go ahead and leave the rest blank and select **Create subscription**.

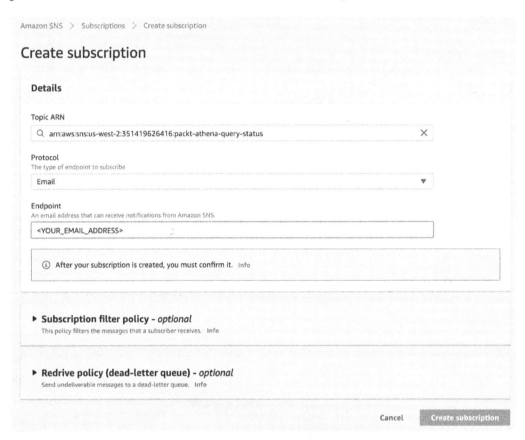

Figure 10.4 – Creating an email subscription for your SNS topic

Before you can move on, go and check your email. You should receive an email that looks like the following. Assuming everything lines up (this is, in fact, the SNS topic you created), go ahead and click the **Confirm subscription** link.

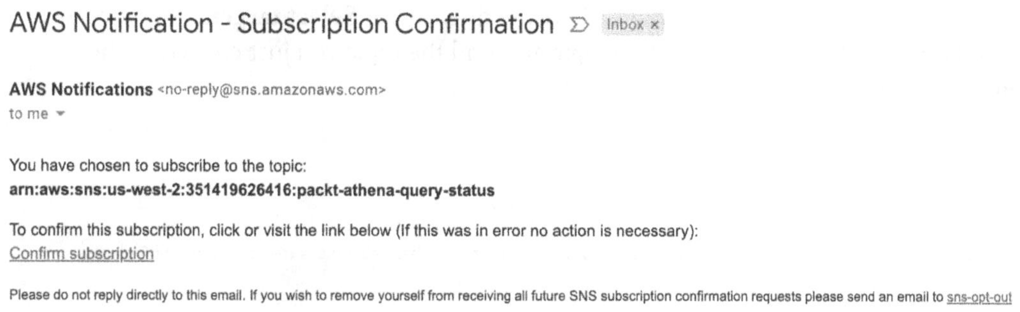

AWS Notification - Subscription Confirmation ∑ Inbox ×

AWS Notifications <no-reply@sns.amazonaws.com>
to me ▾

You have chosen to subscribe to the topic:
arn:aws:sns:us-west-2:351419626416:packt-athena-query-status

To confirm this subscription, click or visit the link below (If this was in error no action is necessary):
Confirm subscription

Please do not reply directly to this email. If you wish to remove yourself from receiving all future SNS subscription confirmation requests please send an email to sns-opt-out

Figure 10.5 – Sample subscription confirmation email

Step 2 – Setting up the EventBridge rule

Head on over to the EventBridge console and find **Events** > **Rules** in the navigation bar. Then, find and click the **Create rule** button. On the **Create** page, give a descriptive name; I called it `packt-athena-emailer`. For the pattern, select **Event pattern** > **Pre-defined pattern by service**. The provider should be **AWS** > **Athena**, and the event type should be **Athena Query State Change**. It should look like the following:

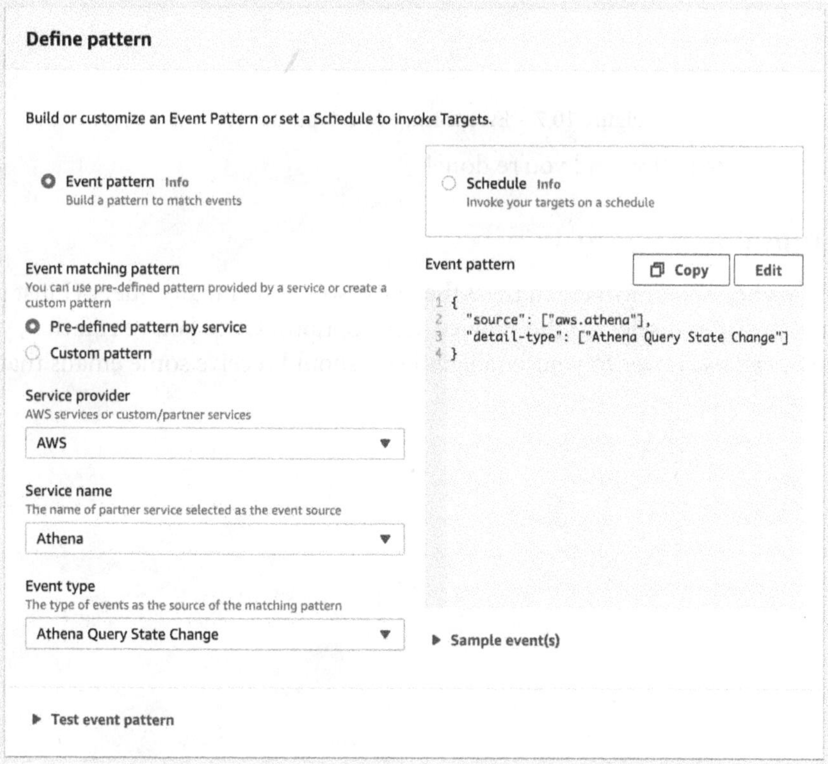

Figure 10.6 – EventBridge Athena query state change event pattern

Skip the event bus section; the defaults there are fine. Under **Select targets**, where it says **Lambda function**, change that to SNS **topic** and find the topic you just created in the previous section.

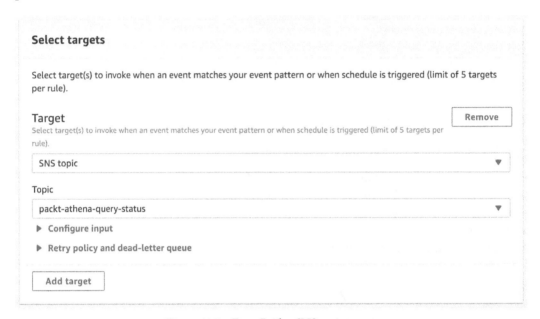

Figure 10.7 – EventBridge SNS topic target

Now, click the **Create** button and you're done!

Step 3 – Running a query

Finally, let's go ahead and navigate to the Athena console and run any query (I just picked a table in my catalog and selected the Preview table option). Once you've run that, you can go ahead and head back over to your email and you should receive some emails that look like the following:

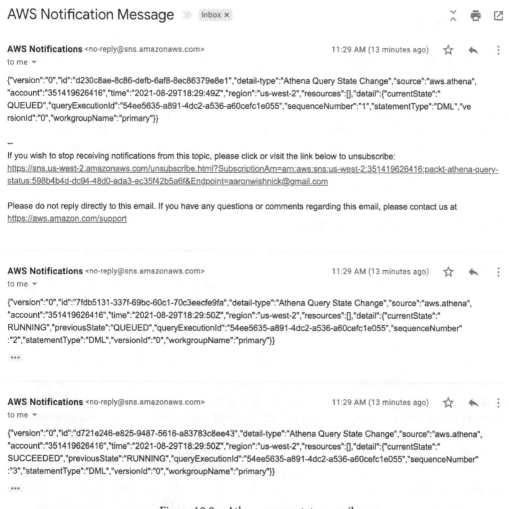

AWS Notification Message Inbox ×

AWS Notifications <no-reply@sns.amazonaws.com>   11:29 AM (13 minutes ago)
to me ▾

{"version":"0","id":"d230c8ae-8c86-defb-6af8-8ec86379e8e1","detail-type":"Athena Query State Change","source":"aws.athena",
"account":"351419626416","time":"2021-08-29T18:29:49Z","region":"us-west-2","resources":[],"detail":{"currentState":"
QUEUED","queryExecutionId":"54ee5635-a891-4dc2-a536-a60cefc1e055","sequenceNumber":"1","statementType":"DML","ve
rsionId":"0","workgroupName":"primary"}}

--

If you wish to stop receiving notifications from this topic, please click or visit the link below to unsubscribe:
https://sns.us-west-2.amazonaws.com/unsubscribe.html?SubscriptionArn=arn:aws:sns:us-west-2:351419626416:packt-athena-query-
status:598b4b4d-dc94-48d0-ada3-ec35f42b5a6f&Endpoint=aaronwishnick@gmail.com

Please do not reply directly to this email. If you have any questions or comments regarding this email, please contact us at
https://aws.amazon.com/support

AWS Notifications <no-reply@sns.amazonaws.com>   11:29 AM (13 minutes ago)
to me ▾

{"version":"0","id":"7fdb5131-337f-69bc-60c1-70c3eecfe9fa","detail-type":"Athena Query State Change","source":"aws.athena",
"account":"351419626416","time":"2021-08-29T18:29:50Z","region":"us-west-2","resources":[],"detail":{"currentState":"
RUNNING","previousState":"QUEUED","queryExecutionId":"54ee5635-a891-4dc2-a536-a60cefc1e055","sequenceNumber"
:"2","statementType":"DML","versionId":"0","workgroupName":"primary"}}

AWS Notifications <no-reply@sns.amazonaws.com>   11:29 AM (13 minutes ago)
to me ▾

{"version":"0","id":"d721e246-e825-9487-5616-a83783c8ee43","detail-type":"Athena Query State Change","source":"aws.athena",
"account":"351419626416","time":"2021-08-29T18:29:50Z","region":"us-west-2","resources":[],"detail":{"currentState":"
SUCCEEDED","previousState":"RUNNING","queryExecutionId":"54ee5635-a891-4dc2-a536-a60cefc1e055","sequenceNumber"
:"3","statementType":"DML","versionId":"0","workgroupName":"primary"}}

Figure 10.8 – Athena query status emails

If you take a look, you will see we received three notifications for the status of our query. First, the execution went into the **QUEUED** state, then **RUNNING**, and finally **SUCCEEDED**. Pretty neat, huh?!

Emails are great and all, but you will probably want your automated system to be able to react to these events. You probably noticed already when you were setting up the rule, but if you didn't, EventBridge has a huge selection of possible targets that you can configure for a rule, so there's a really good chance that there is a target option that will fit nicely into your application. Take a look at the following URL, `https://amzn.to/2UXrj4x`, for the full list of targets.

Securing your application

In the previous section, we talked about some best practices when it comes to connecting to and calling Athena. In this section, we're going to touch a little bit more on that point, but with a focus on security, and then focus on some other mechanisms for using Athena in the safest way possible. In *Chapter 5, Securing Your Data*, we discussed the concepts of the attack surface and blast radius, two metrics by which you can measure how safe your application is, both in terms of preventing a bad actor from gaining access and then minimizing the impact in the event that they do gain access. Some of the stuff we are going to cover is not necessarily specific to an Athena-based application, but it is still very valuable information to keep in mind.

Credential management

Firstly, we're going to take a look at credentials, the entry point for secure communication with AWS. We're going to focus on two specific aspects of it – life cycle management and the distribution/persistence of credentials. Also, whether you are running your application within some standard AWS offerings versus the alternative makes a big difference here, so we'll discuss each separately.

If you are running your application on AWS compute options, such as EC2, ECS, Lambda, and others, then the problem of distributing credentials can be rather simple. In these cases, credentials are distributed to the hosts by way of metadata services (or in the case of Lambda, it is simply the credentials that are being used to execute the function). What this means is that, assuming the credentials distributed to the host are the ones that you need, you can simply rely on the default credentials provider within the AWS Client, and it will know to look for the metadata service. However, if you require credentials that are different from the ones that are distributed directly to the host, then you have a couple of options. The first is the case where you need IAM user credentials, and we're going to cover that in the next section when we discuss on-premises (out of AWS) applications. The second option is that you use IAM roles, which is the recommended approach in any case. In these cases, the credentials that are automatically distributed need the `iam:AssumeRole` permission on whatever role you want to actually assume. Then you can call the STS service to retrieve temporary credentials for that role and then instantiate the desired client with those credentials. In Java, this looks like the following:

```
AWSSecurityTokenService stsClient =
                AWSSecurityTokenServiceClientBuilder.
standard()
                                          .build();
AssumeRoleRequest roleRequest = new AssumeRoleRequest()
                        .withRoleArn(roleARN)
```

```
.withRoleSessionName(roleSessionName);
AssumeRoleResult roleResponse = stsClient.
assumeRole(roleRequest);
Credentials sessionCredentials = roleResponse.getCredentials();

BasicSessionCredentials awsCredentials = new
BasicSessionCredentials(
                sessionCredentials.getAccessKeyId(),
                sessionCredentials.getSecretAccessKey(),
                sessionCredentials.getSessionToken());

AthenaClient athenaClient = AthenaClient.builder()
    .withCredentials(new
AWSStaticCredentialsProvider(awsCredentials))
    .build();
```

Roles make it such that you don't have to concern yourself with credential rotation at all; they are temporary credentials that you can just get new ones of whenever you need. The benefit of this is that if somehow role credentials get leaked, a bad actor will only be able to use them for a short period of time, thereby reducing the blast radius.

If you are in a situation where there is no automatic distribution of credentials handled by AWS, such as in an on-premises solution, then it is, of course, your responsibility to solve that. Many organizations in these cases end up building their own solutions, often referred to as **credential stores**. Credential stores are far preferable to the other option, where you store credentials on disk, such as in the AWS ~/.aws/credentials file, or even worse, in your code repository, the reason being, again, because we want to reduce the blast radius in the event a bad actor gains access to the host. If the credentials are stored elsewhere, then the actor will not necessarily be able to access them, but if they are on the disk, then now the actor has access to whatever resources those credentials have access to. Since these credential stores are often very custom and involve significant integration with whatever enterprise authentication mechanism that is being used, we're going to focus more on what to consider when following this approach. You still want to utilize roles as much as possible.

The credentials you are managing should primarily be utilized to access those roles so that in the event the credentials are leaked, you can quickly revoke those permissions, and any future attempts to use the role will fail. This is called **least privilege**, the idea being that any actor within a system has exactly the permissions it requires to perform its duty and no more, which is with the aim, again, of reducing the blast radius.

The other key consideration is automatic credential rotation. You should ensure that credentials get rotated so that in the event any credentials are leaked, they cannot be used indefinitely. By default, IAM will not rotate your credentials, which means that they will live on forever. IAM has a very helpful pattern for setting an automatic rotation system that you can use or at least reference here: `https://amzn.to/3gRDHLk`. In general, the system is a three-step process:

1. Generate new keys.

2. A short amount of time later, deactivate the old keys.

3. A short amount of time later again, delete the deactivated keys.

The idea here is that after *step 1*, the system should pick up the new keys. After *step 2*, any systems still reliant on the old keys will start to fail, but if need be, you can reactivate until you switch, and then, in *step 3*, you completely eliminate the old keys once you've confirmed it's OK.

At the end of the day, doing your absolute best to keep credentials secret is, of course, the primary goal. But the other tools we've discussed are invaluable in reducing the blast radius in the event that credentials are in the hands of someone who shouldn't have them.

Network safety

The next thing we're going to focus on is ensuring that the communication between your application and Athena is as protected as possible. By default, all communications between the AWS SDK and an AWS service are encrypted via HTTPS and signed using AWS's Signature Version 4 signing process (more info on that here: `https://amzn.to/3DvAg6I`). These mechanisms do an excellent job of ensuring that any message sent to AWS is tamper-proof. So, this does a good job of minimizing your attack surface; however, if you haven't properly configured your network, then you are still requiring access to the **public internet** to communicate with AWS since, by default, all communication goes to AWS's public endpoint. The public internet, in this case, refers to anything accessible to anyone via the internet without requiring any additional network configuration (note: *accessible* means that the IP address will resolve, not that they have the necessary credentials). So, the implications of that are that there is a larger potential blast radius (for example, the bad actor, having gained access to your hardware, could call out to the public internet to retrieve a nasty script that they've already prepared in advance for wreaking havoc). I would guess by now you are seeing a pattern; AWS has an answer for this problem as well!

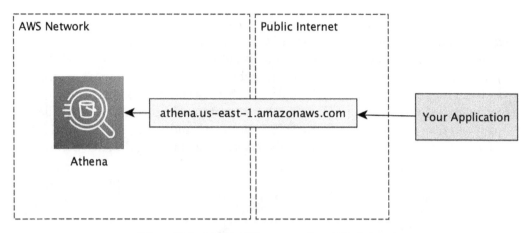

Figure 10.9 – Calling Athena over the public internet

VPCs, or **Virtual Private Clouds,** provide you with the ability to create isolated networks and are consistently one of the most recommended security features within AWS. VPCs enable fine-grain control over network traffic in and out of them and also within the VPC itself. There is much, much more to VPCs, but that sufficiently covers what we need to worry about here.

So, great, we can configure rules to allow traffic in and out of our private network. But we still have to communicate with AWS, so we still need access to the public internet to talk to AWS's public endpoint. This means that our VPC rules must allow for that traffic out of our network. Or… do we? (Hint: we don't.)

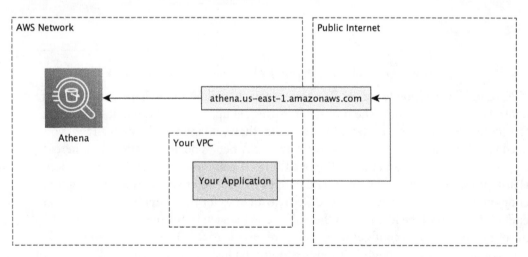

Figure 10.10 – Calling Athena over the public internet from inside your own VPC

AWS PrivateLink and **VPC endpoints** exist to solve this exact use case. A VPC endpoint is a resource you can provision inside your VPC, which can be communicated with by way of a private IP address, meaning that the IP address exists only in your VPC. A private IP address is explicitly separate from the public internet, meaning anyone outside of the VPC, if they tried to access that IP address, either it would exist on the public internet, pointing to a completely different resource, or it would simply not resolve. The VPC endpoint then routes your traffic to AWS PrivateLink. And finally, AWS PrivateLink allows for direct communication with an AWS service without leaving the AWS network!

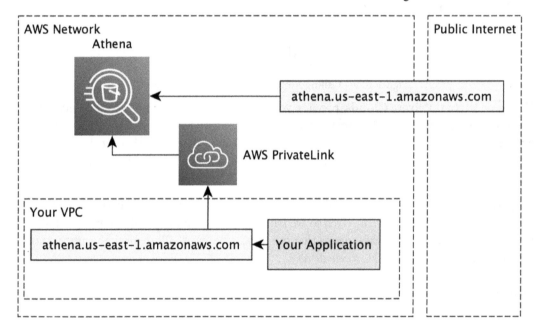

Figure 10.11 – Calling Athena using a VPC endpoint and PrivateLInk

VPCs and VPC endpoints are very powerful tools that allow you to have much finer-grain control over your network. The process of getting set up, while not difficult, requires a few more steps than we have time for. So, rather than walking you through all that you need to, I'm going to suggest you head on over to Athena's documentation on it here: https://amzn.to/3aifWrE. The one thing I'll point to from that documentation is just how easy it is once your VPC and endpoint are set up to actually start using the endpoint. In fact, if you enable private DNS hostnames for the endpoint you create, the endpoint to call will be identical to the public one (as seen in the preceding diagrams), meaning no additional configuration will be required. Your SDK will just automatically start communicating through PrivateLink to Athena instead of over the public endpoint.

Optimizing for performance and cost

Switching gears one last time, let's orient ourselves to optimizing our use of Athena. Again, remember that what is optimal differs depending on what your greater concerns are; either reducing your overall dollar costs or having the fastest possible experience, that's going to depend on your priorities. By the end of this section, you'll have a good starting point for achieving them.

Workload isolation

In *Chapter 3, Key Features, Query Types, and Functions*, we covered **workgroups** and how to leverage them to isolate workloads. Just to reiterate, workgroups allow you to splice up your Athena usage in such a way that you can specify who gets access to which data and how much of that data they can access through the WorkGroup resource and resource policies. Another huge benefit of workgroups is that you can visualize aggregated query metrics at the workgroup level. The way you can do this is when you create a workgroup, you make sure to enable publishing query metrics to AWS CloudWatch (see *Figure 10.17*). Note that this is disabled by default because there are additional charges associated with using CloudWatch.

Metrics

Metrics   ☑ Publish query metrics to AWS CloudWatch ⓘ

Figure 10.12 – Workgroup CloudWatch query metrics option

Once you've enabled that, you should be able to head over to the workgroup and you can view the **Metrics** tab on your workgroup page, and you'll see some pretty handy metrics there! Now, if you are thoughtful in terms of how you break data up into different workgroups, you can leverage these metrics to determine which workgroups (and therefore which sets of data) are resulting in the worst performance and highest costs (in other words, most data reads). We'll take a closer look at how best to think about this soon, but for now, just observe and enjoy.

That relates to monitoring WorkGroup usage, but what about optimizing your workloads? I know I'm being super repetitive, but this is an important thing to keep in mind; what are you trying to optimize? WorkGroups have a nice feature for that where you can specify how much data a given WorkGroup can process, either in a given query or in a given time period. Remember that data processed is the metric utilized by Athena for billing. So, this feature allows you to tune what individual workgroups will cost you over time, so that's what we're optimizing for in this case.

And again, on the other hand, we have performance optimizations. As of the time of writing this book, in most regions, Athena allows you to have 20 active DML queries at a time. An attempt at running a query beyond that will result in a "too many queries" error. So, let's say you have two workloads, one is fast and frequent, while the other is slow and infrequent. And let's say that this fast and frequent workload is pretty consistently at or near 20 active DML queries at a time. What do we think is going to happen with that slow and infrequent query? Well, either it's going to frequently hit the "too many queries" error, or it's going to occupy an active query slot for the fast query for a long period of time, impacting the amount that can be executed in parallel with the fast one. The issue here is that these two workloads have very different scaling vectors, and none of that is based on the amount of data processed, and so WorkGroups won't really help in this case, since these limits are at the account level. So now it's time to look elsewhere for a solution. In these cases, it's a good idea to start considering branching out workloads into separate AWS accounts that can scale completely independently of one another. **AWS Organizations** is an excellent offering that makes it really easy to create AWS account trees where billing is all aggregated at the root of the tree, but the accounts still effectively act as independent entities.

Application monitoring

By now, you've split out your various workloads into separate WorkGroups and maybe even separate AWS accounts. But software and its use is a living thing; it's forever growing and, just as importantly, changing. So, monitoring the status of your application is extremely important.

By default, Athena logs all API calls plus the associated request parameters to **AWS CloudTrail**. AWS CloudTrail is a service aimed at empowering customers to audit all actions that are taken within their accounts. Actions, in this case, are API calls made against all services that log to CloudTrail (which should be most, if not all, of them). The data logged by Athena within CloudTrail includes the request parameters, such as the query string, and other valuable data such as the caller.

To get started with diving into your CloudTrail usage, we're actually going to use Athena to gain insights into our Athena usage. To get started, if you haven't done this already, you need to go to the CloudTrail console and create a **trail**. On the first page of the trail creation, give the trail a name, something descriptive. For your encryption settings, if you plan on using this in a production environment, you will want to turn this on, but keep in mind that KMS has costs associated with it. Each **customer-managed key** (**CMK**) is $1/month and then you pay based on your usage of the key as well (there is a free tier for this part). If you are just doing this for testing purposes, it's your decision whether or not you want to turn that on. Just don't forget what you decide if you end up continuing to use it (or not).

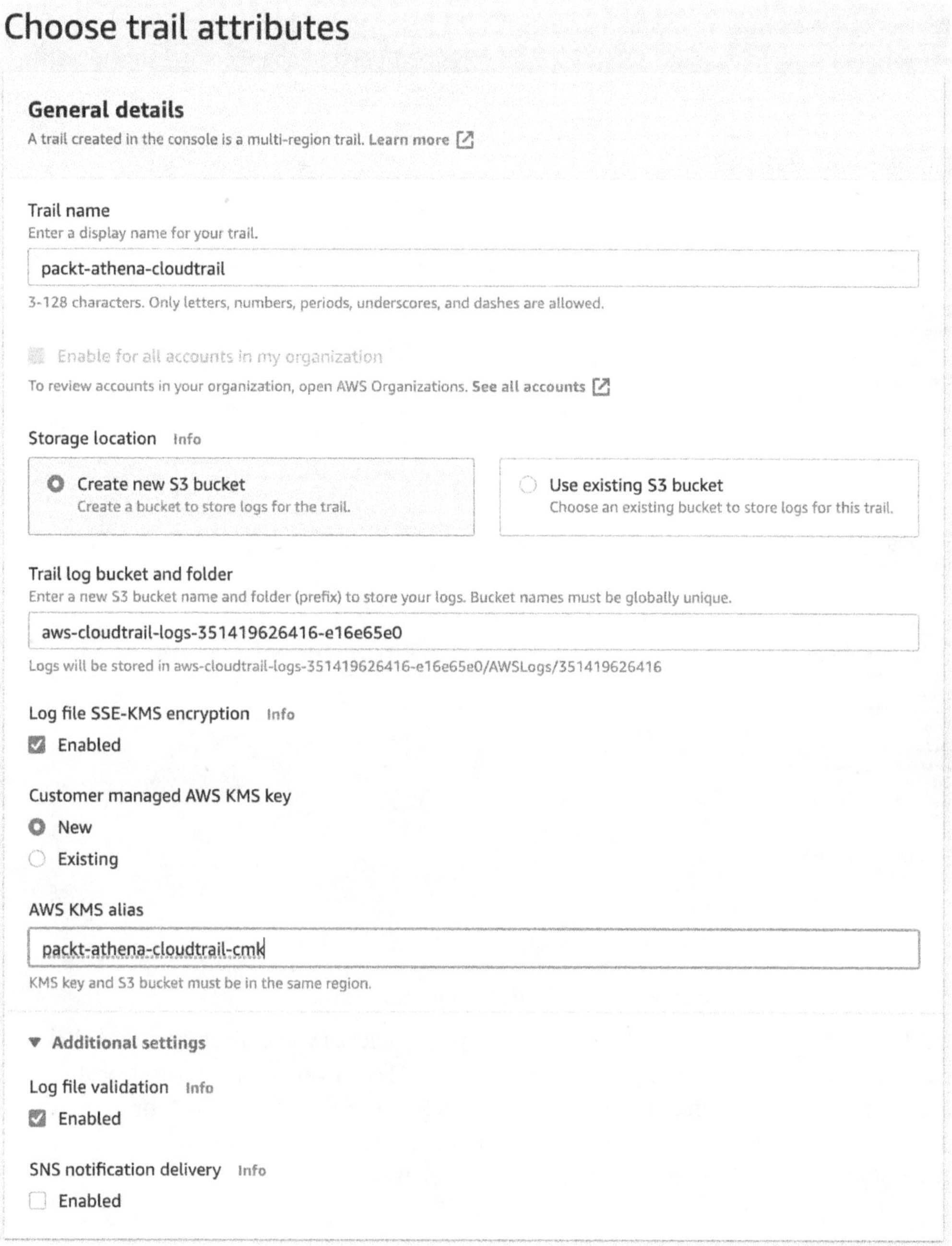

Choose trail attributes

General details

A trail created in the console is a multi-region trail. Learn more [↗]

Trail name

Enter a display name for your trail.

```
packt-athena-cloudtrail
```

3-128 characters. Only letters, numbers, periods, underscores, and dashes are allowed.

☐ Enable for all accounts in my organization

To review accounts in your organization, open AWS Organizations. See all accounts [↗]

Storage location   Info

◉ **Create new S3 bucket**
Create a bucket to store logs for the trail.

○ **Use existing S3 bucket**
Choose an existing bucket to store logs for this trail.

Trail log bucket and folder

Enter a new S3 bucket name and folder (prefix) to store your logs. Bucket names must be globally unique.

```
aws-cloudtrail-logs-351419626416-e16e65e0
```

Logs will be stored in aws-cloudtrail-logs-351419626416-e16e65e0/AWSLogs/351419626416

Log file SSE-KMS encryption   Info

☑ Enabled

Customer managed AWS KMS key

◉ New

○ Existing

AWS KMS alias

```
packt-athena-cloudtrail-cmk
```

KMS key and S3 bucket must be in the same region.

▼ **Additional settings**

Log file validation   Info

☑ Enabled

SNS notification delivery   Info

☐ Enabled

Figure 10.13 – CloudTrail trail attributes

Note the bucket name being used for the trail. The rest you can leave unmodified, and now move on to the next page. On this page, you don't need to change anything; all of the defaults apply. The management event type just refers to general AWS API calls.

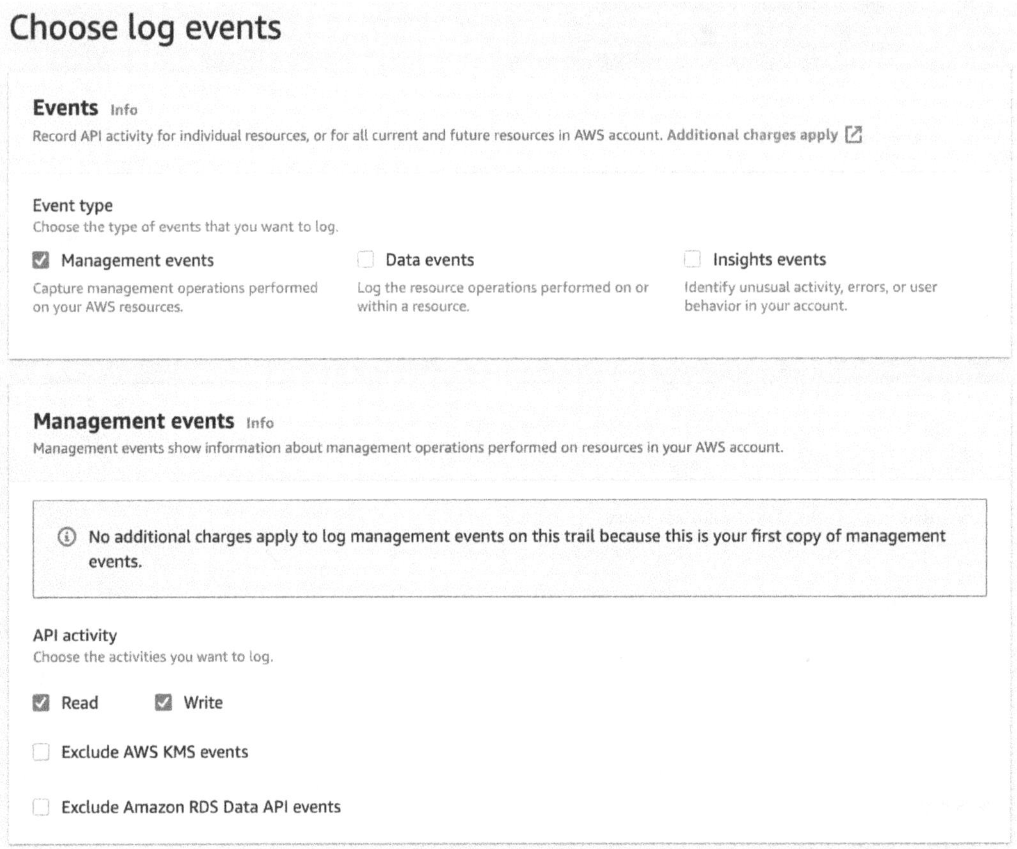

Choose log events

Events Info

Record API activity for individual resources, or for all current and future resources in AWS account. Additional charges apply

Event type
Choose the type of events that you want to log.

☑ **Management events**
Capture management operations performed on your AWS resources.

☐ **Data events**
Log the resource operations performed on or within a resource.

☐ **Insights events**
Identify unusual activity, errors, or user behavior in your account.

Management events Info

Management events show information about management operations performed on resources in your AWS account.

ⓘ No additional charges apply to log management events on this trail because this is your first copy of management events.

API activity
Choose the activities you want to log.

☑ Read   ☑ Write

☐ Exclude AWS KMS events

☐ Exclude Amazon RDS Data API events

Figure 10.14 – CloudTrail log events selection

Now, click **Next** again, verify that the summary page looks as you expect it to, and click **Create trail** at the bottom. Navigate over to the CloudTrail Event history page, locate the button that says **Create Athena table**, and then click that. Select the bucket corresponding to the trail we just created and then scroll down and click **Create table**. And now you're ready to start gaining insights from your CloudTrail events!

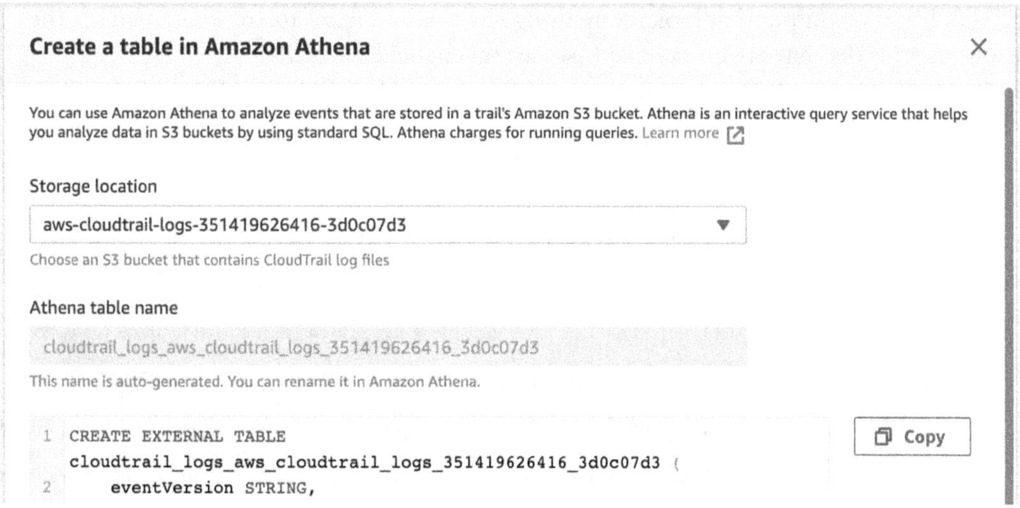

Create a table in Amazon Athena   ✕

You can use Amazon Athena to analyze events that are stored in a trail's Amazon S3 bucket. Athena is an interactive query service that helps you analyze data in S3 buckets by using standard SQL. Athena charges for running queries. Learn more ↗

Storage location

aws-cloudtrail-logs-351419626416-3d0c07d3   ▼

Choose an S3 bucket that contains CloudTrail log files

Athena table name

cloudtrail_logs_aws_cloudtrail_logs_351419626416_3d0c07d3

This name is auto-generated. You can rename it in Amazon Athena.

```
1  CREATE EXTERNAL TABLE                                  □ Copy
   cloudtrail_logs_aws_cloudtrail_logs_351419626416_3d0c07d3 (
2      eventVersion STRING,
```

Figure 10.15 – CloudTrail Athena table creation dialog

Move over to Athena and switch to the default data catalog and you should see your newly created table there. Go ahead and preview it and take a quick look at what the data looks like. Three columns worthy of highlighting that we are particularly interested in right now are eventsource, eventname, and requestparameters. If you look at some samples of these, you'll see that eventsource corresponds to the service or caller that triggered the particular event, eventname is the API that was called, and requestparameters contains the values provided for that API call in the form of a JSON object (on many occasions, services will redact sensitive fields). So now let's try to derive some more useful information from here. Try running the following query. (If you just set up CloudTrail, you'll want to run a few random queries first before you run this one, otherwise you will get no results):

```
SELECT json_extract(requestparameters, '$.queryString') AS
queryString
FROM "default"."<CLOUD_TRAIL_TABLE_NAME>"
WHERE eventsource = 'athena.amazonaws.com' and
       eventname = 'StartQueryExecution'
```

You should get an output that looks something like that of *Figure 10.16*, where you see the various queries that have been executed (since you enabled the trail):

Results

▲	queryString ▽
1	"SELECT * FROM \"packt_serverless_analytics\".\"chapter_7_counties\" limit 10;"
2	"SELECT * FROM \"packt_serverless_analytics\".\"taxi_ridership_data\" limit 10;"
3	"CREATE EXTERNAL TABLE cloudtrail_logs_aws_cloudtrail_logs_351419626416_3d0c07d3 (\n event
4	"SELECT * FROM \"default\".\"cloudtrail_logs_aws_cloudtrail_logs_351419626416_3d0c07d3\" limit 10;
5	"SELECT * FROM \"default\".\"cloudtrail_logs_aws_cloudtrail_logs_351419626416_3d0c07d3\" limit 10;
6	"SELECT * FROM \"default\".\"cloudtrail_logs_aws_cloudtrail_logs_351419626416_3d0c07d3\" WHER
7	"SELECT * FROM \"default\".\"cloudtrail_logs_aws_cloudtrail_logs_351419626416_3d0c07d3\";"

Figure 10.16 – Query strings from the CloudTrail Athena table

Now, at this point, you might be thinking that this is a difficult dataset to analyze, and you'd be right! Especially if you imagine that you've expanded your use of Athena to a massive amount. This is why it's so important that you use all of the various things we've discussed here together.

Let's say we've got a data warehouse for our coffee shop from earlier in the chapter. This warehouse contains data on transactions that have occurred over the past year and also data on what we have in our stockroom. For the stockroom, to begin with, we really only care about what is in there at any given time. So, we create a workgroup for checking that information. Essentially, our data is a daily snapshot of the items in stock. Our other workgroup contains the transaction data that we've got nicely partitioned by month and contains information about all the transactions that occurred each day. One day, one of our data analysts (yes this is a tech-coffee shop) runs a query to try and correlate transactions that are occurring and how they relate to how frequently perishable stock is being returned to determine what we need to buy less of. This ad hoc query turns out to be super useful, so it gets added as a regular job that gets run. But no one told the data engineer! Over time, the data engineer is checking on the metrics of the workgroup and notices that the performance for the stockroom workgroup has degraded significantly over time. The data engineer decides to query the CloudTrail logs for the table in that workgroup and notices a large number of queries that are running over a range of time instead of just a single day (the latest day), and because the table is not partitioned, it requires the entire table to be scanned. They now determine that this is a valuable dataset to have and create a new table that adds month-based partitioning on the stockroom so that it aligns with the transactions table.

I hope that, with the help of my silly little coffee shop example, you can see the power of combining all of these monitoring tools to ensure that you are always operating in the most optimal manner.

CTAS for large result sets

The last topic we are going to briefly discuss is not so much a best practice but just a nice trick to have in your back pocket in case you ever need it. Sometimes, you have queries that you run that produce very large numbers of results. As usual, you call GetQueryResults to get them and notice that you are spending a really long time on this part. The reason for this is that Athena stores all results in a single CSV file. And so GetQueryResults is, in turn, slowly reading through that line by line. In *Chapter 3, Key Features, Query Types, and Functions*, we learned about the CTAS (Create Table as Select) clause, which allows you to run a select query and rather than return the results to you directly, it puts those results into a new table in your catalog. So, one option to consider instead of reading through your large numbers of results in a single thread is to temporarily store the results in a separate table using CTAS and partition that table in such a way whereby you can leverage parallel reads by reading different partitions at the same time!

Summary

In this chapter, we covered a really broad array of topics, all focused on giving you the right concepts to consider when building an application that leverages Athena (though many topics would benefit you no matter what you are building).

We discussed your different options for connecting to Athena and how to decide which one is right for you, whether it is using the AWS SDK, the JDBC driver, or the ODBC driver – deciding between the convenience of implementation, especially if you are already familiar with the JDBC/ODBC frameworks, versus the flexibility of having direct access to the SDK.

Then we continued the discussion of connecting to Athena, but with a focus on best practices. Firstly, we covered making sure you are leveraging idempotency tokens (in Athena's case, ClientRequestTokens) to make sure you are safely retrying on unclear failures, which is a feature you get for free with the SDK! And then we looked at how best to track the status of queries, moving away from the standard model of polling GetQueryExecution until the query completes, and instead utilizing the push model by working with AWS EventBridge.

Next, we looked at being secure! We discussed how best to manage credentials, particularly when your application is not running with an AWS environment, and then, when you are in an AWS environment, how best to manage your network traffic to and from your application by leveraging VPCs and VPC endpoints.

Finally, we took a look at the various options you have for optimizing your application, whether it be for minimizing cost or maximizing performance. In this section, we reiterated from *Chapter 3, Key Features, Query Types, and Functions*, the idea of leveraging WorkGroups as a mechanism to isolate workloads both from an access and cost perspective. We also looked at how you can leverage WorkGroup-aggregated CloudWatch metrics for analyzing the overall performance and cost of workloads. Then, we saw where WorkGroups may not be able to help, which is when you have workloads with significantly different scaling vectors that you don't want to impact one another, and in that case, we recommended that you consider separating those into different AWS accounts under a single AWS organization. Continuing with the theme of monitoring, we discussed how you can leverage AWS CloudTrail in addition to well-defined workloads by WorkGroup to discover common access patterns that need to be optimized. Finally, we took a look at a trick you can do to speed up queries with very large result sets by leveraging CTAS to take advantage of the multi-file upload capability of CTAS.

Of course, there is so much to consider when building an application and we've only scratched the surface, but these topics should take you a long way by creating a solid foundation from which to get started. In the next chapter, we will check out operational excellence, in other words, how to monitor and optimize Athena for various uses.

11

Operational Excellence – Monitoring, Optimization, and Troubleshooting

In this chapter, we will focus on operational excellence. Operational excellence in this chapter has three components: monitoring Athena to ensure it is healthy and running normally, optimizing our usage of the system for cost and performance, and, lastly, how to troubleshoot issues when they occur.

When monitoring systems, it is essential to know what to monitor and what steps to take when something goes wrong. This information is valuable because when the system is not operating correctly, the data will give you clues on possible issues, which reduces investigation time. You can also act before problems occur, preventing calls from users on why things are not working. We will look into processes that can be put in place to ensure that Athena and our usage of it are normal and efficient. When there are issues, we will know how to fix common problems.

We also want to get the most out of Athena. To run optimally and cost-effectively, we will optimize our use of Athena by going through best practices on how to store our datasets and best write queries. Following these best practices can significantly reduce your monthly bills and keep your users happy, with low query times.

Lastly, we will look at how we can troubleshoot failing queries. We will dive deep into the most common problems users encounter, what they mean, and how to address them.

In this chapter, we will learn about the following:

- How to monitor Athena to ensure queries run smoothly
- How to optimize for cost and performance
- How to troubleshoot failing queries

Technical requirements

For this chapter, if you wish to follow some of the walk-throughs, you will require the following:

- Internet access to GitHub, S3, and the AWS Management Console.
- A computer with a Chrome, Safari, or Microsoft Edge browser installed.
- An AWS account and accompanying IAM user (or role) with sufficient privileges to complete this chapter's activities. For simplicity, you can always run through these exercises with a user that has full access. However, we recommend using scoped-down IAM policies to avoid making costly mistakes and to learn how to best use IAM to secure your applications and data. You can find a minimally scoped IAM policy for this chapter in the book's accompanying GitHub repository, listed as `chapter_11/iam_policy_chapter_11.json` (https://bit. ly/3hgOdfG). This policy includes the following:

  - Permissions to read, list, and write access to an S3 bucket
  - Permissions to read and write access to the AWS Glue Data Catalog databases, tables, and partitions:

- You will be creating databases, tables, and partitions manually and with Glue crawlers.

- Access to run Athena queries

Monitoring Athena to ensure queries run smoothly

Monitoring your usage of Athena is essential to ensuring that your users' queries continue to run uninterrupted without issues. Many issues can be addressed before they impact users and applications. This section will look into the metrics that Athena emits to CloudWatch Metrics and the metrics that should be monitored and alarmed on, so that actions can be taken before users reach out to their administrators. Before we do, let's take a look at which metrics are emitted by Athena.

Query metrics emitted by Athena

Athena emits query-level metrics for customers to be able to monitor and alarm on. These metrics exist in CloudWatch Metrics under the namespace "AWS/Athena" and three dimensions, *QueryType*, *QueryState*, and the *Workgroup* name. *QueryType* can be DML (INSERT/SELECT queries) or DDL (metadata queries such as CREATE TABLE). *QueryState* can be SUCCEEDED, FAILED, QUEUED, RUNNING, or CANCELED. The *Workgroup* dimension aggregates metrics within the Athena workgroup that the query executed in.

The metrics that are emitted are listed here:

- **TotalExecutionTime** – in milliseconds. The entire execution time of the query from when the query is accepted by Athena to when it reaches its final state (SUCCEEDED, FAILED, or CANCELED).

- **QueryQueueTime** – in milliseconds. This is the time a query spent waiting for resources to run on. This measures the time after Athena has accepted a query for execution and before it is sent to the execution engine for execution.

- **EngineExecutionTime** – in milliseconds. This is the amount of time taken when the query is received by the execution engine to when it completes executing it. This metric includes the *QueryPlanningTime* metric. This applies to both DML and DDL queries.

- **QueryPlanningTime** – in milliseconds. This is the amount of time the execution engine (that is, PrestoDB) took to parse the query and create its execution plan. This includes operations such as retrieving partition information from the metastore, optimizing the execution plan, and so on. This applies to DML queries only.

- **ServiceProcessingTime** – in milliseconds. This is the time from when the query has finished in the execution engine (that is, PrestoDB) and the time Athena uses to read the results and push them to S3. This applies to both DML and DDL queries.

- **ProcessedBytes** – in megabytes. This is the amount of data that the execution engine processed for DML (that is, `SELECT`) queries. This can be used as an approximation for billing.

With these metrics, we can build dashboards and alarms. The process to create the alarms will be included in the following sections.

Monitoring query queue time

To protect available resources for all customers, Athena allows a certain number of queries to be run at any given time from a single AWS account. When a query is submitted for execution, Athena will check how many queries the submitting account is executing. If it exceeds the account limit, or if there are not enough resources, say at peak times during the day, then the query will be queued until both of those conditions are met. When clients submit their queries and notice that their queries are not running or taking a significant time to run even a small query, it is likely because they are queued.

Monitoring and actioning when Athena queue time occurs is essential to prevent users' queries from being constantly queued. Since queue time metrics are emitted to CloudWatch metrics, alarms can be created and actioned against. We have included a sample script that can be used as templates to monitor and email when thresholds are exceeded, which can be found in this chapter's GitHub location at `https://bit.ly/3j7Nzly`. The script can be adjusted for your use case. It will create four alarms, two for DML queries and two for DDL queries. Each set will generate a warning threshold and one for when production impact occurs. The split between DML and DDL queries is due to the query types having their own queues, and one will not impact the other.

Once the alarms are created, then the CloudWatch alarms dashboard may look like the following.

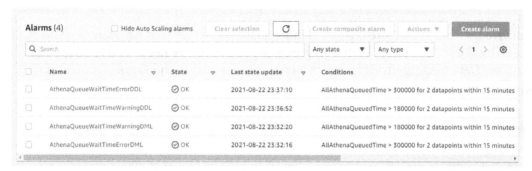

Figure 11.1 – A sample alarm dashboard in Amazon CloudWatch for alarms

When users' queries start to get significantly queued, three actions can be taken. First, reduce the frequency that queries are submitted or spread them out throughout the day. This can be done by asking users or by disabling low-priority users. This is not an ideal solution but can be a short-term solution to prioritize applications or users that need to execute queries. The second action that can be taken is to submit a support case and ask for the AWS account to increase concurrent query execution for your AWS account. This generally happens automatically but is not done when there is a sudden increase in sustained usage. Requests to increase query concurrency need to be considered by AWS and may be approved or disapproved.

The last solution is to split queries among AWS accounts. This approach has the additional benefit of isolating applications and users from impacting each other. It is best practice to use an AWS account for SLA-sensitive applications and users, and other ad hoc queries from users doing data exploration in another account.

Let's now look at monitoring Athena's costs.

Monitoring and controlling Athena costs

No one wants to be emailed or especially called because their team's Athena costs are significantly higher than initially expected. Unexpected increases in costs can be due to a single user running a query without a partition filter that unintentionally scans terabytes of data or an application with a bug that makes an unexpectedly high number of calls. There are mechanisms that Athena provides to prevent these scenarios when using Athena workgroups.

To prevent the scenario of a single user running every large and expensive query, workgroups can be configured to cancel individual queries that exceed a quantity of data read. This is configurable for each workgroup, depending on the need. To set this feature, go to the **Data usage controls** tab when editing a workgroup in the Athena console, as shown in the following figure:

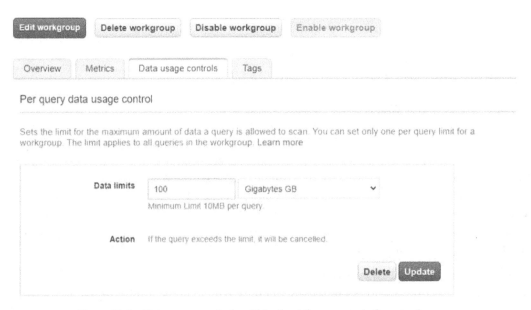

Figure 11.2 – Data usage controls within the Athena console for a workgroup

Within this tab, you can set data usage limits at a query level. By selecting this limit, Athena will cancel queries that exceed the query limit. It is recommended that this limit be configured to a value that prevents legitimate queries from being interrupted but is low enough to identify issues.

In addition to setting limits at a query level, you can also set **ProcessedBytes** limits at a workgroup level. CloudWatch alarms can be created in CloudWatch or through the Athena console in the workgroups tab, as shown in the following figure:

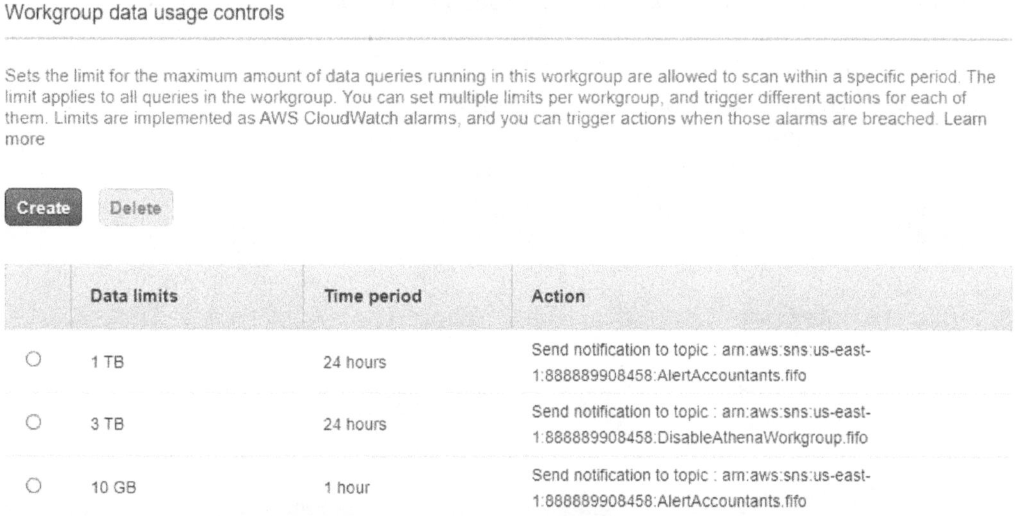

Workgroup data usage controls

Sets the limit for the maximum amount of data queries running in this workgroup are allowed to scan within a specific period. The limit applies to all queries in the workgroup. You can set multiple limits per workgroup, and trigger different actions for each of them. Limits are implemented as AWS CloudWatch alarms, and you can trigger actions when those alarms are breached. Learn more

Create   Delete

	Data limits	Time period	Action
○	1 TB	24 hours	Send notification to topic : arn:aws:sns:us-east-1:888889908458:AlertAccountants.fifo
○	3 TB	24 hours	Send notification to topic : arn:aws:sns:us-east-1:888889908458:DisableAthenaWorkgroup.fifo
○	10 GB	1 hour	Send notification to topic : arn:aws:sns:us-east-1:888889908458:AlertAccountants.fifo

Figure 11.3 – Workgroup data usage controls section below per query limits

The previous figure shows an example in which three alarms notify different Amazon SNS topics in which different actions can be taken. This is an example where warning emails can be sent out to interested parties if the usage and cost exceed certain thresholds. The other two thresholds can disable the workgroup. Going above 3 TB of usage would be considered highly unusual for this use case, and an investigation should be done.

The above example alarms were created by clicking the **Create** button in *Figure 11.2*. As seen in *Figure 11.3*, the window that pops up allows data limits and a time period to be entered, and an optional SNS queue to send a notification to. Each alarm created is backed by a CloudWatch alarm and can be tweaked through the CloudWatch alarm console.

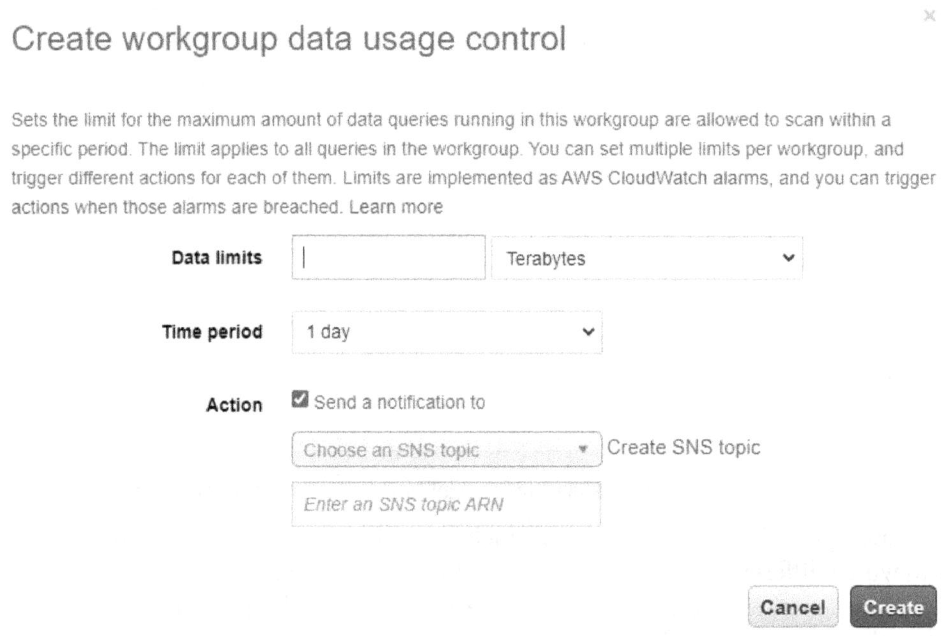

Create workgroup data usage control

Sets the limit for the maximum amount of data queries running in this workgroup are allowed to scan within a specific period. The limit applies to all queries in the workgroup. You can set multiple limits per workgroup, and trigger different actions for each of them. Limits are implemented as AWS CloudWatch alarms, and you can trigger actions when those alarms are breached. Learn more

Data limits		Terabytes ⌄
Time period	1 day ⌄	
Action	☑ Send a notification to	
	Choose an SNS topic ▾	Create SNS topic
	Enter an SNS topic ARN	

Cancel   Create

Figure 11.4 – Creating a workgroup data usage alarm

Now that we have seen how to monitor and put limits on query usage, let's look at how we can optimize using Athena to reduce costs and query runtime.

> **Important Note**
>
> If you are using the federated connectors, you will incur costs associated with the connectors, such as the cost of launching Lambda functions and the resources used when running Lambda.

Optimizing for cost and performance

When optimizing performance for any execution engine, two goals should always be kept in mind: read as little data from your storage as possible, which reduces costs and reduces query time, and make sure that your query engine does as little work (processing) as possible, which reduces query time.

This section will look similar to the AWS Big Data Blog post titled *Top 10 Performance Tuning Tips for Amazon Athena* (https://amzn.to/2VIFv1y) that I wrote. Still, we will provide some additional details that the blog post did not offer. Many customers bookmark this page and refer to it, and I recommend visiting it often to improve its view count.

> **Important Note**
>
> The recommendations in this section are generalizations and may not apply to all circumstances. Everyone's data, data structure, and queries are different, so not all of these recommendations may drive an improvement. Testing and prototyping are highly recommended when going through the process of optimizing usage.

Let's get started by looking at some optimizations on how to efficiently store data.

Optimizing how your data is stored

It is essential to consider how your data is stored when being read by execution engines such as Athena. How your data is stored usually has the most significant impact on how queries perform and how much they cost. Also, if you need to regenerate data when your system is live, it is much more expensive than doing it upfront. Changing queries is much easier and cheaper. With this in mind, some planning and prototyping are highly recommended.

Let's look at how file sizes and count impact performance.

File sizes and count

The size and number of files have a pretty significant impact on the performance of your Athena queries.

> **Important Note**
>
> The general recommendation is that your file sizes are between 128 Mb and 1 GB.

There are many reasons for this:

- S3 list operations are expensive. If you have a high number of files for a table, more S3 list operations need to be performed to get the list of files to read for a dataset.

- For each file, the engine needs to perform many S3 operations to consume it. It will first need to open the file by running a `HeadObject()` function to get the file size, encryption keys, and any other information necessary to start reading the file. This operation is expensive. Next, it will need to call a `GetObject()` function that returns a pointer to the data and the first data block. Ideally, you want to minimize the overhead of calling the `HeadObject()` function.

- The smaller your files, the less effective the compression, increasing the total amount of data stored.

- If your files are encrypted using server-side encryption, S3 will need to call the AWS KMS service to get decryption keys. This introduces overhead and **increases costs** because KMS charges for each call. I had a customer where their KMS costs were higher than their Athena costs because the cost of getting KMS keys was higher than reading the data. Having larger files reduces the number of calls needed for encryption keys.

Having appropriate file sizes reduces the amount of work that the execution engine needs to do.

Compression

Using compression makes the engine read less than uncompressed data, reducing network traffic from the data source to the Athena engine. For S3, it reduces storage costs. There is a trade-off of CPU usage as compression requires extra work to decompress data. Still, this cost is most of the time outweighed by making fewer calls to S3, and most queries do not exhaust CPU resources.

> **Important Note**
>
> Always compress your data when using text-based file formats, such as JSON and CSV. If your file sizes are larger than 512 Mb, use a compression algorithm that allows for files to be splittable. When using Apache Parquet or Apache ORC, compression should be applied within the column blocks (not to be confused with compressing the entire file).

When trying to decide on a compression algorithm, there are two aspects to consider. First, there is a trade-off between higher compression ratios and higher CPU usage. Second, whether the compression algorithm allows a query engine to read different parts of a file without reading the entire file. This is called if the file is splittable. If the algorithm produces splittable files, then multiple readers can read the file simultaneously, which increases parallelism. If a file is not splittable, a single reader must read the entire file, reducing parallelism.

For text-based file formats, such as JSON and CSV, it's always recommended to compress them because the compression ratios are generally very high. For columnar formats such as Apache ORC and Apache Parquet, they support compressing column data blocks. Because compression works best when groups of similar values exist, compressing all the values for a column in a single block usually leads to better compression ratios. Typically, Parquet and ORC are configured to compress column blocks by default.

File formats

Data formats impact the amount of data that a query engine reads and the amount of work the engine needs to do to read the data contained in the files. If your data is not in an optimal format, then transforming the data may reduce your overall cost if the cost of transformation is less than the cost of querying the data.

> **Important Note**
>
> For datasets that are read frequently, use Apache Parquet or Apache ORC. For data that is not likely to be queried or is queried infrequently, any compressed file format should be used. Datasets stored in CSV or JSON that will be queried frequently should be transformed into Parquet or ORC.

Let's dive into some common file formats.

- **Row formats**: CSV and JSON formats are the most common file formats used today but are the most inefficient. They are text-based, which is less efficient than storing the data in a binary format. For example, the int data type can be stored in binary using 4 bytes of data, but the number 1234567890 uses 10 bytes to store as a string. Add the delimiter for CSV and the field names in JSON, and they can take a substantial amount of space and memory. Also, when the file parser reads a number, it first needs to read the number as a string and then convert it to a number.

- **Columnar formats**: Columnar formats store data differently than row-based formats. With columnar formats, the data is grouped by the columns and stored in column blocks. A columnar file is then created by storing all the column blocks. When a reader wants to read the file, it will read each column block and generate a row by putting together all the columns by the index in the block. There are many reasons why this is cheaper and faster:

  - Field values are stored in binary instead of text. This reduces storage size and eliminates conversion from strings to numeric types, reducing the engine's read amount and work.

  - If a query only contains a subset of the columns, the execution engine will only read those columns, reducing the amount of data read.

- Compression on blocks of data that contain similar values is generally more efficient. This reduces the amount of data needed to be read, which reduces the cost and work demanded from the engine.

- Both Parquet and ORC support predicate pushdown, also known as predicate filtering. Parquet and ORC store statistics about each column block that can help skip reading entire blocks by pushing a filter to the reader and evaluating the filter on these statistics. If it is determined that the filter value is not in the data block, it is skipped. The statistics include ranges of values in the block and, for string data types, a Bloom filter. This reduces the cost and work demanded from the engine.

Parquet and ORC are better formats in almost every way than text-based formats. Let's take a look at how partitioning and/or bucketing your data can improve performance and costs.

Partitioning and bucketing

Partitioning and bucketing are two different optimizations that can lead to significant improvement in cost and performance. These features require an understanding of the data's usage patterns or how users or applications will query the datasets. Depending on the queries that will mainly be executed, it will inform the best way to leverage partitioning and bucketing.

Let's look at both features in a bit more detail:

- **Partitioning**: Partitioning your table separates rows into separate directories based on a column value. When a query contains a filter on a partition column value, only the partitions that meet the filter will be read, reducing the amount of data read. We talked about partitioned tables in *Chapter 4, Metastores, Data Sources, and Data Lakes*.

> **Important Note**
> Partitioning can significantly improve the performance and cost of your queries. A general recommendation for partitioning is to keep the number of partitions for a table under 100,000 while maintaining file sizes within the partitions.

Choosing partition columns can be a challenge, but here are some best practices. When looking at the queries executed against the table, the columns in WHERE clauses are great candidates to look at. An example would be a dataset containing a transaction date. The date is in the WHERE clause very frequently because users only need the most recent data.

The next best practice is to keep in mind the number of partitions your table has. The more partitions your table has, the smaller the files may be in each partition, which goes against file size best practice. Additionally, there is overhead when using partitioning, but if the partition column is chosen wisely, the overhead will be insignificant compared to the performance and cost savings. When Athena reads a partitioned table, it will need to fetch partition information from the metastore, and the greater the number of partitions, the more partitions it will need to fetch. A general recommendation is not to exceed 100,000 partitions, but this number depends on the upper bound of the query execution time and the amount of data in the dataset.

One unique feature that Athena has that could help read tables with many partitions is partition projection. It allows tables to specify the partition columns and the expected values that those columns may take within the table properties. When Athena queries the partitioned table, it generates the partitions on the fly instead of going to the metastore to retrieve the partition information. This works for tables that store their partitions on S3 in a consistent directory structure, with partition columns whose values can be specified in a list or a range. Partition projection supports integer and string data types and supports date formats as well. You can see examples of partition projection using various datasets in this book at `https://amzn.to/38a2FAC`.

Although not yet supported in Athena, one last optimization is indexing your Glue Data Catalog tables' partition data. When this feature is supported, it will significantly improve partition retrieval performance within Athena and reduce query time. Keep an eye out for when this is available.

- **Bucketing**: Bucketing is similar to partitioning because it groups rows with the same column values in a file within a partition. You specify the number of buckets you want at table creation time and the column to bucket on. The engine will then hash the bucket column values and put the rows with the same hash value in the same file. When a query engine has a filter for specific values for the bucketed column, it can then run the hash on filter values and determine which files it needs to scan. This could lead to entire files being skipped.

> **Important Note**
>
> Bucketing can significantly improve the performance and cost of your queries. However, bucketing adds complexity and should be employed by advanced users for the most time-sensitive workloads.

The following diagram shows what the NYC taxi dataset may look like if bucketing is employed. The sample CREATE TABLE statement is located at https://bit.ly/3kjd4j9.

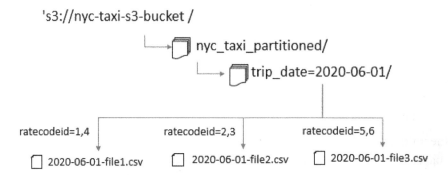

Figure 11.5 – An example of bucketing on the NYC taxi dataset

The dataset is partitioned by the trip_date value but bucketed on the ratecodeid column. All rows that contain the value in the ratecodid column of 1 and 3 will go into 2020-06-01-file1.csv, 2 and 3 will go into the 2020-06-01-file2.csv file, and 5 and 6 would go into the last file. If the query SELECT * FROM nyc_taxi_partitioned where ratecodeid = 3 is executed, Athena will determine that ratecodeid only existed in the 2020-06-01-file2.csv file and hence can skip the other two files. However, if the query SELECT * FROM nyc_taxi_partitioned where ratecodeid > 3 is executed, Athena will read all the files because it does not know the complete list of possible values.

There are some limitations to discuss. The current version of PrestoDB that Athena uses only supports tables that were bucketed using Hive, without the ability to insert data after a table or partition has been created. Once Athena offers a newer version of PrestoDB, this limitation may be removed and support Apache Spark's bucketing algorithm. Also, once the number of buckets is chosen for the dataset, it cannot be changed unless the entire dataset is regenerated. For these reasons, it is recommended that only advanced users attempt to leverage bucketing.

Now that we have gone through the optimization techniques to lay out our datasets on S3, let's look at some optimization techniques when writing queries.

Optimizing queries for performance

Although how data is stored can make the most significant impact on the performance of Athena queries, how queries are written is also important. In this section, we will go through some best practices when optimizing your queries.

Explain plans

Athena recently released a new feature that allows you to look at the execution plan of your queries. The execution plan is the set of operations that the engine performs to execute the query. It is not a requirement to read and understand the execution plans to optimize, but if we know how to read them, they can give us a valuable tool to dive deep into how queries are being executed. If you are not able to follow the technical details, it is okay. The other sections for optimizing your query will provide general recommendations that anyone can follow.

Let's take a quick look at an example of the information that EXPLAIN provides. If we take the simple query EXPLAIN SELECT SOURCE_ADDR, COUNT(*) FROM website_clicks GROUP BY source_addr, we get the following logical execution plan (edited to simplify the output):

```
Query Plan
- Output [SOURCE_ADDR, _col1] => [[source_addr, count]]
    - RemoteExchange [GATHER] => [[source_addr, count]]
        - Project [] => [[source_addr, count]]
            - Aggregate (FINAL) [source_addr] [$hashvalue] =>
[[source_addr, $hashvalue, count]]
                - LocalExchange [HASH] [$hashvalue] ("source_
addr") => [[source_addr, count_8, $hashvalue]]
                    - RemoteExchange [REPARTITION] [$hashvalue_9]
=> [[source_addr, count_8, $hashvalue_9]]
                        - Aggregate (PARTIAL) [source_addr]
[$hashvalue_10] => [[source_addr, $hashvalue_10, count_8]]
                            - ScanProject [table
schemaName=packt_serverless_analytics_chapter_11,
tableName=website_clicks, analyzePartitionValues=Optional.
empty}] => [[source_addr, $hashvalue_10]]
                                LAYOUT: packt_serverless_
analytics_chapter_11.website_clicks
                                source_addr := source_
addr:string:1:REGULAR
```

This can look daunting at first, so let's break it down. The plan from the top down goes backward from the order of operations. The operation executed is `ScanProject`, which does the reading of our source data, our `website_clicks` table. The second operation is `Aggregate`, which does a partial `GROUP BY` function on the local node before sending it to the `RemoteExchange` operation. `RemoteExchange` shuffles data between the nodes of the partially aggregated data based on a hash code so that rows that contain the same `GROUP BY` columns go to the same node. `LocalExchange` shuffles data within a worker node. Then, a final `Aggregate` operation aggregates all the rows with the same `GROUP BY` values. The `Project` operator removes the hash code column and then performs the last `RemoteExchange` operation to a single node, to output the results using the `Output` operator.

To graph a visual representation of the plan, you can specify the format to `GRAPHVIZ` and use an online conversion tool to convert the output to an image. The one that is used within this chapter is `https://dreampuf.github.io/GraphvizOnline/`. The converted image for the query `EXPLAIN (FORMAT GRAPHVIZ) SELECT SOURCE_ADDR, COUNT(*) FROM website_clicks GROUP BY source_addr` is located at `https://bit.ly/3yOMFzD`.

If the type of execution plan is not specified, such as the previous example, a logical plan is provided. But Athena supports three other types of execution plans. They are `VALIDATE`, `IO`, and `DISTRIBUTED`, which can be specified in the query. For example, to validate whether a SQL statement is valid before executing it, you can run `EXPLAIN (TYPE VALIDATE) <SQL STATEMENT>`. It will return a `true` or `false` value, depending on whether Athena can parse and execute the query. The IO execution plan outputs the input and outputs of the query. An IO plan for the previous example can be seen at `https://bit.ly/2VYnzjh`.

The `DISTRIBUTED` plan provides fragments of the execution plan that is executed across different nodes. Each fragment is performed on one or more nodes depending on the type of the fragment. There are several fragment types, including `SINGLE`, `HASH`, `ROUND_ROBIN`, `BROADCAST`, and `SOURCE`. The `SINGLE` type of fragment executes only on a single node. The `HASH` type executes on a fixed number of nodes where the data is distributed among the nodes, based on a `HASH` code derived from one or more column values. For example, the `source_addr` column would be hashed for the previous query because it is in `GROUP BY`. To perform the `GROUP BY` function, rows with the same `source_addr` value need to be on the same node to do the aggregation. The `ROUND_ROBIN` type means that data is sent in a round robin to multiple nodes for operations such as transformations. The `BROADCAST` type means that the input of the fragment is the same across one or more nodes. This type is sometimes used with joins if a table is small enough to send to all nodes to do the join, which can significantly improve join performance.

Lastly, the SOURCE type specifies a fragment that reads from a source data store. In each fragment in the plan, the input data is determined by the *RemoteSource[FragmentNumber]* value, where *FragmentNumber* is the source fragment. To see the distribution plan for the previous query example, visit `https://bit.ly/3g58Lqw`.

Now that we have a basic understanding of how to read execution plans, let's look at some of the optimizations we can make to our queries, starting with optimizing joins.

Optimizing joins

The order in which tables are expressed in a join operation can significantly impact your query performance.

> **Important Note**
>
> When joining tables, order the largest tables on the left to the smallest tables on the right.

You may ask why the ordering of tables matters. Athena does not have access to statistics yet to reorder joins optimally as other database systems do. This may change in the future, but it is up to the user to perform this ordering for now.

For those interested in the technical details of why ordering matters, we need to understand how Athena performs joins. In summary, both tables get read and shuffled to a join operator to perform the join. However, there is an extra operation for the table on the right side. If the right-side table is smaller, the extra operation will be cheaper than if the operation occurred on the larger table. If we look at the explain plan for a query that performs a join, `EXPLAIN (Type DISTRIBUTED) SELECT larger_table. table_data FROM larger_table LEFT OUTER JOIN smaller_table on larger_table.table_key = smaller_table.table_key`, we see the following subsection of the distributed plan (the full explain plan can be found at `https://bit. ly/37MH0hG`):

```
Fragment 1 [HASH]
    Output layout: [table_data]
    Output partitioning: SINGLE []
    Stage Execution Strategy: UNGROUPED_EXECUTION
    - LeftJoin[("table_key" = "table_key_0")][$hashvalue,
$hashvalue_9] => [[table_data]]
            Distribution: PARTITIONED
        - RemoteSource[2] => larger_table
```

```
            - LocalExchange [HASH]
                - RemoteSource [3] => smaller_table
```

The `LocalExchange` operator reshuffles the data within the worker. If the join order was reversed, the reshuffle would occur on the larger table, which would require more work and would cause the query to run longer.

Now let's look at optimizing the `ORDER BY` operator.

Optimizing ORDER BY

You will often need to order your results to get the top *N* number of results to generate reports or look at a subset of data when exploring a dataset. However, doing an `ORDER BY` operation on a large dataset can be a costly operation.

> **Important Note**
>
> When performing `ORDER BY` operations, using `LIMIT` can dramatically reduce query time if only the top *N* results are needed.

We need to understand why performing a global ordering requires a single worker to get the entire result set and perform a global sort, even if the input is sorted from many workers. Performing a global sort is very memory- and CPU-intensive. By limiting the number of results in the output, workers pushing rows to the global sort of an operator can limit the number of rows to it. The global sort can be done on a much smaller set of data. Let's look at the execution plan when `LIMIT` is specified for the query `EXPLAIN SELECT total_amount FROM nyc_taxi where payment_type = 1 ORDER BY trip_distance LIMIT 100` (the full explain plan can be found at https://bit.ly/2VSMvc4):

```
Query Plan
- Output [total_amount] => [[total_amount]]
    - Project [] => [[total_amount]]
        - TopN [100 by (trip_distance DESC_NULLS_LAST)]
            - LocalExchange [SINGLE]
                - RemoteExchange [GATHER]
                    - TopNPartial [100 by (trip_distance DESC)]
                        - ScanFilterProject [table = nyc_taxi]
```

Without the `LIMIT` operator, the `TopNPartial` operator would not be in the plan. All results would go to the `TopN` operator. Performing the local sort before performing the `RemoteExchange` operation limits the amount of data shuffled, saving time and bandwidth.

Let's now look at the next best practice.

Selecting only the columns that are needed

This recommendation should be self-evident, but I have seen many customers not do this.

> **Important Note**
> Only select the columns in your query that are required as the output of your query.

There are many reasons why this can save both time and cost. For columnar data types, less data is read, which reduces cost. Another reason is that there is less data that needs to be shuffled between workers and outputted.

Let's now look at our last best practice.

Parallelizing the writing of query results

When Athena executes a `SELECT` query, the query's output is written by a single worker. If there is a huge result set, the amount of time to write the results from the single worker can be significant.

> **Important Note**
> For queries that produce a large number of results, use `CTAS`, `INSERT INTO`, or `UNLOAD` to parallelize the writing of the output.

Troubleshooting failing queries

When Athena works, it is excellent. It queries data in S3 without having to worry about servers or installing and maintaining software. But when Athena fails to execute a query, it can be tricky to know how and where to start looking. Issues can include how you wrote your query, problems with your metadata, or your data. In this section, we will go through some common failures and how to approach them. However, this list is not exhaustive. Athena's documentation publishes many error messages that customers see and how to deal with them, so bookmark it and refer to it when needed.

When your query starts failing, here is a list of actions that you can take:

- If your queries were working previously but are failing now, determine what has changed. Source control for queries that applications submit can help keep track of code and queries that have changed. If the queries have not changed, then most of the time, the issue is due to the loading of new data that it cannot process or metadata that was changed. This question is usually the first one that AWS Support would ask.

- Retry your query after a few minutes. Some failures with Athena are transient, such as when S3 throttles Athena because the load was too high on a particular S3 partition.

- Go to Athena's troubleshooting documentation, which contains a list of error messages and solutions (located at `https://amzn.to/3kjBuJt`).

- If all else fails, and you have access to AWS Support, then enter a support ticket. When creating a support ticket, the query ID and AWS region should be provided to help with the investigation. Providing a small sample of data is super helpful to AWS Support and the Athena development team to reproduce the issue. Just ensure that it does not contain any sensitive data.

Let's look at some common issues that customers face with Athena.

My query is running slow!

This is the most common issue that customers have when using Athena. Following the recommendations in the optimization section generally solves this issue. Using partitioning, converting to Apache Parquet or Apache ORC, and ensuring queries are optimally written will solve most of the reasons why queries may be running slow. If these do not, the other reason may be that too many concurrent queries are being run, and queries are being queued by the Athena service. You can check this by running your query and running a CLI command to get the amount of time the query spent in the queue. The following shows an example of the CLI command and its results:

```
aws athena get-query-execution --query-execution-id <EXECUTION
ID>
{
    "QueryExecution": {
        "QueryExecutionId": "edea5091-6061-44bb-89ce-
96090098c1b1",
        "Query": "select * from customers limit 10",
        "StatementType": "DML",
```

```
        ... (Section omitted ) ...
        "Statistics": {
            "EngineExecutionTimeInMillis": 511,
            "DataScannedInBytes": 35223,
            "TotalExecutionTimeInMillis": 765,
            "QueryQueueTimeInMillis": 155,
            "QueryPlanningTimeInMillis": 89,
            "ServiceProcessingTimeInMillis": 99
        },
        "WorkGroup": "packt-chapter11"
    }
}
```

Under the statistics section, you will see the `QueryQueueTimeInMillis` statistic. This value shows you the amount of time the query spent in Athena's queue, waiting for resources to execute on. If this value is consistently high, then your query rate is too high. Recommendations on how to monitor and the steps to correct this are in *Monitoring query queue time* in this chapter.

My query is failing due to the scale of data

This is the next most common issue customers face. The amount of data that Athena can scan is only limited to the maximum amount of time the query can run for, which by default is 30 minutes. When Athena performs simple table scans, it can process petabytes of data. However, if you see error messages such as `INTERNAL_ERROR_QUERY_ENGINE, EXCEEDED_MEMORY_LIMIT: Query exceeded local memory limit`, `Query exhausted resources at this scale factor`, and `encountered too many errors talking to a worker node. The node may have crashed or be under too much load`, then it's highly likely that your query contains complex operations, such as joins, aggregations, or windowing functions. These operations are performed by shuffling data around to nodes based on the values of the rows and stored in memory until the operation is completed. If a single node in the execution engine's cluster exhausts its resources, the query will fail.

There are a few strategies to overcome this issue. The first is to reduce the amount of data processed within the query by filtering data as soon as you read a table before complex operations. For example, take the following query:

```
SELECT upper(col1), sum(col1 + col2) FROM
    (SELECT
        table1.key, table1.col1, table2.col2
    FROM table1
    LEFT OUTER JOIN table2
    ON table1.key = table2.key) innerQuery
WHERE col2 > 10
```

This can be rewritten as follows:

```
SELECT upper(col1), sum(col1 + col2) FROM
    (SELECT
        table1.key, table1.col1, table2.col2
    FROM table1
    LEFT OUTER JOIN
        (select * from table2 WHERE col2 > 10)
    ON table1.key = table2.key) innerQuery
```

Filtering data before performing a complex operation can really improve performance and reduce memory requirements. Selecting only the columns that are of interest can help as well. Lastly, splitting up the query into smaller queries that scan a subset of partitions may help.

The other strategy is to find out whether your data has any data skews. Data skews exist when your data is not evenly distributed across a cluster when a complex operation is performed. For example, suppose there was a dataset that tracked all the different types of chairs. You performed a join on the number of legs a chair has to a dimension table, that is, `SELECT * FROM chairs JOIN dim ON chairs.legcount = dim.legcount`. Since most chairs have four legs, there will be significantly more data going to one node to perform the join, exhausting all the available memory. The only way to deal with this is to distribute the joins data across several nodes by joining on more than the `legcount` column or to reduce the number of rows by aggregating the data before the join occurs.

Now that we have gone through some troubleshooting techniques, let's summarize what we have learned in this chapter.

Summary

In this chapter, we went through best practices to get the most out of Athena while making sure it operates smoothly. We went through how we can create alarms to keep track of query queue time and costs, and take action to prevent Athena's unexpectedly high usage from leaving us with unexpected bills at the end of the month. We then went through how to optimize our usage of Athena by looking at best practices on how to store data and the queries we run. To do that, we explored how to look at the explain plans and how to read them to identify possible bottlenecks or issues with the written queries. Lastly, we looked at what to do when a query fails and the common problems users usually encounter.

The next chapter will dive into query federation, which allows you to query almost any data source with Athena.

Further reading

- *Amazon Athena CloudWatch Metrics* – https://docs.aws.amazon.com/athena/latest/ug/query-metrics-viewing.html

- *Top 10 Performance Tuning Tips for Amazon Athena* – https://aws.amazon.com/blogs/big-data/top-10-performance-tuning-tips-for-amazon-athena/

- *Athena Partition Projection* – https://docs.aws.amazon.com/athena/latest/ug/partition-projection.html

- *Athena EXPLAIN documentation* – https://docs.aws.amazon.com/athena/latest/ug/athena-explain-statement.html

- *PrestoDB 0.217 EXPLAIN documentation* – https://prestodb.io/docs/0.217/sql/explain.html

- *Amazon Athena Troubleshooting* – https://docs.aws.amazon.com/athena/latest/ug/troubleshooting-athena.html

Section 4: Advanced Topics

In this section, we will delve into advanced topics, including how to extend Athena with your own custom data sources and functions.

This section consists of the following chapters:

12
Athena Query Federation

Welcome to *Chapter 12*, *Athena Query Federation*, and the beginning of *Section 4*, *Advanced Topics*. In this part, we will cover topics that go deeper into highly customizable, experimental, or emerging areas of development for Amazon Athena. Don't be intimidated by the "advanced" designation as these topics are not necessarily more difficult to understand or use than those we covered earlier. This chapter is all about getting the most out of Amazon Athena by using Query Federation to expand beyond queries over data in S3. We will learn how Query Federation allows you to combine data from multiple sources, such as DynamoDB and Elasticsearch, to provide a single source of truth for your queries.

We'll begin by peeling back the curtain to explain Athena Query Federation's architecture, including how Athena uses AWS Lambda to run its customizable Connectors. Since the entire Athena Federation SDK and its 14 initially released Connectors are all open source, we can quickly run queries across several sources all from one tool. In addition, these same open source Connectors offer a great set of examples, upon which we can build our custom Connectors for proprietary datastores, formats, or simply integrate them with sources that aren't officially supported yet. Before the end of this chapter, you'll deploy and query one of Athena's off-the-shelf Connectors, and we'll also build our very own custom connector from scratch using Athena's Federation SDK.

In this chapter, we will cover the following topics:

- What is Query Federation?
- How Athena Connectors work
- Using pre-built Connectors
- Building a custom connector

Technical requirements

Wherever possible, we will provide samples or instructions to guide you through the setup. However, to complete the activities in this chapter, you will need to ensure you have the following prerequisites available. Our command-line examples will be executed using **Ubuntu**, but most Linux flavors should work without modification, including Ubuntu on Windows Subsystem for Linux.

You will need internet access to GitHub, S3, and the AWS console.

You will also require a computer with the following installed:

- Chrome, Safari, or Microsoft Edge
- The AWS CLI

This chapter also requires that you have an **AWS account** and an accompanying IAM user (or role) with sufficient privileges to complete this chapter's activities. Throughout this book, we will provide detailed IAM policies that attempt to honor the age-old best practice of "least privilege." For simplicity, you can always run through these exercises with a user that has full access. Still, we recommend using scoped-down IAM policies to avoid making costly mistakes and learning more about using IAM to secure your applications and data. You can find the suggested IAM policy for this chapter in this book's accompanying GitHub repository, listed as `chapter_12/iam_policy_ chapter_12.json`, here: `https://bit.ly/3xCi0ow`. The primary additions from the IAM policy that have been recommended for past chapters include the following:

- Adding CloudFormation access for using Serverless Application Repository
- Adding Lambda access for deploying and executing Athena Connectors.

What is Query Federation?

Simply put, **Query Federation** refers to the concept that a query engine such as Athena may enlist the help of multiple datastores, working together, to execute your query. These datastores are usually capable of more than file-level CRUD operations. Most will support row-level scan, filter, and project operations, with some handling full SQL. We've mentioned this concept earlier in this book, typically concerning ETL versus querying in place. Let's take a closer look at the practical difference between a federated query and what we'll call a classic query.

The following diagram shows an example of a tried and true S3 data lake. There are multiple datastores, namely DynamoDB, RDS Aurora, and a generic database, all feeding into S3. Then, Athena, or another query engine, with the aid of Glue Data Catalog, can access all our data. This is a classic query. You submitted the query to Athena, and Athena directly answered your query by reading the table(s) in the data lake. As you've seen throughout this book, S3-backed data lakes offer countless advantages over alternate models. One of those advantages encapsulates the difference between a federated query and the classic query we just described:

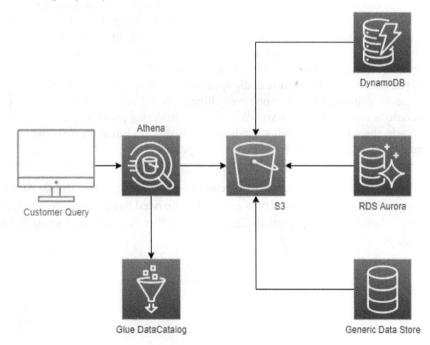

Figure 12.1 – Data lake model

By exporting, copying, or replicating your data into S3, you've decoupled your operational store from your analytics queries. Put another way, if you run a resource-intensive query against your data lake, there is no chance you'll overwhelm your website's MySQL database, which would result in web page timeouts for customers. By decoupling these systems, you've traded consistency for availability. Physics will ensure that the data in your S3 data lake always creates lag in the authoritative system that created the data. In the vast majority of cases, this lag is not a problem, and you should jump at the opportunity to reduce non-critical workloads running on operational stores.

Like many engineers that have come before you, you may be asking yourself whetherthere is a better way. For example, can I avoid the delay or the work of replicating all that data into S3? Query Federation is one way you can avoid copying data to S3 while getting the most up-to-date view possible. However, the second rule of optimization applies here as federating queries across datastores doesn't come without downsides.

The Rules of Optimization

Many a human has fallen foul of the rules of optimization. Optimization is tricky, and our brains are so good at finding patterns that they see patterns that aren't real or are so uncommon that they aren't worth handling. Thus, the first rule of optimization is "don't optimize." The intention is to help avoid adding complexity when a more straightforward approach might be just as good or better. The assumption is that you don't usually know what is worth optimizing until you have a complete picture of the system and its usage patterns. Wait! If we never optimize anything, won't everything be slow?! Yes, and you might even need to redo all that work you did without considering performance. That's why there is a second rule of optimization that pokes fun at the first. The second rule goes as follows: "For experts only, optimize."

The following diagram demonstrates how Athena Query Federation can enable Athena to directly query your operational data stores and avoid the need to copy data to S3. We've deployed three Athena Query Federation Connectors in this example, one for each source we want to query:

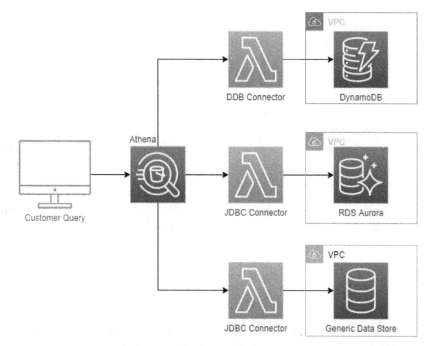

Figure 12.2 – Query Federation model

In the next section, we'll learn how these Connectors work. But for now, you only need to know that they act as translators, turning Athena's representation of rows, columns, and queries into that of the target datastore. So, in this example, when you submit your query to Athena, S3 and Glue Data Catalog do not act as an integration point between Athena at the various systems that house the data you are trying to query. Instead, Athena enlists the help of the underlying datastores to execute your query. Athena's engine begins by analyzing the query and identifies the tables and columns it must read. This step is identical to using a query against data residing in an S3-backed data lake. Then, as Athena attempts to resolve the storage location of each table, it classifies them as being natively supported or residing in a federated source. Finally, for each federated source, Athena determines which of its available Connectors can reach the requested table(s). Together, these pieces of information allow Athena to create a **physical plan** for executing the query.

The resulting physical plan is a blueprint for how Athena will build the results of your query. Much like the construction of a house, some parts of the work will be done by Athena itself, but other details will be sub-contracted to specialists. For example, the contractor that frames your home may not be as proficient at installing a bathroom fixture as a plumber. Athena Query Federation takes the same approach with data stores. Where possible, Athena prefers to delegate portions of the query to the underlying datastore. This is true for all federated sources but is especially prevalent in Athena's JDBC connector. For example, Athena pushes down full SQL fragments to databases such as RDS Aurora so that your query can benefit from any indexes, query caches, or other dark magic that allows Aurora to scale.

Now that we have a better understanding of Query Federation and its differences from classic Data Lake queries, let's look at their unique features.

Athena Query Federation features

Since its announcement in 2018, the Athena team has released open source Connectors for more than 14 databases, storage formats, and live APIs, with dozens more community-built Connectors available in the AWS Serverless Application Repository. At the heart of Athena, Query Federation is an SDK that defines a set of interfaces, as well as an accompanying wire protocol you can implement to enable Athena to delegate portions of its query execution plan to code that you write and deploy. In essence, you can customize Athena's core execution engine with your functionality while still taking advantage of Athena's ease of use and fully managed nature.

Some of the notable features of Athena Query Federation include the following:

- **Lambda-based deployments**: Athena Query Federation allows you to deploy your Connectors and UDFs as AWS Lambda functions, preserving Athena's serverless experience. This ensures your Connectors are easy to manage and scale and can securely access datastores in your VPC, without the need to grant Athena access to your network.

- **Apache Arrow**: Athena has adopted Apache Arrow as a standard format for the data interchange format. Apache Arrow is a widely adopted open source format capable of zero-copy transfers with support from Spark, Python, and AWS Lake Formation.

- **Federated metadata**: It's not always practical or possible to store your table metadata in AWS Glue Data Catalog. As such, Athena Query Federation allows you to choose between AWS Glue Data Catalog and implementing a metastore interface. Athena will use that interface during query planning to look up table and column details.

- **Federated UDFs**: Athena can delegate calls for batchable scalar UDFs to you, allowing you to write custom UDFs.

- **AWS Secrets Manager integration**: If your Connector needs a username, password, or other sensitive information, you can use the SDK's built-in tooling to resolve secrets. This is especially helpful when you're federating to non-cloud-enabled datastores.

- **Federated identity**: When Athena has federated a query to your connector, you may want to perform authorization on the identity of the entity that executed the original Athena query.

- **Parallelized and pipelined reads**: Athena will parallelize reading your tables based on the partitioning information that's returned by your Connector. This allows you to tell Athena how it should divide reads for optimal performance.

- **Predicate pushdown**: Where relevant, Athena will supply your Connector with the associative portion of the query predicate. This enables your Connector to filter results or translate the predicate into the source datastore for maximum performance. It also has the nice side effect of lowering the cost of your Athena query by dropping data before it reaches Athena.

- **Column projection**: Athena supplies your Connector with the columns that need to be projected so that you can reduce the amount of data that's scanned. This is especially useful when you're federating to columnar stores such as Redshift.

- **Congestion Control**: Some datastores may not be as scalable as Athena or run operationally sensitive workloads. Athena automatically detects signs of congestion and supports explicit throttling exceptions from your Connectors. When congestion is suspected, Athena employs an algorithm similar to TCP's flow control to organically "share the road" with other workloads. This makes it easier to scale your usage of Athena while simultaneously protecting sensitive data stores.

With our newfound understanding of federation queries, we can take a deeper look at Athena's implementation of this new query pattern. In the next section, we will unpack the architecture of an Athena connector.

How Athena Connectors work

Unsurprisingly, Athena Connectors follow a similar structure to both Presto and Trinio Connectors, with one notable difference. Athena's Federation SDK is designed to decouple your Connector from Athena's core engine, whereas both Presto and Trinio require Connectors to run within the engine. If you keep this key difference in mind, much of what we are about to describe applies to Athena, Presto, and Trinio. At the heart of each Connector is the Athena Federation SDK, which provides an abstraction over the boilerplate code required to enable Athena to orchestrate your federated query. Every Athena Connector is required to implement the following six functions defined in the SDK (`https://bit.ly/3vXmm9j`):

- `doListSchemaNames(...)`: This function provides Athena with a list of schemas, also known as databases, that the Connector believes are available in the federated source.

- `doListTables(...)`: This function provides Athena with a list of tables in a given schema (that is, a database) that the Connector believes are available in the federated source.

- `doGetTable(...)`: This function provides Athena with the columns, partition, and storage details of a requested table in the federated source.

- `doGetTableLayout(...)`: This function provides Athena with details about how a given table is physically stored as a means to quickly prune portions of the table's physical storage from the query plan. If your source doesn't support such partitioning, you can use Athena's default implementation and mostly skip implementing this function.

- `doGetSplits(...)`: This function influences the level of parallelism that Athena uses to read the data for your federated query. Each **split** represents a chunk of rows that Athena must read from your Connector to complete the query.

- `doReadRecords(...)`: Athena uses this function to read actual rows of data (that is, splits) from your federated source.

Each of these function names begins with the word `do`. This is a common programming convention that hints at the fact that some higher-level abstraction uses these functions to delegate doing some specific thing. The first five functions all supply Athena's query planner with metadata, with the sixth function, `doReadRecords(...)`, being the only one that interacts with the raw data associated with your tables. Later, we'll write a connector from scratch and implement many of these functions to give you an even deeper understanding of what they do.

In true Amazon fashion, the Athena team is always attempting to make life easier for customers. Several of these functions have been further simplified for the most common use cases. Customers can choose to implement them as is for complete control over their connector, or they can opt for default implementations of doGetTableLayout (...) and doReadRecords (...), leaving them to implement a simplified readRecordsWithConstraint (...) function. But we're getting deeper into the code than we need to at the moment. So, for now, the key takeaway is that the SDK offers you a balance of control and customization while also seeking to make common things easy.

The following diagram shows how Athena calls these functions. doListSchemaNames (...) and doListTables (...) are only used when you run a show tables or show databases query, so we will omit them for now. Athena calls the getTable (...) API on your Connector, which is handled by the MetadataHandler class in Athena's Federation SDK for each table in your query:

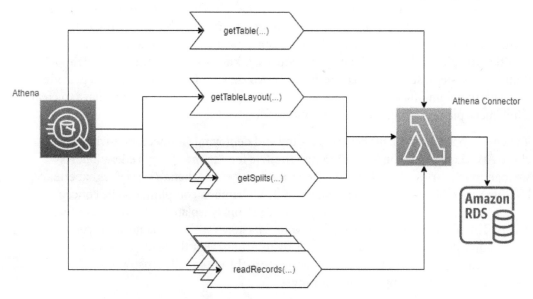

Figure 12.3 – Athena Federation call flow (ordered top to bottom)

MetadataHandler's `getTable(...)` function handles serialization and error checking before calling the `doGetTable(...)` function that your connector implements. Once Athena has retrieved the schema information for your table using the `getTable(...)` API, it uses the `getTableLayout(...)` API to allow your Connector to participate in physical query planning. This is a fancy way of saying your Connector can look at the table's storage characteristics, such as partitioning, and choose the access pattern that best matches the required columns and query predicates that Athena has supplied as part of the API call. A typical example of this would be a table that is partitioned by a column that also appears in the `where` clause. Your Connector can only return partitions that satisfy the `where` clause, potentially pruning large amounts of data without ever having to read those physical partitions.

Next, Athena calls the `getSplits(...)` API for batches of partitions returned by the call to `getTableLayout(...)`. Here, we encounter one of the most confusing aspects of Athena Query Federation. Athena doesn't understand the partition or split objects that your Connector returns. So, why would Athena call our Connector for these values if it doesn't know what they represent? The answer is simple: Athena is acting as the orchestrator of the federated query. While Athena may not understand or care about the partition or split objects your Connector returns, it knows your Connector cares about them in later steps, such as the call to `readRecords(...)`. In this way, Athena provides you with the plumbing and structure to build a Connector so that you don't have to create a full query planner or distributed computing framework.

The desire to simplify the process of building and deploying Connectors is why Athena chose AWS Lambda as the first compute primitive for Athena Query Federation. As we mentioned earlier, AWS Lambda allows us to preserve Athena's serverless experience, but there are other, less obvious reasons why AWS Lambda compliments the concept of federated queries. For example, AWS Lambda strongly isolates functions from their callers. This allows Athena to federate to your datastores without the need for you to open your VPC network or datastore directly to Athena. For those of you who are network-savvy, this also allows everyone to ignore the potentially frustrating task of routing across overlapping IP address ranges.

At a recent re:Invent talk, the Athena team indicated that AWS Lambda would be the first but likely not the last compute option for Athena Federation. They were working to vertically integrate high-performance UDFs and Connectors directly into Athena's engine, without the extra hope of remote calls to Lambda. Still, the choice to use AWS Lambda to process large amounts of data often raises questions about performance and cost. In the next section, we will tackle this topic directly.

Using Lambda for big data

One of the most frequently asked questions the team received after launching Athena Query Federation pertained to the use of Lambda for processing large amounts of data. Customers wanted to understand the performance, scale, and cost implications. The Athena Federation team spent a lot of time designing the APIs, performing serialization, and performing stress tests to ensure that the answers to these questions would be acceptable. In short, Lambda ends up being an excellent option for offloading the compute associated with Athena Connectors in almost every case.

Firstly, Lambda can scale to many thousands of concurrent invocations, ensuring that Lambda concurrency never becomes a bottleneck for running concurrent Athena queries. Second, with all that concurrency, it's a good thing that additional costs associated with Lambda costs are typically less than 5% of the total of a federated Athena query. Lastly, Athena's engine deeply pipelines most operations, such as table scans and aggregations, to reduce latency. Any remaining overhead associated with an extra data transfer hop from Lambda is overcome through sheer parallelism in the form of concurrent Lambda invocations.

So, even though it may seem like an unlikely choice to use Lambda for processing large amounts of data, Lambda offers a powerful compute primitive that, when used appropriately, can process 100 Gbps of data without breaking a sweat. There are, however, a few cases where the use of Lambda makes things a bit tricky. For example, suppose the data source you are federating to does not support parallel scans. In that case, the table scan portion of your federated query will be limited to that of a single Lambda invocation. At the time of writing, a Lambda invocation is limited to a max runtime of 15 minutes. Customers most often encounter this limit when federating to an RDBMS such as MySQL. The other limit pertains to how results are transmitted from Lambda to Athena. A Lambda invocation is limited to a 6 MB response. As we'll see later in this chapter, the Athena Federation SDK hides this limit by spilling data above this limit to S3 in chunks that allow Athena to pipeline reads. As we mentioned previously, the SDK mostly hides this inconvenient truth, so you don't need to do any extra work. It does this by automating the creation and transmission of the spilled data, including using a per-query encryption key that protects the spilled data that is automatically shredded after the query is completed. Your only responsibility is setting up an S3 life cycle policy to periodically delete old files from the spill's location.

Next, we'll cover an often overlooked aspect of Query Federation. The sources you'll want to query will often sit in different networks or VPCs. Thanks to Lambda, Athena allows you and your network engineers to completely sidestep the routing, DNS, and security headaches of presenting a flat network to your query engine.

Federating queries across VPCs

The intricacies of network design, routing, and connectivity have filled many books twice the size of this one. Luckily, you won't need to solve these kinds of problems to configure or use Athena Query Federation. Still, it is worth understanding this class of issues and how Athena solves them for you. Having this basic knowledge will help you when you're evaluating alternatives to Athena. It may also allow you to simplify your network design if it was previously dictated by limitations in the analytics suite you were using.

Suppose our company had three separate VPCs or networks, similar to those shown in the following diagram. VPC-A was our first VPC. We were a young company and just learning about network design, so we ended up using an entire Class A network (10.0.0.0/8) for our VPC. The team figured that this was a good idea to plan for scale upfront, and a Class A network can fit more than 16 million hosts in our VPC, so we'd never outgrow it. Later, we acquired a competitor and inherited VPC-B. The designers of VPC-B had a bit more experience and didn't use an entire Class A network. Instead, they carved out a piece of the 10.0.0.0/8 private space for their VPC. This would leave room for future growth or acquisitions without address overlaps. We'll describe why overlaps can be painful shortly. Until recently, these two VPCs remained independent, but now, there is a push to integrate the two companies, beginning with an analytics solution that can unify siloed data for reporting.

It is at this point that the networking teams realized that the resources in VPC-A and VPC-B have overlapping addresses. Unfortunately, that means we can't connect these two networks because any IP address in VPC-B might also be a valid, in-use IP address in VPC-A. Typically, your network engineers would have to deploy and configure a **Network Address Translation** (**NAT**) capable firewall or edge device to act as a proxy between the two networks. Sometimes, they may even need to introduce a third network, as we've done in the following diagram, because VPC-A and VPC-B resources should remain independent for operational and security reasons. Only the analytics system housed in VPC-C should be able to access the data stores. They never need to access each other:

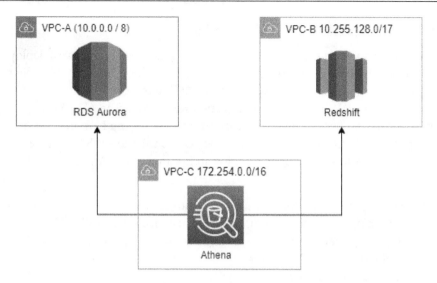

Figure 12.4 – Athena Federation network topology

These problems happen more often than companies would like to admit. Just ask the designers of IPv6. As the successor to the IPv4 address schema used in the preceding diagram, it can support *3.4×1038* addresses while IPv4 offers just *4.2×109*. The IPv6 address space is so large that there isn't even an agreed specification or implementation for IPv6 address translation in the Linux kernel.

Luckily for you, Athena doesn't require your federated sources to be on the same network. Instead, each Athena Connector can be deployed directly into the network housing your datastores via Lambda VPC attach capabilities. When an Athena query attempts to access the RDS Aurora instance in VPC-A, Lambda will quickly attach to that VPC and allow the Connector code to communicate with your RDS instance. From within the same query, you can join against a table housed in a Redshift instance from VPC-B, and Athena will invoke the appropriate Lambda function connected to VPC-B. The networks remain isolated from each other, but your queries are free to span datastores and networks.

This feature is not likely to show up on a flashy presentation or marketing promotion because its existence means you don't need to trouble yourself with these details. It is not until you find yourself in a situation where this capability is missing that you realize it was worth adding to your acceptance criteria. With our introduction to Athena Query Federation completed, we can start our hands-on experience. First, we will deploy and use a pre-built connector before building a connector from the ground up.

Using pre-built Connectors

As part of our first hands-on experience with Athena Query Federation, we'll deploy and query one of Athena's 14+ pre-built Connectors. The Athena team has published these Connectors to the Athena Query Federation GitHub repository (`https://bit.ly/3vXmm9j`) and the AWS Serverless Application Repository. The AWS Serverless Application Repository offers a one-click experience for deploying Lambda-based serverless applications. This section will show you how to use the AWS Serverless Application Repository to search for and deploy an instance of the Athena CMDB Connector.

The Athena CMDB Connector allows you to query various AWS resources using standard Athena SQL. For example, you can `SELECT` all EC2 instances in a specific VPC or search for all S3 objects greater than a particular size and residing in the most expensive storage tier. At the time of writing, the Connector exposes the following AWS services and resources as databases and tables, respectively:

- **ec2**: This database contains EC2-related resources, including the following:

  - **ebs_volumes**: Contains details of your EBS volumes

  - **ec2_instances**: Contains details of your EC2 instances

  - **ec2_images**: Contains details of your EC2 instance images

  - **routing_tables**: Contains details of your VPC routing tables

  - **security_groups**: Contains details of your security groups

  - **subnets**: Contains details of your VPC subnets

  - **vpcs**: Contains details of your VPCs

- **emr**: This database contains EMR-related resources, including the following:

  - **emr_clusters**: Contains details of your EMR clusters

- **rds**: This database contains RDS-related resources, including the following:

  - **rds_instances**: Contains details of your RDS instances

- **s3**: This database contains RDS-related resources, including the following:

  - **buckets**: Contains details of your S3 buckets

  - **objects**: Contains details of your S3 objects (excluding their contents)

Navigate to **Serverless Application Repository** in the AWS console. Then, click on **Available Applications** from the left navigation bar. Check the **Show apps that create custom IAM roles or resource policies** box. Search for `AthenaAwsCmdbConnector` and click on the result labeled as having been published by an AWS-verified author. Alternatively, you can go directly to the CMDB Connector's detail page via this direct link: `https://amzn.to/3x909VH`. Regardless of how you get there, you'll see a page similar to the one shown in *Figure 12.5* and *Figure 12.6*. At the top of the page, you'll see a basic description of the Connector, including where you can find the source code for it, as well as the "AWS verified author" badging, which guarantees that this Connector is officially supported by the Athena team:

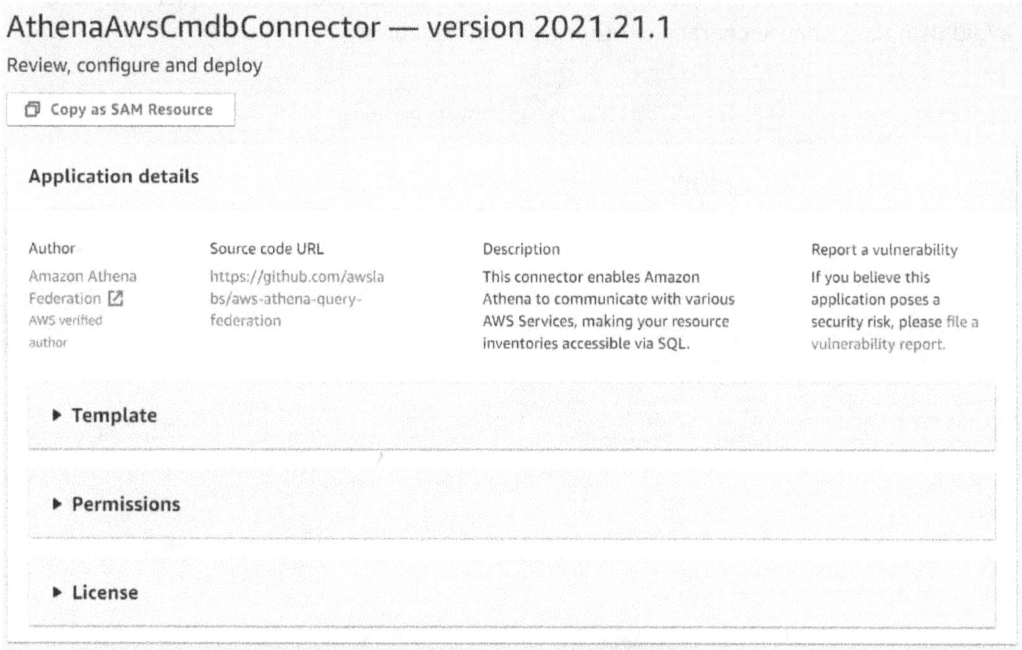

AthenaAwsCmdbConnector — version 2021.21.1

Review, configure and deploy

[Copy as SAM Resource]

Application details

Author	Source code URL	Description	Report a vulnerability
Amazon Athena Federation ↗ AWS verified author	https://github.com/awsla bs/aws-athena-query-federation	This connector enables Amazon Athena to communicate with various AWS Services, making your resource inventories accessible via SQL.	If you believe this application poses a security risk, please file a vulnerability report.

▶ **Template**

▶ **Permissions**

▶ **License**

Figure 12.5 – Athena CMDB Connector summary

For your convenience, you can also click to expand the CloudFormation template and IAM permissions that this Connector will run. The AWS IAM team has released several useful utilities to help you get a handle on your IAM policies, including Access Analyzer. Regardless, it's nice to have the policy that's used by the Connector available before you deploy it. This is the same philosophy that has led us to provide a per-chapter IAM policy in this book's accompanying GitHub repository.

Further down, on the **Connector deployment** page, you'll see the README file for the Connector with more details about its functionality and usage. On the right pane, as shown in the following screenshot, you'll see a form that lets you fill in several settings that will be used to deploy and configure an instance of your Connector. If you are using the IAM policy provided for this chapter, you'll want to ensure that you choose an **application name** that begins with `packt-serverless-analytics-`. This will be used to create the CloudFormation stack that deploys your Connector's Lambda function and IAM role. The other setting you'll need to pay close attention to is **AthenaCatalogName**. Your IAM policy is configured to allow any catalog name that begins with `packt_serverless_analytics`. This will not only be the name of your Lambda function but also the catalog name that you'll use in your Athena query. Be sure to avoid using any special characters other than underscores:

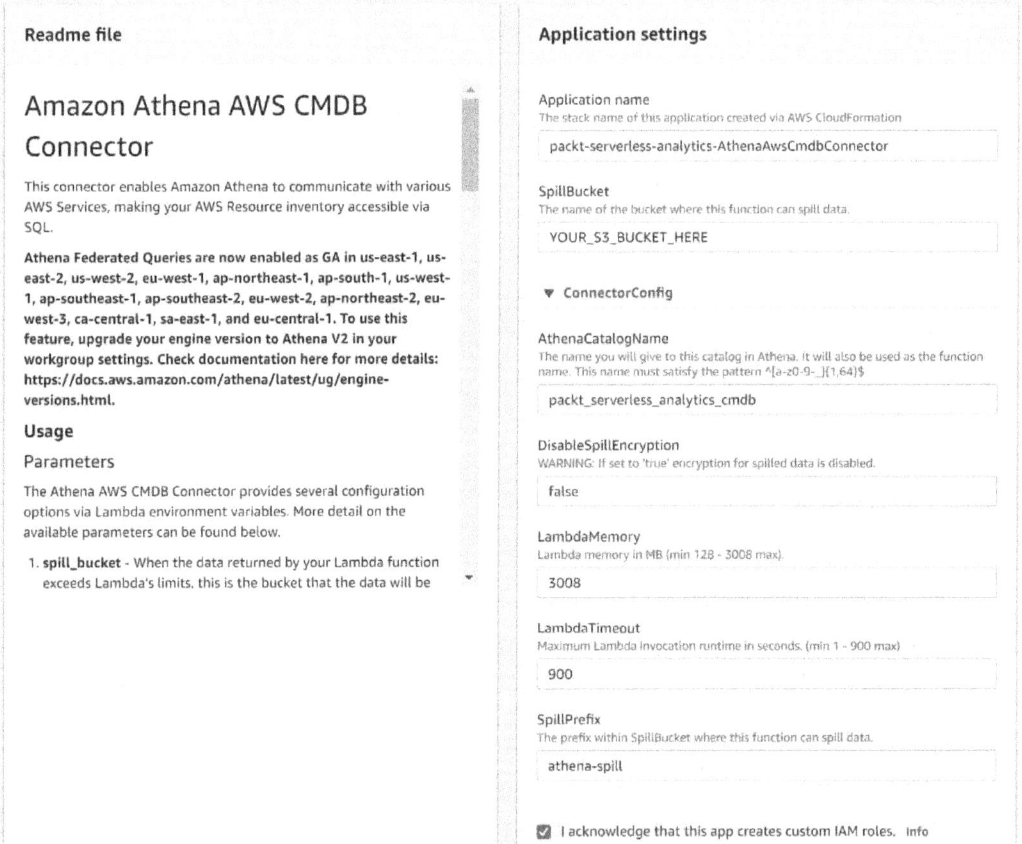

Figure 12.6 – Athena CMDB Connector deployment form

Once you've filled in the other settings, be sure to check the **I acknowledge that this app creates custom IAM roles** dialog box and click **Deploy**. Over the next 5 minutes, the Connectors CloudFormation script will run, creating a Lambda execution role and Lambda function. You can navigate to the CloudFormation console to view the status of each step in the deployment. Once it completes, you can go to the Athena console to run your first federated query!

Our first query will be a basic test of the Connector to ensure that it was deployed correctly. Later, we'll run more interesting queries over our AWS resource inventory. For now, let's see whether we can get our Connector to return a list of schemas it supports. To do this, we'll be using a convenience syntax that Athena offers for federated queries. In the following screenshot, we ran a `show databases in 'lambda:packt_ serverless_analytics_cmdb'` query.

There are three notable aspects to this query. First, it uses backticks around the catalog name. This is one of the few places in Athena's syntax where backticks are used. Second, we can prefix the catalog name with `lambda:`. This is a convenience syntax that tells Athena that this catalog is not registered in Glue Data Catalog or Athena's catalog registry. Instead, treat the rest of this catalog name as a Lambda function. In most cases, you'll want to register the Lambda function as a catalog to make it easier for your customers to discover federated sources. We will use this syntax for now because it lets us get up and running with fewer steps. Lastly, we use the name of our Lambda function, `packt_serverless_analytics_cmdb`. If you used a different name for your Lambda function, be sure to use that in your query. If successful, the query will return a list of schemas, or databases, including `rds`, `s3`, `ec2`, and `emr`:

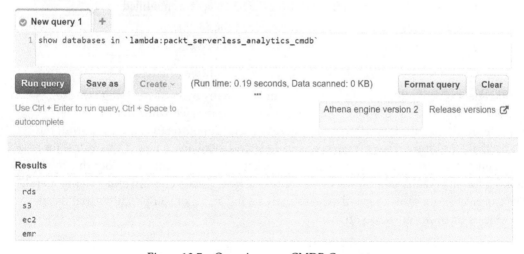

Figure 12.7 – Querying your CMDB Connector

Now that we know our Connector has been deployed successfully and that our IAM user has been configured correctly to interact with the Connector, let's explore one of the more interesting tables that's exposed by this Connector. Running the following SQL from the Athena console will return a description of a virtual table that we can query to get the list of S3 objects in a bucket:

```
DESCRIBE 'lambda:packt_serverless_analytics_cmdb'.s3.objects
```

Then, we can use the full capability of Athena's SQL engine to filter or join this data against other sources. Notice again that we are using the `lambda:` syntax described earlier. This query will return the table schema for the `Objects` table of the S3 database, as shown in the following table. We'll use this schema to craft a query over the Athena results location we've been using in S3 across the various chapters in this book. You can certainly choose to query something else once you get a handle on the schema:

Column Name	Data Type	Description
bucket_name	varchar	The name of the bucket that this object is in. You must specify a specific bucket in the where clause to query this table as that is how the connector will configure its S3 client. It will not list all the objects in all buckets.
key	varchar	The key of the S3 object.
e_tag	varchar	The ETag of the S3 object.
bytes	bigint	The size of the S3 object in bytes.
storage_class	varchar	The S3 storage class of the S3 object.
last_modified	timestamp	The last time the S3 object was modified.
owner_name	varchar	The owner name of the S3 object.
owner_id	varchar	The ID of the S3 object's owner.

Table 12.1 – S3 objects table schema

For this next query, our goal will be to find out how many bytes of S3 storage we are currently using for all the Athena query results we've generated while following the exercises in this book. Accordingly, we will select the sum of the bytes column from the `s3` database and the `objects` table, as shown in the following query. This Connector uses a bit of trickery to determine which S3 bucket you want to query. It does this by looking for an equality condition between the `bucket_name` column and a string literal. The Connector does this to avoid an extremely taxing series of S3 operations that list all the objects in all the buckets you own.

Hence, omitting a `where` clause filter for a specific bucket results in the query failing. Lastly, we will use a regular expression to filter down to keys that begin with `results/`. If your Athena results location is different from this, be sure to update the expression so that it matches the results location you have configured on your Athena workgroup. When you are ready, go ahead and run the query:

```
SELECT sum(bytes)
FROM
    "lambda:packt_serverless_analytics_cmdb".s3.objects
WHERE
    bucket_name = '<YOUR_ATHENA_RESULTS_S3_BUCKET>'
    and regexp_like(key, 'results/.*')
LIMIT 100
```

You can follow the same steps to deploy any of Athena's pre-built Connectors for sources, including Elasticsearch, DynamoDB, Neptune, and more. Now that we've deployed our first Connector and run a few federated queries, we're ready to author a custom Connector. The next section will walk us through integrating Athena with any data source we can imagine, just by writing six Java functions!

Building a custom connector

In addition to the 14 Connectors published by the Athena service team, a growing community of third-party and open source Connector authors is continually adding Connectors to the ecosystem. In most cases, you'll be able to use a ready-made Connector to query your source of interest. However, there may be cases where you'd like to modify or extend one of the existing open source Connectors to better fit your needs. Or maybe your company has a proprietary datastore or storage format that would benefit from a serverless query engine. Whatever the reason, this section will walk you through the key steps of authoring a new Connector and using it with Athena.

Setting up your development environment

To write a new Connector or modify an existing one, we'll need the ability to build, test, and package the code. So, our first task is to ensure we have a development environment with the appropriate builder tools. These tools will include Apache Maven, the AWS CLI, and the AWS SAM build tool. The Apache Foundation describes Maven as a "software project management and comprehension tool." That's a fancy way of saying Maven helps automate dependency management, build orchestration, and a host of related activities that can be added or augmented via plugins. The AWS SAM build tool is how we'll package our Connector so that it can be used with Lambda and Serverless Application Repository. Lastly, the AWS CLI will help us publish our Connector to Serverless Application Repository for deployment.

If you've already got an environment that meets these requirements, you're welcome to use it. If not, don't worry – we'll guide you through the setup for Debian Linux-based systems such as Ubuntu or Ubuntu on Windows Subsystem For Linux. Most of the commands will work on other flavors of *nix with minor modifications. We'll assume you will be using a basic text editor without any fancy builder tool integrations and that you will need a guide to install these other dependencies. Let's get started by cloning this book's accompanying GitHub repository by using the following command:

```
git clone https://github.com/PacktPublishing/Serverless-
Analytics-with-Amazon-Athena.git
```

Inside the chapter_12 directory, you'll find a prepare_dev_env.sh script that you can run to install OpenJDK, Apache Maven, the AWS CLI, and the AWS SAM build tool. Depending on your CPU and disk speeds, the script may take 5 minutes or more to set up your environment. If the script is successful, it will print the installed version of each required tool at the end. The output will look similar to the following, but don't worry if your versions differ slightly:

```
aws-cli/1.19.96 Python/2.7.18 Linux/4.19.128-microsoft-standard
SAM CLI, version 1.24.1
Apache Maven 3.5.4
openjdk version "11.0.11" 2021-04-20
javac 11.0.11
```

Next, we'll start writing the code for our custom Connector!

> **A Note about Java Versions**
>
> While the Athena Federation SDK and its Connectors should work fine with
> Java versions up to and including 11, issues have been reported with Apache
> Arrow with JDK versions beyond 8. The open source project has resolved many
> of these issues and provided configuration workarounds for the remaining
> items. If you run into errors that appear to be related to your JDK version, try
> executing the exercise with OpenJDK 8.

Writing your connector code

For this exercise, we've put together a training Connector with all the structural
boilerplate of an actual Connector taken care of for you. To guide you through the
authoring experience, we've included the working code for each `require` function in
the comments. This allows you to learn by doing while also putting the correct answer
at arm's length so that you won't get stuck or frustrated. To get started, navigate to the
`athena-example` folder in the `chapter_12` directory of the GitHub repository
you cloned in the previous section. You'll want to have a terminal window open in this
directory, and we'll want to open the directory in our favorite text editor or IDE.

The `athena-example` Connector we'll be working with was initially provided by the
Athena team as part of the Athena Federation SDK to teach customers how to write
Connectors. We'll use a fork of that original Connector that reads data stored in S3 using
a custom metadata source and custom file format. To make this exercise more realistic,
the Connector is designed to read fictitious financial transaction data and even provide
column-level masking capabilities. The Connector itself is intentionally simple, so you
can focus on learning how to build a custom Connector instead of how to integrate with
a specific source. In the sections that follow, we'll fill in the missing metadata and record
reading code. We'll also run the included unit tests and use the Athena Federation SDK's
ConnectorValidator to simulate Athena calling our Connector during an actual query.
And of course, we'll end by using our new Connector in an Athena query.

Editing ExampleMetadataHandler

When you open the `athena-example` folder, you'll find several configuration files, a license file, and some sample data in a CSV file. You'll also see an `src` directory that contains the code for your soon-to-be Connector. The first file we'll need to modify is the `ExampleMetadataHandler.java` file in the `src/main/java/com/amazonaws/connectors/athena/example` directory. This class is responsible for providing Athena with metadata about the schemas (that is, databases), tables, columns, and the general layout of your data source. Lastly, this class tells Athena how to break up reads against your data source. This gives you control over the level of performance and parallelism Athena achieves when reading your tables. Let's go function by function in this class, explaining what your code needs to do to complete the exercise. For brevity, we won't include all the code you need in this book. Function signatures, returns, and other boilerplate will be omitted. This is an exercise that requires that you use this book's GitHub repository to get the full effect. If you aren't coding along with us, you can open the appropriate file in this book's GitHub repository in your browser using this link: `https://bit.ly/3iRrHv8`.

Our first function is `doListSchemaNames`. Since this is the first function you will be editing, we've included the entire function here. This function has an elementary responsibility. Athena will call it any time you run a `show databases in 'lambda:<function_name>'` query to get the list of schemas (that is, databases) from your Connector. Looking at the function example here and in the GitHub project, you can see a working solution already included in the function, but it's commented out. The first time you create a new Connector, we recommend reading the commented solution and then uncommenting it to get a working Connector. You can repeat this exercise and make changes once you understand how everything works:

```
public ListSchemasResponse doListSchemaNames(BlockAllocator
allocator,
                                             ListSchemasRequest
request)
    {
        Set<String> schemas = new HashSet<>();
        /**
         * TODO: Add schemas, example below
        schemas.add("schema1");
        */
        return new ListSchemasResponse(request.
getCatalogName(),
```

```
                                        schemas);
    }
```

Our next function, doListTables, is just as trivial. This is the final time that we'll show the entire function body so that you get the hang of completing the exercise. Similar to the previous function, Athena will call this function when your run a show tables in 'lambda:<function_name>'.schema1 query to get the list of tables in the requested schema. Again, your job, when completing this function, is to uncomment the provided solution. In this case, we are returning the same three tables, regardless of which schema was specified in the request object:

```java
public ListTablesResponse doListTables(BlockAllocator
allocator,
                                        ListTablesRequest
request)
    {
        List<TableName> tables = new ArrayList<>();
        /**
         * TODO: Add tables for the requested schema, example
below
        tables.add(new TableName(request.getSchemaName(),
"table1"));
        tables.add(new TableName(request.getSchemaName(),
"table2"));
        tables.add(new TableName(request.getSchemaName(),
"table3"));
         */
        return new ListTablesResponse(request.getCatalogName(),
                                        tables,
                                        null);
    }
```

We've finished two of the six functions that we need to write! But don't get too excited – doListSchemaNames and doListTables were just a warmup. The remaining four functions have some meat to them, and each implements a vital aspect of executing our queries. The doGetTable function is on deck. When Athena is parsing our query, it will call doGetTable to ensure the tables and columns that are referenced in our query are valid and get the types of each column. In our example Connector, we don't bother validating whether the requested table exists, but normally, this is a key part of this function. For now, our example focuses on building and returning the schema of the single table our Connector supports. The function begins with the following code snippet, which specifies the names of the table's partition columns. You may be wondering about the significance of declaring a partition column. From Athena's perspective, this designation means little other than to indicate your data source can prune data, along these dimensions with elevated efficiency. In our case, we have three partition columns called year, month, and day:

```
Set<String> partitionColNames = new HashSet<>();
/**
 * TODO: Add partitions columns, example below.
partitionColNames.add("year");
partitionColNames.add("month");
partitionColNames.add("day");
*/
```

In addition to the partition columns, we also need to return the complete list of columns and their associated data types so that Athena knows what kind of data to expect. You may recall from earlier in this chapter that Athena uses Apache Arrow as its data interchange format. The Apache Arrow specification also provides a way of defining the schema of your data. Accordingly, Athena expects the schema of your federated tables to be defined as an Apache Arrow schema. The Athena Federation SDK provides a convenient SchemaBuilder to make interacting with Apache Arrow's Schema object easier. In the following code snippet, we are using SchemaBuilder to produce an Apache Arrow schema object from our hardcoded schema. That hardcoded schema contains the three partition columns that we declared earlier (month, year, and day). Here, we are defining them as integers.

Next, we must add an `account_id` column defined as a string and a `transaction` column of the `Struct` type with two child columns. Since the Athena Federation SDK is built on Apache Arrow, we can use complex types such as `Struct` and `List`. The transaction struct contains an `id` field of the `Integer` type and a Boolean field named `completed`, indicating whether the system has finished processing this particular transaction:

```
        SchemaBuilder tableSchemaBuilder = SchemaBuilder.
 newBuilder();
        /**
         * TODO: Generate a schema for the requested table.
        tableSchemaBuilder.addIntField("year")
        .addIntField("month")
        .addIntField("day")
        .addStringField("account_id")
        .addStringField("encrypted_payload")
        .addStructField("transaction")
        .addChildField("transaction", "id",
                    Types.MinorType.INT.getType())
        .addChildField("transaction", "completed",
                    Types.MinorType.BIT.getType())
        */
```

We momentarily skipped the string-based `encrypted_payload` field because it's a bit special. This field is intended to demonstrate the level of customization that is possible with the Athena Federation SDK. In our sample data file, we've stored a piece of sensitive information that only certain users should be able to decrypt. We'll use a **UDF**, covered more deeply in *Chapter 13*, *Athena UDFs and ML*, to decrypt this secret field right in our query. This may seem like a cumbersome way to handle row- or cell-level access control, and it is. AWS Lake Formation offers better options for this. We're only doing this to illustrate the level of customization you can achieve when writing a Connector. As with the previous code snippets, uncomment the working solution before moving on to the next function.

Later in this chapter, we'll upload sample data that matches this schema to S3 for our custom Connector to consume. To ensure the sample data works as expected, you should refrain from modifying the example code unless you plan to make appropriate changes throughout the example.

The `getPartitions` function is next, and while technically this function is considered optional to allow for unpartitioned tables, we'll implement it so that you know how it works. Since our example Connector doesn't connect to an actual data store, we will use hardcoded values for our partitions. This can be accomplished in the sample code by using a series of nested `for` loops. The outer loop generates `year` values from 2000 to 2018. The middle loop generates `month` values ranging from 1 to 12. The final loop naively generates `day` values from 1 to 31 without any regard for how many days are in the month. But this isn't the exciting part of the function, which comes next:

```
public void getPartitions(/* arguments omitted */) throws
Exception {
    for (int year = 2000; year < 2018; year++) {
        for (int month = 1; month < 12; month++) {
            for (int day = 1; day < 31; day++) {
                final int yearVal = year;
                final int monthVal = month;
                final int dayVal = day;
```

Based on this `getPartitions` code, Athena queries that use this Connector would always need to process 6,324 partitions. That would be inefficient and slow if the query filtered on a specific partition using a WHERE clause such as `"year = 2001 and month = 1 and day = 1"`! Luckily, Athena will include relevant predicate conjuncts from your query when it calls your Connector. This allows us to use the built-in features of the Athena Federation SDK to implement partition pruning and filter out irrelevant partitions much earlier in the query's execution.

When the Athena Federation SDK calls our `getPartitions` function, it supplies us with an instance of `BlockWriter`, which we can use to add partitions to the API's response. `BlockWriter` is automatically configured with the query predicates that were sent by Athena. Keep in mind that Athena won't send all the query predicates. For example, if you use the result of a function in your WHERE clause, Athena won't send that part of the predicate to you since your Connector may not support that function. In general, Athena only sends associative conjuncts on literal values.

In the following snippet, we are calling the supplied BlockWriter's `writeRows` function for each partition and providing a Lambda expression to set the values of the partition columns:

```
/**
 * TODO: Build partitions
 blockWriter.writeRows((Block block, int row) ->
{
        boolean matched = true;
        matched &= block.setValue("year", row,
yearVal);
        matched &= block.setValue("month", row,
monthVal);
        matched &= block.setValue("day", row,
dayVal);
        return matched ? 1 : 0;});
     */
    }
   }
  }
 }
```

The Athena Federation SDK makes frequent use of this pattern wherever your code needs to interact with blocks of Apache Arrow data. By accepting a lambda instead of giving your function direct access to the `Block` objects, the Athena Federation SDK can handle most nuances of Arrow memory management. This dramatically reduces the chances that the author of a Connector introduces an unintended memory leak or race condition. Experts can still get full access to the Apache Arrow objects, but the default experience is much more curated.

Within the Lambda expression, we can use the `setValue` function of `Block` to enter the values for the `year`, `month`, and `day` columns that uniquely identify our partition. After each call to `setValue`, we record whether the column's value matched the partition pruning predicate sent by Athena. Under the hood, the `BlockWriter` and `Block` constructs are applying the query conjuncts to determine whether the values match the query. However, this is just an optimization since Athena applies its partition pruning over your result to capitalize on any predicates that it could not send to your Connector. This is in addition to the actual data filtering that happens later in the query.

Finally, we've reached our final metadata operation, doGetSplits. This function complements getPartitions. Athena will call doGetSplits for each partition supplied by getPartitions. Even if your Connector does not support partitioning, the Athena Federation SDK needs to return at least one partition if Athena thinks your query was fully partition pruned. The doGetSplits example starts with a bit of boilerplate code that extracts key input parameters from the request object. The only notable part is how we retrieve the details of which partitions Athena is requesting splits from. Each call to doGetSplits receives a batch of partitions in the form of an Apache Arrow block. To retrieve the values from the partitions block, we must use the getFieldReader method for each column we need to read:

```
public GetSplitsResponse doGetSplits(BlockAllocator allocator,
                                     GetSplitsRequest request){
    String catalogName = request.getCatalogName();
    Set<Split> splits = new HashSet<>();
    Block partitions = request.getPartitions();
    FieldReader day = partitions.getFieldReader("day");
    FieldReader month = partitions.getFieldReader("month");
    FieldReader year = partitions.getFieldReader("year");
```

Now that we have a handle for the values in each partition column, we can loop over all the rows in the partition block that Athena sent us. We can do that using a simple for loop from zero to partitions.getRowCount(). The getRowCount function conveniently returns the number of rows in the Apache Arrow block. Because Apache Arrow is designed as an in-memory columnar data format, the FieldReaders we created earlier implement random access to each column. Accordingly, the first thing we must do inside our loop is set the row number on each of our partition column FieldReaders:

```
    for (int i = 0; i < partitions.getRowCount(); i++) {
        //Set the readers to the partition row we area on
        year.setPosition(i);
        month.setPosition(i);
        day.setPosition(i);
```

This next part may be a bit confusing. For each partition, we need to generate one or more splits. What is a split, you ask? It's whatever you want or need it to be. There is only one thing Athena understands about a split. A split represents a piece of your table that needs to be read to complete the query. In this way, we can say that you split up the read into one or more read operations. This is the most critical unit of parallelism for your query. More splits means more opportunities to parallelize. Everything else about a split exists to support your ability to execute the read operation it represents. As such, Athena will call your Connector to read the data associated with a split, and it will supply the split object itself as a parameter to that read operation. We'll cover the `readWithConstraint` function and how it can use a split as input in the next section. For now, let's look at how we can construct a split.

A split is primarily a map of string keys and string values that serve as a place for the developer of a Connector to store arguments that the `readWithConstraint` function will need. When we construct the split, we supply a `SpillLocation` value and an `EncryptionKey` value using helper methods provided by the Athena Federation SDK. When reading a split, if the Lambda function generates more than 6 MB of data, the Athena Federation SDK must spill the data to S3 to avoid Lambda's response size limit. This spill location and optional encryption key are provided to Athena, as well as the eventual call to `readWithConstraint` via the split. Every split must have a unique spill location to avoid duplicate data, throttling, and, in some cases, query failure. The `makeSpillLocation` function ensures no two calls to the method return overlapping spill locations. The `makeEncryptionKey` function supports locally generated AES-GCM keys, as well as AWS KMS-generated AES-GCM keys. We recommend using at least the local key generation as it's free and doesn't meaningfully impact performance.

Lastly, in our example Connector's `doGetSplits` function, we must generate one split for each partition and store the `year`, `month`, and `day` partition field values on the split:

```
        /**
         * TODO: For each partition, create one or more splits.
        Split split = Split.
 newBuilder(makeSpillLocation(request),

                                        makeEncryptionKey())
            .add("year", String.valueOf(year.readInteger()))
            .add("month", String.valueOf(month.
 readInteger()))
            .add("day", String.valueOf(day.readInteger()))
            .build();


        splits.add(split);
```

```
        */
    }
    return new GetSplitsResponse(catalogName, splits);
}
```

The function concludes by accumulating and returning all the splits we generated. This example doesn't utilize the doGetSplit function's ability to support paginated responses. If you plan to produce more than a few hundred splits per call or need to store non-trivially sized parameters on the Split object, you should limit the size of your response by returning a pagination token to Athena. You can find examples of how to do this by looking at the athena-cloudwatch connector in this book's GitHub repository.

Now that we have covered all the metadata functions and five of the size total functions in our Connector, let's move on to the ExampleRecordHandler class and finish the Connector's implementation.

Editing ExampleMetadataHandler

In a few minutes, we'll be ready to package and deploy our new Connector. But before that, we'll need to implement the final piece of code by modifying the ExampleRecordHandler.java file in the src/main/java/com/amazonaws/connectors/athena/example directory. This class is responsible for providing Athena with row data from your data source. If you aren't coding along with us, you can follow along with this book's GitHub repository in your browser by going to https://bit.ly/3vISV9X.

Again, our function starts with a few lines of boilerplate code where we extract the configuration from the request object. In the following code snippet, we've consolidated some of the multi-line statements you'll find in the GitHub example. They are functionally equivalent, but the shortened form is a bit easier to read in book form. These lines are mostly uninteresting as we're extracting the information we stored on the split in doGetSplits:

```
protected void readWithConstraint(/* arguments omitted */){
    Split split = recordsRequest.getSplit();
    int splitYear = split.getPropertyAsInt("year");
    int splitMonth = split.getPropertyAsInt("month");
    int splitDay = split.getPropertyAsInt("day");
```

Next, we can see why we took the time to pass the partition column values of year, month, and day to the readWithContraint method via the split. The example Connector is intended to read financial transaction data from S3. The Connector uses the year, month, and day to determine which S3 file to read! This is similar to how Athena, Spark, Hive, and other analytics engines resolve which files to query in S3. Our Connector uses the Java.lang.String.format function to substitute the year, month, and day values into a hardcoded string representing the S3 key we need to read. In the preceding line, we retrieved the S3 bucket that our data is stored in by reading the data_bucket environment variable via Java's System.getenv(...) facility.

For this example Connector, we could have hardcoded dataBucket too, but this approach lets us demonstrate how to use AWS Lambda's environment variables to pass runtime configuration to your Connector. You can modify these settings at any time from the AWS Lambda console or API without the need to recompile or redeploy your Connector. Once we have the S3 bucket and object key, we must use the provided openS3File helper function to obtain a handle we can use to read the data contained in the same data file that we'll upload to S3 later. If the file doesn't exist, the helper returns null, and our Connector exits without writing any rows:

```
String dataBucket = dataBucket = System.getenv("data_
bucket");
String dataKey = format("%s/%s/%s/sample_data.csv",
                    splitYear,
                    splitMonth,
                    splitDay);

BufferedReader s3Reader = openS3File(dataBucket, dataKey);
if (s3Reader == null)
    return; //There is no data to read for this split.
```

With BufferedReader in hand, our Connector is finally ready to send row data to Athena. Doing so requires that we translate the data from its storage format and type system into Athena's data interchange format, Apache Arrow. In the getPartitions method, we used BlockWriter to simplify writing our partition column values into Apache Arrow's type and storage format. This code is only expected to write a few thousand values of primitive types, such as integers. readWithConstraint is anticipated to write many megabytes of data across hundreds of thousands of cells. Rather than writing Apache Arrow data row by row, we can dramatically improve throughput by adopting Apache Arrow's column-wise approach to data storage.

The Athena Federation SDK's `GeneratedRowWriter` and `RowWriterBuilder` functions provide simplified models for decomposing the steps for translating each column in a row. `GeneratedRowWriter` also automatically applies query constraints that are passed by Athena, saving us the effort of writing code to filter results. The SDK then uses a primitive form of code generation to reduce the number of `if` statements or branches in the critical execution paths. With fewer branches, a CPU's branch prediction logic and Java's code caches can utilize the available resources better, leading to faster queries.

Let's take a closer look at how we can use `RowWriterBuilder`. Be sure to consult this book's GitHub repository for the complete set of column translations as we'll only be covering two columns here. You'll find that the rest are nearly identical to these examples. In the following code snippet, we can see how to create the `Extractor` method for the `year` column. An `Extractor` is what the Athena Federation SDK calls the method that is capable of translating a column value from the source data into Apache Arrow. Extractors contain a single method that accepts an Apache Arrow value *holder* that corresponds to the data type of the column they are capable of extracting from the source data. Since the `year` column is defined as an `Integer` column, its `Extractor` is of the `IntExtractor` type, and it expects `NullableIntHolder` for storing the value it extracts from the source:

```
builder.withExtractor("year", (IntExtractor) (Object
context, NullableIntHolder value) -> {
        value.isSet = 1;
        value.value = Integer.parseInt(((String[])
context)[0]);
    });
```

These *holders* are an essential concept in Apache Arrow. Apache Arrow seeks to eliminate, or, at the very least, limit, memory copies and churn when going from one analytics system to another. A value holder is typically a long-lived object with pre-allocated memory that can be reused many times to avoid frequent, small memory allocations and collections. You'd be surprised how big a difference this can make when you're reading and writing millions of integers. Yes, the code can seem a bit terse, but the throughput associated with this programming model can be seven times that of more naïve approaches such as that used in the `getPartitions` example. And that example isn't even the most naïve!

The other argument that every `Extractor` expects is context. The context is declared to be an `Object` because the framework doesn't know what it is. In this particular case, the context is a single line from the CSV sample data file we will store in S3. Later, in the `readWithContraint` function, we'll see how the context gets populated. This is ultimately why our `Extractor` can cast the context to an array of strings before parsing the first value into an integer.

The `account_id` extractor builds on these concepts and adds a twist. Suppose our organization has decided that account IDs are sensitive **Personally Identifiable Information (PII)**. To comply with regulatory mandates, our organization must mask the `account_id` field whenever it is queried. We could rewrite our entire dataset or produce a sanitized copy, but that can be expensive and difficult to manage. Instead, we can do exactly as our example Connector does here and mask the field while translating it into Apache Arrow:

```
    builder.withExtractor("account_id", (VarCharExtractor)
  (Object context, NullableVarCharHolder value) -> {
        value.isSet = 1;
        String id = ((String[]) context)[3];
        value.value = (id.length() > 4) ?
        value.value = id.substring(id.length() - 4) : id; });
```

The `account_id` column uses a `VarCharExtractor`, which performs a substring on the `account_id` value to ensure it never returns more than the last four characters of the source data's value. In practice, hardcoded masking like this isn't practical. Most customers will be better off leveraging AWS Lake Formation's masking capabilities. However, if you have to apply masking to a non-Lake Formation source, this can be a great option. It also illustrates that Connectors can apply intelligence to their translations. Masking is just one of many possibilities.

The final column translation we'll review belongs to the `transaction` field. Unlike the previous fields, which were primitive types, the `transaction` column was defined as a struct. The Athena Federation SDK does not provide generalized optimizations for translating complex types such as `Struct` or `List`. Instead, `RowWriterBuilder` expects the Connector author to provide a `FieldWriterFactory` for such columns. If your complex types are not deeply nested, the experience will closely resemble the `Extractor` model we just used with the `year` and `account_id` columns. When you start deeply nesting, efficient translation can feel like trying to codify *Inception* but with data analytics instead of dreams.

In the following final column translation snippet, our Connector is building a map corresponding to the child fields of the `transaction` struct. It then uses the `BlockUtils` helper from the Athena Federation SDK to write the map as an Apache Arrow struct. This helper is an extremely convenient tool for dealing with Apache Arrow data of all types. You'll see it used repeatedly in the unit tests. Unfortunately, this convenience comes at a price. Nearly every method on the `BlockUtils` class is an order of magnitude slower than using columnar models for interacting with Apache Arrow resources. This is why you'll rarely see this utility used in Connector code, except for `getPartitions` where the number of rows is almost always too low to affect performance measurably:

```
builder.withFieldWriterFactory("transaction", (FieldVector
vector, Extractor extractor, ConstraintProjector constraint) ->
    (Object ctx, int rowNum) -> {
        Map<String, Object> eMap = new HashMap<>();
        eMap.put("id", Integer.parseInt(((String[])ctx)[4]));
        eMap.put("completed",Boolean.parseBoolean(((String[])
ctx)[5]));
        BlockUtils.setComplexValue(vector, rowNum,
                                FieldResolver.DEFAULT,
eventMap);
        //predicate pushdown not yet supported on complex types
        return true;
});
```

We've nearly completed the `readWithConstraint` function. All we have left to do is read the source data line by line and invoke `RowWriter`, which we just generated. This is how all the extractors we wrote will receive their context objects. In the last section of `readwithConstraint`, our Connector uses `BufferedReader, which it constructed earlier,` to read the S3 object containing the sample data line by line. Each line is then split using commas as separators before calling the provided `BlockSpiller` to write rows to the Athena response. We didn't show it here, but the example code calls the build method of `RowWriterBuilder` to produce the `rowWriter` object, which processes each line in our `while` loop and adds rows that meet the queries filtering criteria to the block. We then return one or zero to tell `BlockSpiller` whether the row we translated was skipped because it didn't pass the queries filter:

```
while ((line = s3Reader.readLine()) != null) {
    String[] lineParts = line.split(",");
```

```
         spiller.writeRows((Block block, int rowNum) ->
             rowWriter.writeRow(block, rowNum, lineParts) ? 1 :
 0);
     }
 }
```

In a genuine Connector, the functions we completed would most likely have read their metadata from a durable store instead of having hardcoded values or relying on naming conventions in S3. But remember, our goal was to familiarize ourselves with writing a Connector, leaving you free to focus on the nuances of your intended data source when the time comes to write a new Connector. Take a moment to enjoy the feeling of completing the most intensive coding exercise in this book. In the next section, we'll see how good of a job we did by deploying and testing the Connector we just wrote.

Deploying and testing your custom connector

If you've been using an IDE to complete this exercise, you have already run a syntax check, possibly even the unit tests. However, if you've been using a regular text editor, let's begin by using the Apache Maven command-line tool to build our Connector code and execute the included unit tests. The easiest way to do this is to open a terminal and navigate to the `athena-example` directory. Once you're there, execute `mvn clean install -Dpublishing=true`. If this is your first time building an Athena Connector on that machine, Apache Maven will take several minutes to download the necessary dependencies. These dependencies include the Athena Federation SDK, Apache Arrow, and many other open source libraries. Once all the dependencies have been satisfied, Maven will build the Connector run unit tests. This one command will catch the majority of common errors long before Athena enters the picture. This ability to iterate quickly and locally makes developing new Connectors easier. Once the build completes, please note any errors, especially unit test errors, and resolve them before moving on.

Conditional Maven Builds

We pass the `"-Dpublishing=true"` flag to indicate to the `athena-example` Connector's build configuration that we've completed the exercise and that the full unit test suite should be applied. This is a bit atypical because we'd expect unit tests to be run as part of any build, but this Connector is an exception. As a result, the code in our GitHub repository is incomplete until you uncomment or implement the missing functionality. Yet, at the same time, we want to have real unit tests to help ensure you don't miss any steps.

With our Connector code built and passing all unit tests, we're ready to package and deploy the Connector. This process involves producing a specially structured ZIP file and accompanying configuration file that Serverless Application Repository can use to deploy our Lambda function. It is also possible to avoid using Serverless Application Repository and instead deploy directly to AWS Lambda. In this exercise, we'll be using the AWS SAM Build Tool to upload our packaged Connector to Serverless Application Repository for a one-click deployment experience. You can use the provided `publish.sh` file in the root of this chapter's GitHub directory to automate the entire process. The script requires an S3 bucket that it can use to upload your Connector code for later use by Lambda, the directory name that contains the Connector code to package, the AWS Region you'd like to publish to, and the partition type (typically, this is `aws`):

```
./publish.sh S3_BUCKET athena-example AWS_REGION aws
```

When executed, this script will print a confirmation of the steps it is about to perform. It begins by looking for a valid set of AWS credentials. So, if you haven't run `aws configure`, you should do so now. This command ensures your AWS CLI and supporting tools are ready to use. The publish script runs several AWS CLI and SAM build tool commands while orchestrating the Connector's deployment. These commands will fail if they can't find your AWS credentials.

As part of publishing your Connector, the publish script will attempt to add an S3 bucket policy to the S3 bucket you supplied to the command. The policy will grant the AWS Serverless Application Repository read access to the S3 bucket so that it can read a copy of the Connector code on behalf of AWS Lambda when doing deployments. If the script sees an existing policy on the S3 bucket, it will skip this step and assume you've manually configured an appropriate policy to avoid overwriting your work. Keep this in mind if you get failures related to Serverless Application Repository being unable to retrieve your Connector code. The script will then package your Connector code by recompiling it and rerunning unit tests before producing a Lambda-compliant ZIP file of the Connector artifacts. Lastly, the script will upload the Connector code artifact to the S3 bucket and call Serverless Application Repository to create a deployable application in your private repository. The resulting serverless application can then be shared with other accounts if you choose, but it will initially be marked as private.

In the end, the script will print a URL, much like the one shown here, that can be used to view the Connector in Serverless Application Repository:

```
https://console.aws.amazon.com/serverlessrepo/home?region=us-
east-1#/published-applications/arn:aws:serverlessrepo:us-east-
1:XXXXXXX:applications~ExampleAthenaConnector
```

Before we deploy our Connector, let's upload the sample financial transaction data that the Connector will use to answer queries. From the `athena-example` directory, you can execute the following S3 put command to upload the sample data. Be sure to replace BUCKET_NAME with the name of the S3 bucket you'd like to use for this exercise. Later, we'll need to enter the name of this S3 bucket in the Connector deployment configuration, so keep it handy:

```
aws s3 cp ./sample_data.csv s3://BUCKET_NAME/2017/11/1/sample_
data.csv
```

Now, we're ready to go to the AWS console and deploy our custom Connector. We'll repeat several of the same steps from earlier in this chapter, where we deployed a pre-built Connector from Serverless Application Repository. Unfortunately, the link from the end of the publish script can't be used to deploy the Connector. You'll need to open Serverless Application Repository in the AWS console and click on **Available Applications** in the left navigation bar. Then, select the **Private Applications** tab, at which point you should see **ExampleAthenaConnector**. From here, the process is nearly identical to deploying a pre-built Connector.

Click on it, and you'll be prompted to fill in any missing configuration details before deploying, as shown in the following screenshot. If you're using the IAM policy for this chapter, be sure to choose an **application name** that begins with `packt-serverless-analytics` and an **AthenaCatalogName** value that begins with `packt_serverless_analyics` to avoid permissions issues. **Application name** corresponds to the underlying CloudFormation stack this process creates, while **AthenaCatalogName** will become the name of the Lambda function:

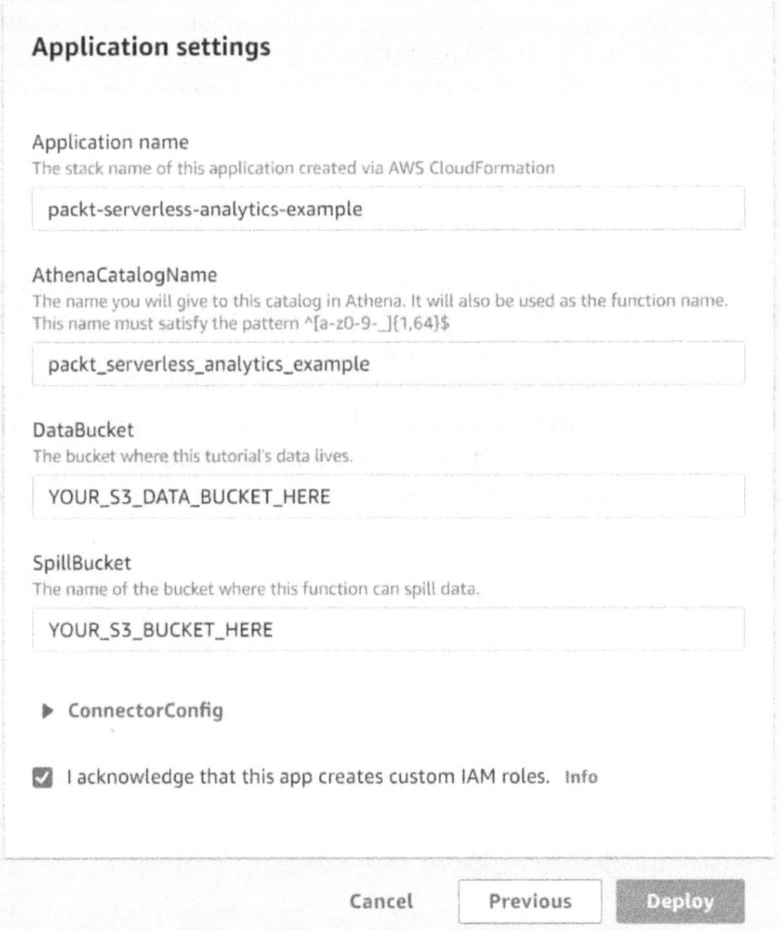

Application settings

Application name
The stack name of this application created via AWS CloudFormation

packt-serverless-analytics-example

AthenaCatalogName
The name you will give to this catalog in Athena. It will also be used as the function name.
This name must satisfy the pattern ^[a-z0-9-_]{1,64}$

packt_serverless_analytics_example

DataBucket
The bucket where this tutorial's data lives.

YOUR_S3_DATA_BUCKET_HERE

SpillBucket
The name of the bucket where this function can spill data.

YOUR_S3_BUCKET_HERE

▶ ConnectorConfig

☑ I acknowledge that this app creates custom IAM roles.   Info

Cancel   Previous   Deploy

Figure 12.8 – Example Connector deployment config

DataBucket should match the S3 bucket where you uploaded the sample data. For simplicity, we recommend using the same S3 bucket for the sample data, **SpillBucket**, and publishing your Connector. Once the deployment completes, you should be able to run the following query in the Athena console. Be sure to replace the Lambda function name with your function name if you didn't follow the suggested naming convention:

```
USING EXTERNAL FUNCTION decrypt(payload VARCHAR )
        RETURNS VARCHAR LAMBDA 'packt_serverless_analytics_
example'
SELECT year,
        month,
        day,
        account_id,
        encrypted_payload,
        decrypt(encrypted_payload) AS decrypted_payload,
        transaction.id AS tx_id
FROM "lambda:packt_serverless_analytics_example".schema1.table1
WHERE year=2017
        AND month=11
        AND day=1;
```

The query will return a few hundred rows of data from the sample data file we uploaded to S3. It's taken us a while, but we've built and deployed a custom Connector! You can use what've you've learned in this chapter to integrate with any data source you may need. Athena Query Federation holds one more secret, and this query is hinting at it. You may have noticed two rather curious columns in the output. The `encrypted_payload` column looks like jibberish, but the `decrypted_payload` column is a human-readable copy of the `encrypted_payload` column that has been postprocessed by an **external** function called decrypt.

Along with our Connector code, this example contained the `decrypt` UDF. In the next chapter, we'll learn more about Athena UDFs, including a special case UDF that allows us to take advantage of SageMaker's machine learning tools within our Athena queries.

> **Troubleshooting Custom Connectors**
>
> Also included in the GitHub repository is the `athena-federation-sdk-tools` module, which provides a Connector validator tool that can be used to troubleshoot malfunctioning Connectors without the need to run Athena queries. You can also use the validator as a form of integration testing. If the preceding query didn't work for you, take a look at the `README.md` file in the tools directory for more details on troubleshooting common errors. Most errors are reported via the Athena console, but some can easily be root causes with the client-side logs that the validator generates.

Summary

In this chapter, we learned the ins and outs of Athena Query Federation, including the differences between a federated query and a "classic data lake query." Then, our journey took us deeper into performance, availability, and the consistency tradeoffs of querying live data via a federated query or a snapshot that's been loaded into S3. We looked at the structure of the Athena Federation SDK and how it relies on Apache Arrow as a memory-compatible columnar format for exchanging data between analytics systems, without the need for multiple performance-robbing serialization steps.

Next, we stepped out of the academic realm and into the thick of things with a hands-on exercise in deploying and querying one of Athena's pre-built Connectors. Our efforts concluded with our most ambitious coding exercise yet, where we built a custom Athena Connector from the ground up using the Athena Query Federation SDK directly. In the next chapter, *Chapter 13, Athena UDFs and ML*, we'll build on the federation concepts we learned here to extend Athena's functionality even further with custom UDFs and machine learning.

13
Athena UDFs and ML

In this chapter, we will continue with the theme of enhancing Athena with our functionality by adding **user-defined functions (UDFs)** using AWS Lambda and AWS SageMaker. In *Chapter 3, Key Features, Query Types, and Functions*, we introduced the built-in functions that are available to you as a user of Athena. But as you build out your data lake and your Athena usage becomes more targeted at specific use cases, you may encounter situations where the built-in functions do not provide the exact functionality that you require. For such scenarios, Athena supports UDFs.

In this chapter, we are going to cover the basics of UDFs and how to create them. By the end, we will learn how we can apply UDFs to non-standard use cases and also to perform machine learning analysis on our data.

In this chapter, we will cover the following topics:

- What are UDFs?
- Writing, deploying, and using UDFs
- Using built-in machine learning UDFs

Technical requirements

Wherever possible, we will provide samples or instructions to guide you through the setup. However, to complete the activities in this chapter, you will need to ensure you have the following prerequisites available. Our command-line examples will be executed using **Ubuntu**, but most Linux flavors should work without modification, including Ubuntu on Windows Subsystem for Linux.

You will need internet access to GitHub, S3, and the AWS Console.

You will also require a computer with the following installed:

- Chrome, Safari, or Microsoft Edge
- The AWS CLI

This chapter also requires that you have an **AWS account** and accompanying IAM user (or role) with sufficient privileges to complete this chapter's activities. Throughout this book, we will provide detailed IAM policies that attempt to honor the age-old best practice of "least privilege." For simplicity, you can always run through these exercises with a user that has full access. Still, we recommend using scoped-down IAM policies to avoid making costly mistakes and learning more about using IAM to secure your applications and data. You can find the suggested IAM policy for this chapter in this book's accompanying GitHub repository, listed as `chapter_13/iam_policy_chapter_13.json`, here: `https://bit.ly/3gnwCSm`. No changes need to be made to the policy from *Chapter 12*, *Athena Query Federation*, so if you completed the exercises in that chapter, you don't need to make any modifications to your role.

What are UDFs?

If it wasn't already obvious before now, it has probably become pretty clear by this point that the world of big data analytics is vast and complex. Athena offers a very wide array of built-in functionality that enables you to analyze your data, but as your use cases grow, you may find that certain situations are not covered. Perhaps your data has a special encoding that can't be converted by Athena, or maybe you want to do some natural language processing to look for general sentiment in some free text fields. Whatever the situation may be, you can turn to **user-defined functions** (**UDFs**) to solve them. UDFs allow us, as users of Athena, to provide custom query behavior that can be used within the queries we are running.

UDFs are not a new concept created by Athena, so if you've been in the data analytics space for a while, you've likely already encountered them. The case of Athena is a bit more unique since you are not managing the query execution hardware, nor are you managing the software installed on that hardware. In traditional, self-managed data warehouse solutions, UDFs are typically registered within or alongside the program itself at startup time. For example, `prestodb` has support for custom functions (`https://bit.ly/36q2Ir5`), which are deployed alongside Presto by simply placing the `.jar` file in a pre-configured plugin directory.

If you read the preceding link on prestodb's support for custom functions, you may have noticed that there are three different types supported by the engine: **scalar** and **aggregate** functions. Scalar functions are used to add custom functionality to data existing in a single row. An example could be `is_null`, where it will simply return a `boolean` indicating whether the provided value is null. Aggregate functions, on the other hand, are used to create behavior across several rows (think `avg`). They require you to use `AccumulatorState`, which is where the aggregation is persisted across rows. At the time of writing this book, Athena only supports scalar functions.

For Athena, UDFs are referred to as **external functions**. In this chapter, we're going to cover the two different options available to you at the time of writing this book. These options are Lambda-based functions and SageMaker endpoint-based functions. Lambda-based functions, as the name implies, utilize a Lambda that gets invoked during the execution of your query. The following diagram shows the flow for Lambda-based UDF execution. If you read *Chapter 12, Athena Query Federation*, then the process of writing and deploying UDFs is going to look very similar to the process of writing and deploying a connector. If you skipped that chapter, then don't worry – we will go over everything again here:

Customer Query   Athena   User Defined Function

Figure 13.1 – Athena Lambda UDF workflow

The preceding diagram shows the flow for SageMaker-based UDF executions. If you completed the exercises in *Chapter 7*, *Ad Hoc Analytics*, some of the SageMaker setup will look familiar. However, we will be using SageMaker to train a model, so there will be some differences here:

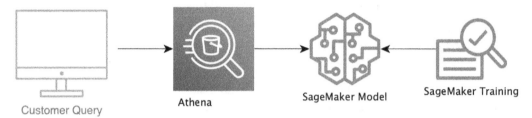

Figure 13.2 – Athena SageMaker UDF workflow

Now that we know what UDFs are, let's create a new one.

Writing a new UDF

So, now that we've gotten a bit of an idea of what UDFs are and when we might want to use them, let's go ahead and create one.

Setting up your development environment

To write a new UDF or modify an existing UDF, we'll need the ability to build, test, and package the code. So, our first task is to ensure we have a development environment with the appropriate builder tools. These tools will include Apache Maven, the AWS CLI, and the AWS Serverless Application Model (SAM) build tool. The Apache Foundation describes Maven as a "software project management and comprehension tool." That's a fancy way of saying Maven helps automate dependency management, build orchestration, and a host of related activities that can be added or augmented via plugins. The AWS SAM build tool is one option for packaging and deploying our UDF for use with Lambda and Serverless Application Repository. And, of course, the AWS CLI will be there for when we'll need to interact with AWS via the command line.

If you've already got an environment that meets these requirements (for example, if you completed the exercises in *Chapter 12, Athena Query Federation*), you're welcome to use it. If not, don't worry. We'll guide you through the setup for Debian Linux-based systems such as Ubuntu or Ubuntu on Windows Subsystem for Linux. Most of the commands will work on other flavors of *nix with minor modifications. We'll assume you will be using a basic text editor without any fancy builder tool integrations and that you need a guide for installing these other dependencies. Let's get started by cloning this book's accompanying GitHub repository by using the following command.

```
git clone https://github.com/PacktPublishing/Serverless-
Analytics-with-Amazon-Athena.git
```

Inside the `chapter_13` directory, you'll find a `prepare_dev_env.sh` script that you can run to install OpenJDK, Apache Maven, the AWS CLI, and the AWS SAM build tool. Depending on your CPU and disk speeds, the script may take 5 minutes or more to set up in your environment. If the script is successful, it will print the installed version of each required tool at the end. The output will look similar to the following, but don't worry if your versions differ slightly:

```
aws-cli/1.19.96 Python/2.7.18 Linux/4.19.128-microsoft-standard
SAM CLI, version 1.24.1
Apache Maven 3.5.4
openjdk version "11.0.11" 2021-04-20
javac 11.0.11
```

Next, we'll start writing the code for our custom UDF!

A Note About Java Versions

While the Athena Federation SDK should work fine with Java versions up to and including 11, issues have been reported with Apache Arrow with JDK versions beyond 8. The open source project has resolved many of these issues and provided configuration workarounds for the remaining items. If you run into errors that appear to be related to your JDK version, try executing the exercise with Open JDK 8.

Writing your UDF code

For this exercise, we'll be taking a closer look at the UDF that we used in *Chapter 12, Athena Query Federation*. Again, don't worry if you skipped that chapter as we will cover everything that you need to know here. To help you learn about writing UDFs, we've included a working example that you can check your work against. However, unlike Query Federation, where Athena provides a pretty wide selection of ready-made connectors, UDFs can be very customer use case-specific, so there isn't currently a large pre-built collection to browse through. So, you must understand the steps to go from nothing to a fully functional UDF. Due to this, I'm going to recommend that you try to avoid just copying and pasting from the working example and leverage that as a way to check and debug your work. To find the working code, navigate to the `udf-example` folder in the `chapter_13` directory of the GitHub repository you cloned in the previous section (if you skipped that section, go ahead and clone it now!).

The UDF we'll be working with was initially provided by the Athena team as part of the Athena Federation SDK to teach customers how to write UDFs. This UDF is intentionally simple so that you can focus on the basics of understanding the components of a UDF rather than having to decipher complex function logic. The function will take in a parameter, decrypt it (using a hardcoded encryption key, which violates every security tenant, so **please** don't do this in production), and return the result of the decryption. We've also included some unit tests to verify the function code. In the end, we will deploy the UDF and use it in a query.

Athena provides an SDK that will handle any of the logic that's necessary for communicating with the main Athena query engine, as well as aiding in interactions with Apache Arrow. The SDK is implemented in Java, so it is recommended that you implement your UDF in Java as well (or Kotlin or Scala if you are feeling adventurous). Since the SDK is fully open source (`https://bit.ly/3vXmm9j`), you can technically write this in any language, so long as you reproduce the behavior, but that is not recommended other than for expert users with language-specific use cases.

Project setup

For this walkthrough, we are going to be using Apache Maven for our dependency management. To get started, let's go ahead and initialize a new Maven project:

```
mvn -B archetype:generate \
  -DarchetypeGroupId=org.apache.maven.archetypes \
  -DgroupId=<YOUR_GROUP_ID> \
  -DartifactId=<YOUR_ARTIFACT_ID>
```

If you aren't familiar with Maven, for this command, all you need to know is that you're going to get a skeleton application that we'll take a little bit of a closer look at in a second. Remember to replace YOUR_GROUP_ID with something that makes sense for you (for example, the AWS group ID; that is com.amazonaws) and then replace YOUR_ARTIFACT_ID with the project name (for example, udf-example).

After running that command, you should see a new directory with the name that you used for YOUR_ARTIFACT_ID. Inside that directory, you should see two things: an src/ directory and a file named pom.xml. The POM file (https://bit.ly/3xDLd2y) is the file where you declare your dependencies for your project. It is also where all of your build configurations go. A quick search on Amazon reveals seven pages of books on Maven, so we're not going to delve any deeper than we need to, but needless to say, it's a very powerful tool. Taking a quick look in the src/ folder, you'll see main/ and test/. Within each, there is some sample code that you can go ahead and delete.

The POM file

Next, we are going to update the POM file. We'll only have to do this once. The POM file is quite large, so rather than taking up two pages, I am going to recommend that you go to this book's GitHub repository and follow along and/or copy-paste (https://bit.ly/3msAs0x). I will cover some important sections that are worth understanding here:

```
<parent>
  <artifactId>aws-athena-query-federation</artifactId>
  <groupId>com.amazonaws</groupId>
  <version>1.1</version>
</parent>
```

The <parent> tag tells Maven that we want to merge our POM file with the POM file of the referenced parent artifact. In this case, that is the POM file for the aws-athena-query-federation artifact, the POM of which you can find in the open source repository (https://bit.ly/2U4IErJ). The result of the merged POM is to ensure that all the dependencies are together:

```
<properties>
  <maven.compiler.source>1.8</maven.compiler.source>
  <maven.compiler.target>1.8</maven.compiler.target>
  ...
</properties>
```

As the note on Java versions states, Apache Arrow has sometimes been reported to have issues with JDK versions beyond 8, so we are forcing the compiler to use JDK 8 for our build, just to be on the safe side:

```xml
<dependency>
  <groupId>com.amazonaws</groupId>
  <artifactId>aws-athena-federation-sdk</artifactId>
  <version>${aws-athena-federation-sdk.version}</version>
</dependency>
```

We are going to be extending a class from the Athena Federation SDK, so we need a dependency on that. Note that we are referencing the ${aws-athena-federation-sdk.version} variable. We did not declare that in our POM; we are getting that value from <parent>:

```xml
  <build>
    <pluginManagement><!-- lock down plugins versions to avoid
using Maven defaults (may be moved to parent pom) -->
      <plugins>
        <plugin>
          <groupId>org.apache.maven.plugins</groupId>
          <artifactId>maven-shade-plugin</artifactId>
          . . .
        </plugin>
      </plugins>
    </pluginManagement>
  </build>
</project>
```

Finally, we are using maven-shade-plugin. This instructs Maven, when it produces the .jar file for our package, to also include all of our dependencies to create an uber-jar (it can also help with renaming packages in case there are conflicts). An uber-jar is one of two ways to deploy your JVM-based application to AWS Lambda, with the other being just a ZIP file containing your code and all dependencies. Both work fine (fun fact, a .jar file is pretty much just a ZIP with a little extra Java-y information).

UserDefinedFunctionHandler

We're just about ready to start writing our UDF code, but before we do that, we're going to take a quick detour and peek at the code provided in the Athena Federation SDK that will aid us in creating our UDF. The SDK contains an abstract class called `UserDefinedFunctionHandler` that we will be extending. This class handles the deserializing messages that are sent from Athena's main engine and then delegates them down to the proper function handler. We're not going to delve super deep into what this handler is doing, but let's take a look at a few notable code pieces. You can see the full implementation here: `https://bit.ly/3riRfTQ`.

The first thing to notice is that the class implements `RequestStreamHandler`. This class comes from the AWS Lambda Java SDK and has a single method, called `handleRequest`, that you have to implement to have a Java-based Lambda function. The contract is very straightforward: you are given an `InputStream` containing the input values to your function, an `OutputStream` where your function will write its results, and a `Context` that contains mostly Lambda metadata about the function itself.

Next, scroll down until you find `extractScalarFunctionMethod`. We haven't discussed how to use UDFs in Athena queries yet, but this method is important for that. Generally speaking, at query time, we will tell Athena the name of the method to execute, and then this logic will use Java Reflection to find the implementation of that method in your UDF code. If you aren't familiar with Java Reflection or just aren't fully following what's happening here, that's okay – it isn't critical that you understand this logic. It can just sometimes be helpful to understand how everything pieces together.

The last bit we'll take a look at is the following block of code, which is located inside the `processRows` method:

```
for (Field field : inputRecords.getFields()) {
  FieldReader fieldReader = inputRecords.getFieldReader(field.
getName());
  ArrowValueProjector arrowValueProjector = ProjectorUtils.
createArrowValueProjector(fieldReader);
  valueProjectors.add(arrowValueProjector);
}
```

As we mentioned previously, Athena leverages Apache Arrow to represent the data in transit between the query's execution and the UDF function. This logic is taking the fields (with `field` being a column in our table) returned in the query, creating an `ArrowValueProjector` for each field. These projectors are put there to make writing UDFs easier so that you, as the function writer, do not have to interact with or even understand Apache Arrow. Instead, you are given Java objects to operate on.

> **A Note About ArrowValueProjector's Performance**
>
> The convenience of ArrowValueProjector's comes at a cost. Data must be copied from the Arrow objects to the Java objects, and any data copying is always going to introduce some degree of latency, which, when magnified over potentially many thousands (or more) of rows of data, can add up. If you are noticing an unacceptable degree of latency introduced from your UDF, you can consider overriding the `processRows` method inside of `UserDefinedFunctionHandler` and operate directly on the Apache Arrow objects instead of converting from Arrow into Java.

UDF code

Now, we're ready to write our UDF! As we mentioned previously, the function is going to decrypt a parameter from our dataset using an encryption key that we've hardcoded. Again, **do not do this in production** – this is just to keep things simple to illustrate how to write a UDF.

To get started, let's go ahead and create a new class in our Java package and call it `UdfExample`. As we covered previously, we are going to be extending `UserDefinedFunctionHandler`, which has a constructor that requires a `String` parameter called `sourceType`. The value you assign to this isn't super important to you as it's primarily used for Athena's internals; just pick something descriptive. I chose `"Packt_UdfExample"`.

Now, we're going to add our function code. Let's go ahead and create a new `public` method and call it `decrypt`. In this case, the return type of our method will be `String`, but in the general sense, the return type should map to whatever type we want the value to be in our query. For our input, we will take in a `String` as well, but again, we are not limited to strings, and we are also not limited to a single input. We can use as many as we want and whatever types we want, based on the types that our columns are stored as in our dataset.

At this point, you should have something that looks like the following:

```
package com.amazonaws;


import com.amazonaws.athena.connector.lambda.handlers.
UserDefinedFunctionHandler;


public class UdfExample extends UserDefinedFunctionHandler {
    private static final String SOURCE_TYPE = "Packt_
UdfExample";
```

```
    public UdfExample() {
        super(SOURCE_TYPE);
    }
    public String decrypt(String encryptedColumnValue)
    {
        return null;
    }
}
```

This is the minimum you would need to be able to register a UDF called `decrypt`! Of course, your function wouldn't perform any decryption, you would just get nulls back, but still, pat yourself on the back – you've created a UDF!

Now, we'll quickly go over the decryption logic. This isn't super important, since it's been created to demonstrate developing UDFs, so if you want, feel free to skip this portion and just copy the logic from the repository at `https://bit.ly/3AZQUsR`.

We are using what's called symmetric encryption. This means that the same key is used to encrypt and decrypt. This is in contrast to asymmetric encryption, where one key is used to encrypt (generally referred to as the public key) and another one is used to decrypt (the private key). We will be using AES as our encryption algorithm and Java's built-in cryptography library:

```
    @VisibleForTesting
    protected String symmetricDecrypt(String text, String
secretKey)
            throws NoSuchPaddingException,
NoSuchAlgorithmException, InvalidKeyException,
BadPaddingException,
            IllegalBlockSizeException
    {
        cipher cipher;
        String encryptedString;
        byte[] encryptText;
        byte[] raw;
        SecretKeySpec skeySpec;
        raw = Base64.decodeBase64(secretKey);
        skeySpec = new SecretKeySpec(raw, "AES");
        encryptText = Base64.decodeBase64(text);
```

```
        cipher = cipher.getInstance("AES");
        cipher.init(cipher.DECRYPT_MODE, skeySpec);
        encryptedString = new String(cipher.
doFinal(encryptText));
        return encryptedString;
    }
```

The preceding code is performing the decryption. Let's look at the code in bold in more detail. First, both the key string and the encrypted values are Base64-encoded, which is used to turn bytes into ASCII. Finally, at the bottom, we are creating a cipher, which is essentially the implementation of the AES algorithm. So, again, we're getting the raw bytes for both the key and the encrypted text, passing them both through the AES cipher, and getting back our decrypted bytes, which we are converting back into strings (which we know is safe to do, because we know the decrypted value is just a string, though it could, in theory, be more non-human-readable bytes).

> **Don't Forget to Test Your Code!**
>
> As with any code base, make sure you clearly define the contracts of your code and verify them with tests. We've included some test code in our sample as well, which I'd recommend you at least read through to understand what it's doing and then copy it over.

Building your UDF code

Before we can deploy our code, we need to build and package it. Thankfully, since we did all that nice setup earlier on in our POM, this is very easy to accomplish. Simply run one of the two commands:

```
mvn clean install
# If you want to run the tests as well run the following (this
is not standard to Maven, it's just how we happened to set this
package up)
mvn clean install -Dpublishing=true
```

Once this completes, you should have a newly generated directory named `target`. Inside of it, there should be a JAR file called `udf-example-2021.33.1.jar`.

Quick Note on Maven Shading

In the same target/directory, you may also see a file that looks like
`original-udf-example-2021.33.1.jar`. This is the original
`.jar` file that was produced by Maven. However, as we mentioned previously,
we need to provide a `.jar` that contains all of our dependencies (the uber
`.jar`). Again, the plugin responsible for that is called Maven Shade, and it
actually moves the original `.jar` to a file called `origin-[JAR_FILE_`
`NAME]`, and then creates a new `.jar` with the same `[JAR_FILE_NAME]`
that contains all the dependencies. If you look at the size of each of the files,
you'll notice that `udf-example-2021.33.1.jar` is quite a bit larger
than `original-udf-example-2021.33.1.jar`. In my case, it's
23 MB versus 4.6 KB.

Deploying your UDF code

We're ready to deploy our code! The process of deploying your function is no different
than any other Java-based Lambda function, so if you are already familiar with that
process, we aren't going to be introducing any new concepts. There are two primary
mechanisms that we are going to cover to make direct calls to the AWS Lambda APIs
and AWS SAM.

Direct calls to AWS Lambda APIs

In this section, we are going to directly call a Lambda by using the AWS CLI. This is a
simpler and quicker way to get started but I wouldn't recommend it when maintaining
your UDF in the long term.

Before we can register the Lambda, we need an execution role. The AWS Lambda docs
provide a good overview of creating execution roles (`https://amzn.to/3ign39a`)
but to get started quickly, you can just run the following commands:

```
aws iam create-role --role-name udf-example-role --assume-
role-policy-document '{"Version": "2012-10-17","Statement": [{
"Effect": "Allow", "Principal": {"Service": "lambda.amazonaws.
com"}, "Action": "sts:AssumeRole"}]}'
aws iam attach-role-policy --role-name udf-example-
role --policy-arn arn:aws:iam::aws:policy/service-role/
AWSLambdaBasicExecutionRole
```

The first command creates a role called udf-example-role with no permissions attached to it and a trust policy saying that AWS Lambda is allowed to assume it. The second command attaches the AWSLambdaBasicExecutionRole managed IAM policy to the newly created role.

Now that we've created our role, let's go ahead and create our function. To accomplish this, we are going to call the create-function API within AWS Lambda:

```
aws lambda create-function \
  --function-name UdfExample \
  --runtime java8 \
  --role arn:aws:iam::1234567890123:role/udf-example-role \
  --handler com.amazonaws.UdfExample \
  --timeout 900 \
  --zip-file fileb://./target/udf-example-2021.33.1.jar
```

And that's it! You've deployed a Lambda function.

Using your UDF

The time has finally come for us to use our shiny new UDF inside an Athena query! Registering a UDF is done at query execution time by way of the USING EXTERNAL FUNCTION clause, before your SELECT statement. The syntax for that looks like this:

```
USING EXTERNAL FUNCTION UDF_name(variable1 data_type[,
variable2 data_type][,...])
RETURNS data_type
LAMBDA 'lambda_function_name'
SELECT  [...] UDF_name(expression) [...]
```

First, let's get set up with some sample data. For that, we've provided some data (https://bit.ly/3gjYfvK) and a CREATE statement (https://bit.ly/3sGpmWz) that you can use in our repository. Upload the sample data and run the CREATE statement in Athena. Make sure that you replace <S3_BUCKET> in the CREATE statement with the name of the S3 bucket where you placed the sample data.

Now that we've got that set up, let's go ahead and try running a query using our UDF! Your results should match those shown in the following table:

```
USING EXTERNAL FUNCTION decrypt(encryptedData VARCHAR)
RETURNS VARCHAR
LAMBDA 'UdfExample'
```

```
SELECT year, month, day, encrypted_payload, decrypt(encrypted_
payload) as decrypted_payload
FROM "packt_serverless_analytics"."chapter_13_udf_data"
limit 5.
```

This results in the following output:

Year	Month	Day	Encrypted Payload	Decrypted Payload
2017	11	1	0UTIXoWnKqtQe8y+BSHNmdEXmWfQalRQH60pobsgwws=	SecretText-1755604178
2017	11	1	i9AoMmLI6JidPjw/SFXduBB6HUmE8aXQLMhekhIfE1U=	SecretText-747575690
2017	11	1	HWsLCXAnGFXnnjD8Nc1RbO0+5JzrhnCB/feJ/EzSxto=	SecretText-1720603622
2017	11	1	lqL0mxeOeEesRY7EU95Fi6QEW92nj2mh8xyex69j+8A=	SecretText-1167647466
2017	11	1	C57VAyZ6Y0C+xKA2Lv6fOcIP0x6Px8BlEVBGSc74C4I=	SecretText-1854103174

Figure 13.3 – Decryption results

Just in case you think there is some wizardry going on here and I'm trying to trick you, I've gone ahead and included a class in the repository that sanity checks the results (https://bit.ly/2XCBgFu). You can give it an encrypted payload and it will return the expected output, so you can double-check it against the preceding values:

```
mvn compile exec:java -q -Dexec.mainClass="com.
amazonaws.ResultSanityChecker" -Dexec.
args="0UTIXoWnKqtQe8y+BSHNmdEXmWfQalRQH60pobsgwws="
Encrypted payload: 0UTIXoWnKqtQe8y+BSHNmdEXmWfQalRQH60pobsgwws=
Decrypted payload: SecretText-1755604178
```

A couple of other things to point out are that the signature of decrypt matches that of the method signature for decrypt in our function code, including its input types and return type. Though again, you'll notice that the types are VARCHAR versus String, which we have in Java – that's Apache Arrow and ArrowValueProjectors at work. And then, the Lambda function's name is just the name that we gave to the Lambda we created in the last step.

And that's it! Congratulations – you've just written your first UDF!

> **Maintaining Your UDF**
>
> Now that you've gotten your UDF running, let's talk briefly about how we're going to maintain the function. As we mentioned previously, there are two ways to deploy it; we chose the simpler way to get started, which was just to call the Lambda APIs from our Terminal, but that's not a maintainable way of doing that long term. For regular maintenance of your UDF, a very good place to turn to is AWS SAM. SAM is an open source framework provided by AWS that includes a ton of super handy functionality for building serverless applications. As an optional exercise, I recommend that you run through the documentation SAM provides on getting set up with SAM and a CI/CD tool of your choice: `https://amzn.to/3kfQVlW`. Alternatively, the walkthrough in *Chapter 12, Athena Query Federation*, also shows you how to utilize SAM to deploy an Athena Federation Connector.

Using built-in ML UDFs

In the previous section, we learned how we can create UDFs using Lambda. In this section, we're going to learn how to use Athena's built-in functionality to create UDFs that delegate down to a ML model. We're not going to delve too deeply into the ML aspects of things, though we will cover some basics just so you know what's happening. If you read *Chapter 7, Ad Hoc Analytics*, then some of this should be familiar.

Before you get started, note that you may incur some SageMaker charges during this. Particularly for the portion where we are training our models, we don't want to be waiting around forever, so we are leveraging the recommended cost/power instance type of `ml.c5.xlarge`. Total charges should be no more than a few dollars.

Pre-setup requirements

Before we are ready to head on over to SageMaker, there's a couple of things we need to put in place. First up is our favorite resource, an IAM role. By now, you're probably a pro at creating IAM roles, but in case you skipped directly to this chapter, we'll cover the creation process again. You can do this by navigating to the IAM Console, selecting the **Roles** section, and clicking the **Create role** button. Once you've done that, you'll be presented with the dialog shown in the following screenshot. Be sure to select **AWS Service** as the type of trusted entity and **SageMaker** as the entity:

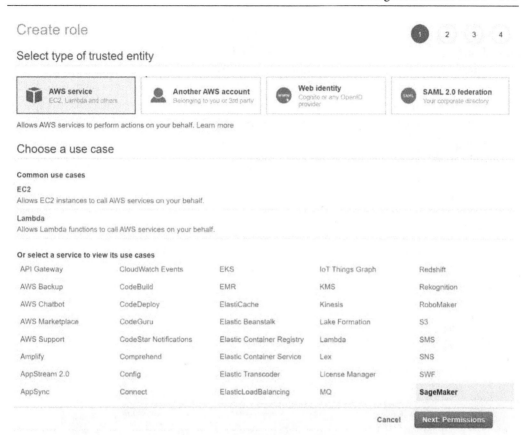

Figure 13.4 – Create role dialog

The settings shown in the preceding screenshot indicate that we are creating a role that can be assumed by SageMaker, allowing SageMaker to perform the actions associated with the role inside your account. This helps scope down both the types of activities the IAM role can perform and the contexts from which it can be assumed. In the next step, you'll have the opportunity to add the specific policies for the activities we plan to perform using this IAM role. We recommend adding the `packt_serverless_analytics` policy that we have been enhancing throughout this book and used earlier in this chapter. As a reminder, you can find the suggested IAM policy in this book's accompanying GitHub repository, listed as `chapter_13/iam_policy_chapter_13.json`, here: `https://bit.ly/3gnwCSm`.

Once you've added the policy, you can move on to the **Add Tags** step. Adding tags is optional, so you can skip that for now and go to the final step of giving your new IAM role a name. We recommend naming your new IAM role `packet-serverless-analytics-sagemaker` since this chapter's IAM policy already includes permissions that will allow you to create and modify roles that match that name without added access. If everything went as expected, your IAM role summary should match what's shown in the following screenshot. If you forgot to attach the `packt_serverless_analytics` policy, you can do so now using the **Attach Policies** button:

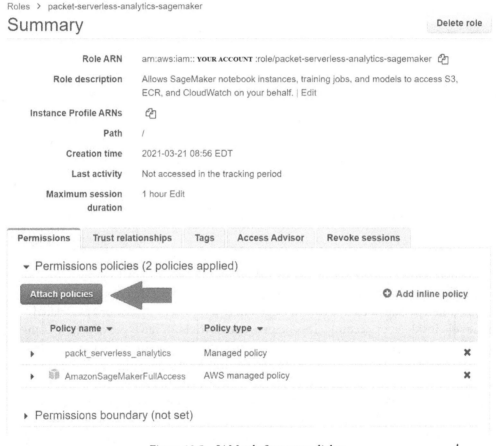

Figure 13.5 – IAM role Summary dialog

Next, go ahead and copy the following CSV file into an S3 bucket of your choosing. Make sure to note where you put it for later:

```
wget -O taxi_ridership.csv https://bit.ly/3kblw45
aws s3 cp taxi_ridership.csv s3://<S3_BUCKET>/packt-serverless-
analytics-chapter-13/ml-example/input/
```

Setting up your SageMaker notebook

Now, it's time to create our SageMaker Jupyter notebook; we're almost ready to start training! Head on over to the SageMaker console, find the **Notebook > Notebook Instances** section and select it. From there, you can click **Create notebook instance** to open the dialog shown in *Figure 13.6* and *Figure 13.7*.

Using our notebook to train a model

Your notebook instance should be ready to use at this point. We're going to cover what we're doing in depth, but first, we will provide a quick overview of the steps we're going to take:

1. Connect our notebook instance to Athena.
2. Create a table in Athena using the CSV file we copied into our S3 bucket earlier.
3. Read the contents of the table.
4. Run a training job on our table data using the Random Cut Forest algorithm.
5. Deploy our trained model to an endpoint.

If you'd like to skip ahead or need added guidance on writing the code snippets we'll be using to train our model, you can find a prepopulated notebook file in this book's GitHub repository at `chapter_13/packt_serverless_analytics_chatper_13.ipynb`, here: `https://bit.ly/3sAErZV`. GitHub nicely renders the notebook file so that you can see it right from the link. Unfortunately, that makes downloading it so that you can upload it to your SageMaker notebook instance later a bit tricky. To get around that, click on the **Raw** view, and then click **Save As** from your browser.

Connecting our notebook instance to Athena

From the SageMaker Console, go ahead and click the **Open Jupyter** link, as shown in the following screenshot. This will open a new browser tab or window connected to your Jupyter Notebook instances. Behind the scenes, SageMaker is handling all the connectivity between your browser and what is your Jupyter Notebook server:

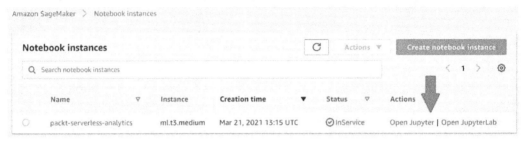

Figure 13.6 – Open Jupyter Notebook

As shown in the preceding screenshot, you'll want to click on **New** and select **conda_python3** for the notebook type. The value may appear at a different position in the dropdown than it does in the preceding screenshot, so don't be afraid to scroll to find it. This setting determines how our notebook will run the data exploration tasks we are about to write. By selecting **conda_python3**, we are telling Jupyter that it can run our code snippets using Python. Once you pick the notebook type, yet another browser tab will open that contains your new notebook. The new notebook file will be named `Untitled.ipynb`, so our first step will be to give it a helpful name by clicking on **File** and then **Rename**:

Figure 13.7 – Creating a new Notebook file

Now that your notebook is ready to use, we'll connect it to Amazon Athena by installing the Athena Python driver. To do this, we'll write the following code snippet in the first **cell** of the notebook. Cells are represented as free-form text boxes and can be executed independently, with subsequent cells having access to variables, data, and other states produced by earlier cells. After executing a cell, its output is shown immediately below it. You can edit, run, edit, and rerun a cell as often as you'd like. You can also add new cells at any time. Let's put this to practice by running our first cell. Once you've typed the code into the cell, you can either click **Run** or press *Shift + Enter* to run the cell and add a new cell directly below it:

```
import sys
!{sys.executable} -m pip install PyAthena
```

This particular cell will take a couple of minutes to execute, with the result containing a few dozen log lines detailing which software packages and dependencies were installed. Now, add another cell and paste the following code into it, make sure to replace <OUTPUT_S3_BUCKET> with the bucket you want the output data to be placed in, as well as <OUTPUT_S3_BUCKET_REGION>:

```
from pyathena import connect
import pandas as pd
import sagemaker

#TODO: Change the bucket to point to an s3 bucket to use for
model output and training data
bucket = <OUTPUT_S3_BUCKET>
output_location = 's3://' + bucket + '/chapter_13/ml_output/'
# Connect to Athena
connection = connect(s3_staging_dir=output_location, region_
name=<OUTPUT_S3_BUCKET_REGION>)
```

Now, we're connected to Athena!

Preparing our training data

Next, we are going to take the data that we copied into our <S3_BUCKET> in the pre-setup steps and create a table for it in Athena. Go ahead and add another cell and insert the following code into it:

```
create_table = \
"""
CREATE EXTERNAL TABLE 'packt_serverless_analytics'.'chapter_13_
taxi_ridership_data'(
    'time' string  ,
    'number' int)
ROW FORMAT SERDE
    'org.apache.hadoop.hive.serde2.OpenCSVSerde'
WITH SERDEPROPERTIES (
    'separatorChar'=',')
STORED AS INPUTFORMAT
    'org.apache.hadoop.mapred.TextInputFormat'
OUTPUTFORMAT
    'org.apache.hadoop.hive.ql.io.HiveIgnoreKeyTextOutputFormat'
LOCATION
    's3://<S3_BUCKET>/packt-serverless-analytics-chapter-13/ml_
input/'
"""
```

Now, add one more cell to execute the CREATE statement:

```
## Create a new Athena table holding data we will use to
predict anomalies
pd.read_sql(create_table, connection)
```

Finally, let's go ahead and read the contents of the table into an object so that we can use it later to train our model. And... you guessed it, add another cell!

```
results = pd.read_sql("SELECT * FROM default.taxi_ridership_
data", connection)
```

Time to train!

This is the last step for our SageMaker notebook: training our model. So, we need one more of our favorite things, a cell!

```python
from sagemaker import RandomCutForest

prefix = 'athena-ml/anomalydetection'
execution_role = sagemaker.get_execution_role()
session = sagemaker.Session()

# specify general training job information
rcf = RandomCutForest(role=execution_role,
                      instance_count=1,
                      instance_type='ml.c5.xlarge',
                      data_location='s3://{}/{}/'.
format(bucket, prefix),
                      output_path='s3://{}/{}/output'.
format(bucket, prefix),
                      num_samples_per_tree=512,
                      num_trees=50,
                      framework_version="2.54.0",
                      py_version="py3")

# Run the training job using the results we got from the Athena
query earlier
rcf.fit(rcf.record_set(results.number.values.reshape(-1,1)))

print('Training job name: {}'.format(rcf.latest_training_job.
job_name))

rcf_inference = rcf.deploy(
    initial_instance_count=1,
    instance_type='ml.c5.xlarge',
)

print('\nEndpoint name (used by Athena): {}'.format(rcf_
inference.endpoint_name))
```

You should see a whole bunch of output for this last cell, but if you scroll to the bottom, you should see the following output. The value we are particularly interested in is the endpoint name, so make sure that you save that for later:

```
Training job name: randomcutforest-2021-08-22-03-07-26-016
-------------!
Endpoint name (used by Athena): randomcutforest-2021-08-22-03-10-43-029
```

Figure 13.8 – Training output

The Random Cut Forest model

In this section, we trained a model in SageMaker using the **Random Cut Forest** (**RCF**) algorithm. Since it's a neat algorithm, we'll briefly cover how it works. RCF is what's known as an unsupervised algorithm. These are often used to detect anomalous data points within a dataset. An unsupervised algorithm means that it does not require additional assistance (sometimes referred to as data labeling) from a human to train the model. If you ever saw the tech talk of the engineer who trained his cat door to be able to detect when his cat brought in a "gift" from outside, that is what's known as a supervised algorithm. In that case, the engineer would physically indicate to the model whether a given image contained a "gift" or not. RCF works by taking in a target value (known as a tree) – in our case, that's the number of riders for a given period – and then comparing it against all of the other values (known as the forest) using random "cuts" through the forest, until it identifies a section containing only the single tree we are looking for. The fewer cuts that are required, the more anomalous the value is determined to be. If you visualize what it would be like to take a literal forest and create slices until you find a specific tree, it's going to take way fewer slices to find the lonesome tree away from all other trees, as opposed to one that is in a very tight cluster of trees.

Using our trained model in an Athena UDF

With our fancy taxi ridership model all trained and ready to do some work, let's find some unexpected ridership amounts! Just like in the walkthrough regarding custom UDFs, we are going to use the USING EXTERNAL FUNCTION clause and then a SELECT statement to utilize the new function. The syntax looks very similar for the ML-based UDFs:

```
USING EXTERNAL FUNCTION ml_function_name (variable1 data_type[,
variable2 data_type] [,...])
RETURNS data_type
SAGEMAKER 'sagemaker_endpoint'
SELECT ml_function_name()
```

There are a few main differences. The biggest difference is that unlike in the Lambda-based UDFs, where the function name was mapped to something in our UDF code, the `ml_function_name` value is just any random identifier that we want to create – it doesn't correspond to anything that we did in our SageMaker Notebook. Then, instead of the type being LAMBDA plus a Lambda function name, it's now SAGEMAKER plus the endpoint name that was output at the end of our Notebook's execution. So, our final query ends up looking something like the following. Notice that we are querying against the same data we used to train our model, which makes sense since we are comparing a single value within our dataset against the entire dataset as a whole:

```
USING EXTERNAL FUNCTION detect_anomaly(b INT) RETURNS DOUBLE
SAGEMAKER 'randomcutforest-2021-08-22-03-10-43-029'
SELECT time, number as number_of_rides, detect_anomaly(number)
as score
FROM "packt_serverless_analytics"."chapter_13_taxi_ridership_
data"
ORDER BY score desc
LIMIT 5
```

You should get the following results:

Time	Number of Rides	Score
11/2/14 1:00	39,197	4.59422851305562
11/2/14 1:30	35,212	4.11130198098128
9/6/14 23:00	30,373	3.1411160633939024
9/6/14 22:30	30,313	3.1235797654023556
1/1/15 1:00	30,236	3.1022678031766535

Figure 13.9 – Top 5 most anomalous half hour periods for NYC taxi ridership

Thus, we have used SageMaker notebooks to train a ML UDF model.

Summary

In this chapter, we walked through a couple of different examples of how Athena allows you to inject custom functionality, known as user-defined functions, into your queries. We started by looking at fully custom UDF behavior through Lambdas. We created and deployed our own Lambda, and then took a closer look at how we can keep a healthy, well-maintained Lambda-based UDF. After that, we took a look at the built-in UDF functionality that Athena provides for integrating your queries with SageMaker ML models. We used this to determine if taxi ridership was anomalous during a specific time.

We've only scratched the surface of the power of UDFs, but this should serve as a solid reminder for when you encounter a business use case that you can't solve perfectly with the functionality provided out of the box. In the next and final chapter, we will summarize some advanced functions that Athena provides and conclude our book!

14
Lake Formation – Advanced Topics

You've reached the final chapter in our journey through *Serverless Analytics with Amazon Athena*. Some authors like to start each chapter with a thought-provoking quote. The pressure to find good, relevant quotes from well-known people was too much for us, so we opted not to employ that pattern until now. I recently came across a quote from Stephen King that does a great job distilling this chapter:

"Sooner or later, everything old is new."

– Stephen King

You see, many of Lake Formation's "new" features are a reimagining of well-known database technologies from the 70s and 80s but scaled up for modern data lakes. As a Lake Formation launch partner, Athena is often one of the first services to support new Lake Formation features. In this chapter, we will learn about Lake Formation's newest features, including row-level security and a new Amazon S3 table type that supports ACID transactions. AWS Lake Formation transactions provide for atomic, consistent, isolated, and durable queries via snapshot isolation, regardless of how many tables your query uses or how many concurrent queries you run. To complement this new table type, Lake Formation also introduced an automatic storage optimizer that continually monitors your tables and reorganizes the underlying storage for optimal performance.

Each of these features has been available in most traditional databases systems for decades. However, these capabilities frequently reduced performance or scalability. The early days of data lakes and their accompanying query engines, such as Athena, shed many of these auxiliary features in the name of scaling. As these systems and their usage evolved beyond solving scaling problems in traditional databases, the need for advanced features such as ACID transactions and row-level security have reemerged.

As many of these features are not generally available yet and should "just work" for your existing queries by toggling a setting, this chapter will focus less on exercises and more on what use cases these capabilities enable. Depending on your AWS Region of choice, these features may not be available to you yet or may still be in preview. Lastly, you may be wondering why we have repeatedly discussed Lake Formation in a book about Athena. Regardless of the analytics engine you choose, AWS looks to Lake Formation as the tide that raises all ships. Put another way, Lake Formation is increasingly where new and foundational data lake features are being built so that customers can seamlessly transition between any of the AWS analytics offerings that support Lake Formation.

In this chapter, we will cover the following topics:

- Reinforcing your data perimeter with Lake Formation
- Understanding the benefits of governed tables

Reinforcing your data perimeter with Lake Formation

We were first introduced to AWS Lake Formation in *Chapter 3*, *Key Features, Query Types, and Functions*, where we explored Lake Formation's ability to go beyond S3 object-level IAM policies to offer fine-grained access control for tables. While security is a focal point for the Lake Formation product, you may not realize that its ambitions extend far beyond this one essential element of data lakes. As we will see later in this chapter, Lake Formation's mandate is to make every aspect of building and managing data lakes simpler, faster, and cheaper. This has led the Lake Formation team to focus on the most frustrating parts of operating a data lake, such as access control.

Before discussing the most significant changes to Lake Formation since it went GA in 2019, let's make sure we genuinely understand how things worked before these new features. The following diagram illustrates the high-level interactions between Athena, Lake Formation, Glue Data Catalog, and S3 during the execution of a simple query:

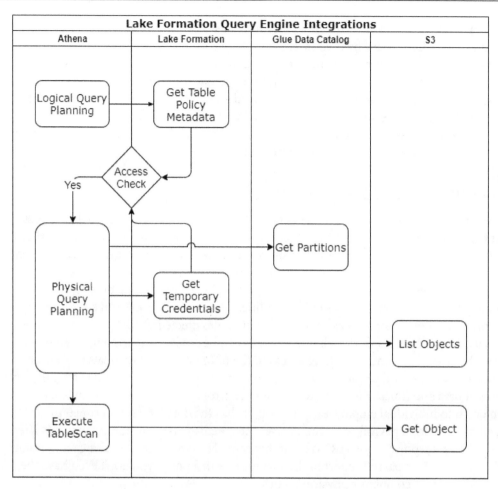

Figure 14.1 – Lake Formation

As with all Athena queries, the process begins with Athena's engine parsing the query and forming a logical plan. This logical plan contains a list of tables and columns that need to be read and a sequence of operators to apply to the resulting data. During the planning process, Athena calls Lake Formation to obtain policy metadata for each referenced table. This metadata, along with column projections from the query, is used to affect access control. Assuming the access check passes, Athena moves on to forming a physical query plan, where it gathers partitioning information for each table from Glue Data Catalog. Before starting the actual query execution, Athena needs to call Lake Formation to obtain scoped-down temporary credentials for reading the required S3 objects. The Lake Formation API calls to get temporary credentials are the second place where an access enforcement check occurs. At this point, Athena has everything it needs to execute the query.

Much of the control flow shown in the preceding diagram is unsurprising, but a couple of nuances may have snuck by if you weren't looking closely. Firstly, Lake Formation is involved in both temporary credential vending and metadata operations, such as getting the list of columns in a table. The initial iteration of Lake Formation's fine-grained access control mechanisms enabled fully managed engines such as Athena and Redshift Spectrum to improve permission management. While this was a marked improvement over the previously available solutions, many customers still found themselves contorting their data models to create effective **data perimeters**.

Establishing a data perimeter

You've undoubtedly heard many vendors talk about democratizing access to data across your organization. We explored this topic by looking at some hands-on exercises as part of *Chapter 7, Ad Hoc Analytics*, but we avoid a pervasive issue by increasing access to your data. As we improve the accessibility of data, so too must we elevate our understanding of **data perimeters**. The word perimeter has historically referred to the outer edges of a company's physical assets, such as office buildings. When the internet and e-commerce revolutionized how business was conducted, companies erected virtual perimeters using firewalls. These concepts work well if your assets can be easily compartmentalized from those who should and shouldn't have access to them. In practice, the threats to your data are not always clear and certainly not always external to your company. There are different classes of data and times where you will need to control access at a department level and even down to individual employees. For example, has the data left your perimeter if an HR employee runs a payroll report and leaves intermediate data on storage, which is later accessed by someone outside HR? What if that same HR employee is working from home and downloads that payroll report to their laptop? At this point, you don't even have the protection of your company's physical security.

In these cases, data lake security is more important than ever. Lake Formation can help companies balance security and compliance needs with their growing desire to share data across departments, groups, and individuals. In many cases, safely sharing data across individuals with different job functions requires making redacted copies of the data. Aside from the additional storage and compute costs to ETL these copies, the organization had to manage consistency and correctness across a web of dependencies. We routinely help customers who have dozens of important datasets but somehow find themselves with thousands of derivative datasets, simply for accommodating different levels of access. Until recently, this was the state-of-the-art approach to creating a robust data perimeter because you get fine-grained control over which use cases and entities need access to specific slices of your data. Paradoxically, this approach created so many subtle variations of the original data that customers feared making mistakes that could lead to unintended information disclosure.

It's probably already pretty clear that security is far from easy to define, let alone build. It can be even more challenging when basic computational building blocks we all depend on seemingly stop playing by the rules. We'll dig deeper into this topic as part of understanding how customers often overlook their part in shared responsibility models.

Shared responsibility security model

Simply put, a shared responsibility security model refers to the basic idea that the customer and the service must work together to ensure any given workload is *secure*. We're using the word secure a bit tongue in cheek here because most security-conscious individuals will recoil at the thought of distilling all the complex nuances of security into one word. Security is rarely binary, meaning it's uncommon for any application to be described as secure or not secure. It's more common to think of these things as a gradient or, even better, in terms of specific threats and mitigations.

For example, one use case may require that data be encrypted when stored at rest. The reasons vary, but a typical example is that the underlying storage does not encrypt data replication traffic that's generated when the storage nodes failover. Another application may run workloads from multiple internal teams on shared infrastructure to improve costs. Since these workloads are all internal, the business valued utilization above protecting internal workloads from one another. If that same application started running workloads from external entities on that same, shared infrastructure, the definition of *secure* might change.

We've already called out that fully managed engines such as Athena and Redshift Spectrum avoid the disclaimer of a shared security model. Still, the reason has less to do with being fully managed and more to do with the level of control or abstraction these services offer. Both Athena and Redshift Spectrum essentially operate over SQL, whereas EMR and Glue ETL offer far more customer control. An EMR or Glue ETL customer can choose to run arbitrary code in their jobs. If you've ever used spark-submit or a Jupyter notebook with EMR, then you've executed arbitrary code on your EMR cluster. So, why the big fuss over arbitrary code? Well, the ability to run arbitrary code provides fairly low-level access to the machines that run your workloads.

Suppose you are running your analytics applications in a shared Spark cluster. During a Spark job, the state of any given node is represented as shown in the following diagram, with your arbitrary Spark code running side by side with the arbitrary code from some other workload:

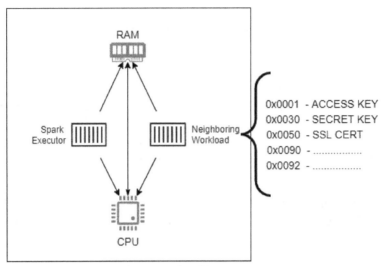

Typical Analytics Node (e.g. Spark)

Figure 14.2 – Process-level isolation of Spark workloads

Running each workload in separate processes that run as a different user improves the security posture by limiting how neighboring workloads can interact. If your organization is mainly concerned with avoiding accidental data leakage from bugs or typos, this level of isolation may be sufficient. But how do you know whether you've set it up correctly? If you depend on process-level isolation, it becomes increasingly important to ensure your customers cannot tamper with the operating system or Spark itself. Ensuring only administrators have root access to the host is a good start, but it isn't always easy to know whether that is enough.

Now, let's suppose that you'd like to go a step further and ensure customers can only access data they are authorized for. You might choose a tool such as Apache Ranger for access control. With Apache Ranger, policy enforcement takes place within Spark, alongside your workload. This pluggability makes it easy to get started with Apache Ranger for Spark, but what level of protection does it provide? For example, what prevents someone from running a Spark job that hijacks the Java classpath and injecting their copy of the `RangerHiverAuthorizerFactory` class? The `RangerHiverAuthorizerFactory` class plays a central role in data access policy enforcement. If an attacker can replace this class with one they control, the workload can bypass access control policies. Because their workload includes arbitrary code and has access to the Java class loader, such attacks become possible.

An analogy may be helpful here. This mitigation is akin to the lock on the front door of a house. It will keep most people from entering your home without permission, but it won't stop a determined adversary. There is a steep difference between keeping honest people honest and mitigating attacks from sophisticated attackers such as nation states. If you aren't using a managed service, your organization must play a more significant role in deciding where to draw the line.

This is one of the big distinctions between a service such as Athena and Glue ETL, which offers fully managed runtimes and lets you run highly customizable environments using your own Spark cluster or EMR. The attack surface is much different, so the customer shares responsibility in the security model.

It may be hard to believe, but we've only discussed the obvious examples that feed into the shared security model so far. Next, we will discuss the more insidious examples that have contributed, in part, to Lake Formation's release of governed tables. In recent years, the computing world as a whole has learned that processor design is not immune from security flaws. While we've seen exploits in software for decades, many of us had been spoiled by the reliability of hardware security controls. When Spectre and Meltdown were announced to the world on January 3, 2018, our ability to depend on previously trusted operating system process-level isolations was shaken. Researchers had managed to use timing variations in memory cache reads to extract information from mispredicted code branches. There is a lot to unpack in that one statement, and while this is not a book on security or processor design, this is a topic worth understanding a bit more deeply. Recognizing the fundamental issues at play here will also help you understand the motivation for several of the new Lake Formation features we'll be discussing shortly.

The following diagram shows two possible ways an x86 process could order the instructions your query engine may need to perform while enforcing column-level access policies. As you read this, please keep in mind that we have greatly simplified what your processor and an attacker would do during an exploit. We recommend reading online to learn more about side-channel exploits such as Spectre and Meltdown:

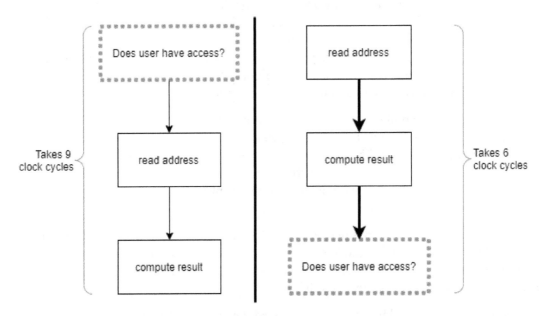

Figure 14.3 – Speculative execution example

On the left-hand side of the preceding diagram, we can see the instruction ordering that's been requested by our query engine. Naturally, it begins with checking whether the caller has access to read the column. Assuming that conditional passes, the query engine then attempts to read the data and compute a result. The right-hand side of the diagram shows how your CPU likely executed these instructions. Notice that the order changed! Some of these pseudo instructions take more time, often measured in clock cycles compared to others. Modern x86 CPUs can work on multiple instructions each clock cycle. While one instruction is fetching its operands from the cache, another instruction might be using a floating-point unit to calculate the result of a division operation. Coordinating which instructions are utilizing each part of the CPU is often referred to as pipelining.

The deeper the pipeline, the more efficient a CPU can be, and the faster customers will perceive the CPU. The trick is keeping all the parts of the CPU busy by guessing what instructions might be run in the future when earlier instructions take too long and stall execution. Your CPU is making a bet. It can remain idle while waiting for the earlier instruction to finish, or it can guess at what it will be asked to do next. Naively waiting has a 100% probability of wasting CPU cycles. Guessing is highly likely to perform better than waiting. Chasing this opportunity is what has driven modern x86 CPUs to reorder instructions and, at times, speculatively execute instructions.

In our access check example shown in the preceding diagram, the memory read and compute result steps have to wait for the access check branch to decide which path to take. While that branch is being evaluated, the memory dispatcher is idle, despite having an impending memory read. Your CPU has a surprisingly large surface area of the physical chip that's dedicated to branch prediction so that it can guess whether read operations will be required. So, your CPU will start reading and maybe even calculating the result while waiting to find out whether the branch will need those instructions to be carried out. This might seem like a bad idea, especially when instructions have side effects such as writing to memory. Luckily, your CPU can unwind mispredicted branches so that they have no materialized side effects.

Unfortunately, Spectre and Meltdown highlighted subtle side effects in the form of changing the cache's state. Imagine that I can fool your CPU into speculatively executing a conditional read of a memory address I don't own – maybe even the address where you are storing an encryption key. Later, I can run a similar operation and use the timing of when the instruction was completed to tell me whether the value was already in the CPU cache. If the value was in the cache, I can infer the result of the conditional check and thus learn about the value that was stored at an address I don't have access to – all because the CPU cache state wasn't rolled back. In this example, the cache created a side channel between the erased world of the failed branch prediction and the resumed execution path.

With this primitive memory gadget, an attacker can steal a few bits of memory from a neighboring process at a time. In practice, this class of vulnerability has been used to crack cryptographic keys that are used for SSH, SSL, and credential storage. Many organizations lack the deep security expertise to identify or worry about these kinds of vulnerabilities. Luckily, Lake Formation can help you stay a step ahead in the race to securing your data lake.

> **What Is a Gadget?**
>
> In the context of malicious code exploits, a gadget is a utility that can be used to exploit a known vulnerability. Most gadgets are small, typically comprised of a few dozen code lines, and appear pretty innocuous on their own. When a malicious actor initially accesses a system that intends to compromise, either through legitimate means or via an initial vulnerability, they often begin constructing gadgets that allow them to elevate their privilege or extract information from the target system.

How Lake Formation can help

At re:Invent 2020, the AWS Lake Formation team announced a preview release of Lake Formation's next-generation security features. Among these new features were a set of APIs for reading and writing data to Lake Formation-managed tables, with the ability to enforce row-level access. The following screenshot shows how to grant access to US customer data in a table containing data from customers around the globe:

Figure 14.4 – Row-level access control

These two features can be combined to address many of the challenges we discussed earlier as part of the shared responsibility security model and data perimeters. The new APIs essentially offload the TableScan operation from your analytics engine into Lake Formation's secure filtering fleet. By doing so, Lake Formation can make strong security guarantees, regardless of the analytics engine you are using. Since Lake Formation's read and write APIs apply policy enforcement remotely to any arbitrary and potentially untrusted code within your workload, the attack surface is much smaller. You no longer need to worry about side channels or admin access to the underlying host. This model also makes it easier to build multi-tenant analytics applications. Its built-in filtering capability also allows Lake Formation to enforce previously impossible row-level access control policies without the need to ETL redacted copies of your dataset.

This functionality is slated to become generally available in late 2021, alongside Lake Formation's new ACID-compliant governed table type.

Understanding the benefits of governed tables

The entire AWS analytics suite of services, including Athena, EMR, Glue, Redshift, and Lake Formation, continually makes building and managing data lakes on S3 easier. What used to take months with traditional data warehouses can be accomplished in days using these tools with S3. Despite all the advances in these services, customers still face difficult choices when it comes to the following:

- Ingesting streaming data such as **Change-Data-Capture (CDC)**, click data, or application logs
- Complying with new regulations such as GDPR and CCPA
- Understanding how your data changes over time
- Adapting table storage to meet evolving usage and access patterns

In addition to the security-oriented features we discussed earlier in this chapter, Lake Formation's new governed table type takes several steps toward addressing these common sources of data lake frustration. Governed tables are a new Amazon S3 table type that supports **atomic, consistent, isolated, and durable (ACID)** transactions and automatic storage optimization. To the uninitiated, this may seem like a home run of marketing buzzwords, but governed tables are poised to change how we build everything, from ETL pipelines to interactive analytics applications. Next, we'll look at a common problem that governed tables and their ACID transactions can help us overcome.

ACID transactions on S3-backed tables

Have you ever queried multiples data lake tables in the same query, perhaps via a join clause? If different source systems or ETL jobs populated those tables, there is a significant probability that any query against them reads inconsistent data. The picture becomes even bleaker when you factor in partial failures, which can be just as challenging to identify as they are to repair. This might be a good time for an example.

Suppose we work for an advertising company and routinely track the performance of different advertising campaigns by joining three tables. The first table contains details about all the campaigns, including their total budget, start date, end date, and sponsor. This table is relatively stable, changing only when new campaigns are booked. Next, the impressions table contains a row representing every time we served an ad placement from this campaign. This table changes rapidly, with new entries appearing in near-real time. The final table contains conversion data that identifies which impressions resulted in an ad click or, better still, a purchase! This table doesn't change as often as we like, but it is far from static and mostly populated with data from third-party systems.

When you open your Athena console and run your company's conversion rate reporting query in preparation for a client meeting, you are rolling the dice that the result you get is an accurate representation of the world. Suppose the impression table has fallen behind because of a traffic surge leading up to the holiday season. The conversion table has a much lower flow and doesn't encounter any issues. Even if your query uses set date ranges, you may still find yourself pulling more conversion data than impression data, resulting in an overly optimistic view of how well the campaign is doing. The opposite can also be true when an unexpected issue causes the third-party source data to be late or incomplete. In that case, you may be scrambling to make up for an inexplicably underperforming campaign and give unnecessary concessions to an important client.

In our experience, all data lake use cases fall into one of three buckets concerning consistency:

- **Consistency is irrelevant**: The data is typically historical (backward-looking), immutable, or consistency is inherent due to the records containing correlation IDs that self-identify consistency issues.

- **Consistency is unknown**: The producers and consumers do not know or understand the implications of datasets being used together. The organization spends many hours chasing phantom data quality *heisenbugs* that seem to resolve themselves when investigated.

- **Consistency is required and designed for**: Producers and consumers take steps to ensure that the data in the lake is consistent. This often includes publishing metadata alongside the data that describes its currency. Many organizations also produce snapshot datasets that simplify consumers by treating data as immutable at the expense of increased ETL compute and storage costs.

Heisenbug

This is one of our favorite pieces of computer science jargon that plays on the famous observer effect of quantum mechanics that Werner Heisenberg first described as the Heisenberg Uncertainty Principle. The theory asserts that the act of observing a quantum particle changes its behavior and reduces the reliability of multi-variable measurements. Naturally, frustrated software engineers rallied behind this theory, which accurately describes a class of bugs that are usually timing-related. In such cases, a new log line is added or a debugger is attached to observe how the bug changes how the system behaves and causes the bug to disappear. In practice, the typical mechanisms that are used to observe a bug also change the speed or timing of program execution, which has a real effect on timing bugs resulting from race conditions.

Now that we have a better understanding of data lake consistency, we can look at an example of how to use transactions against Lake Formation governed tables to simplify how we produce and consume data. At the time of writing, Athena can read governed tables but has not released its specification for writes to governed tables yet. Since most of the interesting consistency work is taken on by the producer or writer, we'll use an Apache Spark example from Glue ETL instead.

In the following code block, we are creating a Glue Spark context and then calling Lake Formation's new `begin_transaction` API. This API returns a transaction identifier that represents a specific point in time within our data lake, commonly called an epoch. With this single API call, we've established a point of observation that will be applied to all reads and writes that are performed within this script. This is important enough that it warrants repeating. No matter what any other reader or writer does to any table in our data lake, we are guaranteed a view of the world as soon as we start the transaction, thanks to the snapshot isolation mechanism offered by governed tables.

The script then uses the transaction ID to configure a Spark sink that points to our impressions table in the ads database. This is primarily boilerplate and is no different from non-governed table use cases, except for passing the transaction ID to the creation function:

```
glueContext = GlueContext(SparkContext.getOrCreate())
txid1 = glueContext.begin_transaction(read_only=False)
sink = glueContext.getSink(connection_type="s3",
                path="s3://my_bucket/ads/impressions/",
                enableUpdateCatalog=True,
                updateBehavior="UPDATE_IN_DATABASE",
                transactionId=txid1)
sink.setFormat("glueparquet")
sink.setCatalogInfo(catalogDatabase=ads,
            catalogTableName=impressions)
```

Once the sink has been created, the script uses it to write new and updated impression data into the data lake via a DataFrame that we loaded offscreen from a third-party source. In the following code block, the script uses a try-except block to ensure that it either commits or aborts the transaction, depending on the success of the write operation. As the developer of the script, you can choose when to call commit_transaction or abort_transaction. For extra protection, you may choose to query the newly written data to ensure it's available before declaring the write successful and committing the transaction. Since governed tables support read-your-own-write semantics, you can easily add this quality check and simplify operations by automatically rolling back the errant or partial data without human intervention:

```
try:
    sink.writeFrame(new_and_updated_impressions_dataframe)
    glueContext.commit_transaction(txid1)
except:
    glueContext.abort_transaction(txid1)
```

There are many other use cases where having transactional capabilities is helpful. Combining Lake Formation's new data read and write APIs with ACID transactions enables compliance with data protection laws such as GPDR, which were previously hampered by the immutable nature of S3 objects.

Despite S3 objects being inherently immutable, organizations have been modifying data in their data lakes for years. Most customers are familiar with adding new data as it arrives, but some must also apply backfills or restate past values by rewriting entire files or tables. With all these modifications flying around, we often find ourselves wondering, "what did that table contain when this job ran?". Your compliance officer might even mandate that specific tables be versioned, even though few, if any, tools exist to automate reading past versions of essentially random S3 objects. The same machinery that Lake Formation uses to create ACID transactions enables reading your data lake through any committed transaction. This is the basic building block of time-travel capabilities, which we will discuss in the next section.

Time-traveling queries

To resolve transaction conflicts and support rollbacks, more ACID-compliant transaction managers maintain a transaction log of some kind. The ledger records every change, addition, or deletion that occurred as part of each transaction. With this information, the system can rebuild the system's state before or after each transaction. Normally, this aids in error recovery or transaction rollback when you call the abort_transaction API. Lake Formation extends the utility of the transaction log to offer time-traveling capabilities.

When activated, time travel allows queries against one or more governed tables to read from a consistent snapshot of the data lake, as of the specified time or transaction ID. The following code block shows how to run an Athena query against the advertising impression table from the previous section. Despite what 80s movies may have taught you, you won't need a Delorean or 1.21 Gigawatts of power to calculate the number of impressions for our advertising campaign as of 30 days ago. We can simply specify a SYSTEM_TIME value that Athena will use to set the read point in the transaction log:

```
SELECT campaign_id,
   count(*) as total_impressions,
   avg(linger_time_ms) as avg_impression_duration
FROM
   lakeformation.ads.impressions
WHERE
      campaign_id = 87348519457
FOR SYSTEM_TIME AS OF datesub(day, 30, now())
GROUP BY campaign_id
```

We can use such queries to debug updates to a dataset, observing when and how data changed. If an update was done incorrectly, then the transaction that caused the data quality issue can be rolled back. For example, if you have impression data that gets updated regularly and a customer suggests that the data is incorrect, using time-travel queries can pinpoint the time when the inaccurate data was updated.

As you might imagine, the underlying storage for transaction tables is more complex than a basic list of S3 objects. Luckily, governed tables are supported by Lake Formation's new storage optimizer.

Automated compaction of data

We first covered the role of the physical table layouts as part of *Chapter 2, Introduction to Amazon Athena, and Built-In Functions*. This subject resurfaced in *Chapter 11, Operational Excellence – Maintenance, Optimization, and Troubleshooting*, where partitioning and file formats became a focal point of operating Athena workloads at scale. The size and arranging S3 objects into partitions and tables dictates both the performance and cost of your analytics queries. When customers ask why their queries are not running as quickly as expected, file size is one of the first things we must check. Most of the time, the files being read are tiny, 10 KB to 10 MB. Small files can be detrimental to query performance because there is overhead associated with each object in the form of metadata, connection time, and data roundtrips from the underlying storage. This overhead can account for as much as 80% of the overall time taken to read the data for small objects.

When enabled for your governed tables, Lake Formation monitors the file sizes and read performance to identify opportunities where reorganizing the data would improve performance. The first such optimization comes in the form of small file compaction. If you've ever processed a data stream from the likes of Kinesis or Kafka, you'll likely have dealt with an accumulation of thousands or millions of small files. Lake Formation will automatically rewrite the small files into more appropriately sized ones, according to the given format's recommended file size. Since these compaction operations happen as part of an ACID transaction, they all occur seamlessly, without your producers or consumers needing to be aware of the activity.

While this is the final Lake Formation feature we'll cover, it is far from the least, given the proliferation of self-managed compaction jobs that many customers run.

Summary

In this chapter, we concluded our exploration of Athena by looking at upcoming Lake Formation features. AWS is increasingly positioning Lake Formation as their one-stop shop for data lake creation and management. If they succeed in making Lake Formation a foundational component of AWS data lakes, customers could expect increased interoperability across the various AWS analytics engines.

It may not be the flashiest feature, but we expect to see many applications mimic Lake Formation's new security features. Using dedicated data access APIs to decouple policy enforcement from workload execution is like an easy button for reducing your attack surface. The addition of ACID transactions with the new governed table type will open a host of new possibilities such as time travel. Look for these features to reach general availability in late 2021.

If you'd like to learn more, consult the *Further reading* section and consider signing up for the public preview of these features.

Further reading

In this section, we've gathered links to additional materials that you may find helpful in diving deeper into some of the primary sources for topics mentioned in this chapter:

- AWS Big Data Blog: Getting Started with Governed Tables: `https://amzn.to/3AsSjYX`

- AWS Big Data Blog: Creating Governed Tables: `https://amzn.to/3s9pJJ1`

- AWS Big Data Blog: Using ACID Transactions on Governed Tables: `https://amzn.to/2VtEV87`

- AWS Big Data Blog: Implementing Cell-Level and Row-Level Security: `https://amzn.to/3CtrkhB`

- AWS Big Data Blog: Securing Data Lakes: `https://amzn.to/2X9QGkv`

- Side-channel a-ttacks: Spectre and Meltdown: `https://bit.ly/3Cu0pSR`

Packt.com

Subscribe to our online digital library for full access to over 7,000 books and videos, as well as industry leading tools to help you plan your personal development and advance your career. For more information, please visit our website.

Why subscribe?

- Spend less time learning and more time coding with practical eBooks and Videos from over 4,000 industry professionals

- Improve your learning with Skill Plans built especially for you

- Get a free eBook or video every month

- Fully searchable for easy access to vital information

- Copy and paste, print, and bookmark content

Did you know that Packt offers eBook versions of every book published, with PDF and ePub files available? You can upgrade to the eBook version at packt.com and as a print book customer, you are entitled to a discount on the eBook copy. Get in touch with us at customercare@packtpub.com for more details.

At www.packt.com, you can also read a collection of free technical articles, sign up for a range of free newsletters, and receive exclusive discounts and offers on Packt books and eBooks.

Other Books You May Enjoy

If you enjoyed this book, you may be interested in these other books by Packt:

Learn Amazon SageMaker

Julien Simon

ISBN: 9781800208919

- Create and automate end-to-end machine learning workflows on Amazon Web Services (AWS)
- Become well-versed with data annotation and preparation techniques
- Use AutoML features to build and train machine learning models with AutoPilot
- Create models using built-in algorithms and frameworks and your own code
- Train computer vision and NLP models using real-world examples

- Cover training techniques for scaling, model optimization, model debugging, and cost optimization
- Automate deployment tasks in a variety of configurations using SDK and several automation tools

Scalable Data Streaming with Amazon Kinesis

Tarik Makota, Brian Maguire, Danny Gagne, Rajeev Chakrabarti

ISBN: 9781800565401

- Get to grips with data streams, decoupled design, and real-time stream processing
- Understand the properties of KFH that differentiate it from other Kinesis services
- Monitor and scale KDS using CloudWatch metrics
- Secure KDA with identity and access management (IAM)
- Deploy KVS as infrastructure as code (IaC)
- Integrate services such as Redshift, Dynamo Database, and Splunk into Kinesis

Packt is searching for authors like you

If you're interested in becoming an author for Packt, please visit `authors.packtpub.com` and apply today. We have worked with thousands of developers and tech professionals, just like you, to help them share their insight with the global tech community. You can make a general application, apply for a specific hot topic that we are recruiting an author for, or submit your own idea.

Share Your Thoughts

Now you've finished *Serverless Analytics with Amazon Athena*, we'd love to hear your thoughts! Scan the QR code below to go straight to the Amazon review page for this book and share your feedback or leave a review on the site that you purchased it from.

`https://packt.link/r/1-800-56234-9`

Your review is important to us and the tech community and will help us make sure we're delivering excellent quality content.

Index

V

W

Z